Mechanical Design

Peter R. N. Childs
BSc. (Hons), D.Phil, C.Eng, M.I.Mech.E.
University of Sussex, UK

A member of the Hodder Headline Group
LONDON • SYDNEY • AUCKLAND
Copublished in North, Central and South America by
John Wiley and Sons Inc., New York • Toronto

First published in Great Britain in 1998 by Arnold
a member of the Hodder Headline Group,
338 Euston Road, London, NW1 3BH
http://www.arnoldpublishers.com

Copublished in North, Central and South America by
John Wiley & Sons, Inc.
605 Third Avenue, New York, NY 10158

British Library Cataloguing in Publication Data
A catalogue record for this book is available from the British Library

Library of Congress Cataloging-in-Publication Data
A catalog record for this book is available from the Library of Congress

ISBN: 0 340 69236 7
ISBN: 0 470 32740 5 (Wiley)

1 2 3 4 5 6 7 8 9 10

Publisher: Matthew Flynn
Production Editor: James Rabson
Production Controller: Rose James
Cover designer: Mouse Mat Design
Cover photography: Mr K. Hunt, University of Sussex

Typeset in 10/11pt Times by Academic & Technical, Bristol
Printed and bound in Great Britain by The Bath Press, Bath

This book is dedicated to Fiona, Hazel and Tabitha

Contents

Preface

The overall aims of this book are to introduce the subject of holistic or total design and to introduce the design and selection of various common mechanical engineering components. These provide 'building blocks' with which the engineer can practise his or her art. The primary focus is the design of rotating machinery.

The approach adopted for defining design follows that developed by the SEED (Sharing Experience in Engineering Design) programme and reported by Pugh (Pugh, S., 1990, *Total Design*, Addison Wesley) where design is viewed as the 'the total activity necessary to provide a product or process to meet a market need'. Within this framework the text is concentrated on developing detailed mechanical design skills in the areas of bearings, shafts, gears, seals, belt and chain drives, clutches and brakes. Where standard components are available from manufacturers, the steps necessary for their specification and selection are developed.

The framework used within the text has been to provide descriptive and illustrative information to introduce principles and individual components and to expose the reader to the detailed methods and calculations necessary to specify and design or select a component. To provide the reader with sufficient information to develop the necessary skills to repeat calculations and selection processes, detailed examples and worked solutions are supplied throughout the text.

This book is pitched toward the latter part of year one, year two and preliminary year three undergraduate level. Prerequisite modules include year one undergraduate mathematics, fluid mechanics and heat transfer, principles of materials, statics and dynamics. However, as the subjects are introduced in a descriptive and illustrative format and as full worked solutions are provided, it is possible that readers without this formal level of education could nevertheless benefit from this book. The text is specifically aimed at mechanical engineering degree programmes and would be of value for modules in mechanical, mechanical engineering design, design and manufacture, design studies and tribology, as well as modules and project work incorporating a design element requiring knowledge about any of the content described.

The aims and objectives described are achieved by a short introductory chapter on total design followed by six chapters on machine elements covering: bearings, shafts, gears, seals, chain and belt drives, clutches and brakes. The subject of tolerancing from a component to a process level is introduced in Chapter 8. The last chapter covers the total design process in more detail and sets the context of the detailed design aspects covered within the book. The design methods where appropriate are developed to national and international standards (e.g. ANSI, ASME, AGMA, BSI, DIN, ISO).

Acknowledgements

I would like to thank the engineers and individuals who have contributed towards my own education, particularly Prof. Alan Turner, Dr Christopher Long and Prof. Fred Bayley. Also I wish to acknowledge the industrial sponsors of the research and design projects I have been involved with, especially European Gas Turbines and Rolls-Royce and the respective managers Mr J. Hannis and Mr J. Millward. Without the opportunity to practise engineering design and see the generation of the final product, engineering design can be abstract and frustrating. By direct involvement, the exciting world of engineering opens up. It is my hope that this book is a step towards opening up that exciting world for the reader. I would also like to thank Arnold for the opportunity to publish this book.

Foreword

There is considerable evidence that companies are increasingly adopting a global approach to product development and marketing and are recognising the critical part design plays in overall competitiveness. However, industry also recognises that to meet these challenges a more sophisticated, rigorous and systematic design process is required. The engineering design process is extremely complex, involving much plan and decision making at many levels. The modern Total Design process as proposed by Stuart Pugh and widely adopted by UK educationalists is rightly employed by Peter Childs in this text. However, after many years of engineering design management practice even Stuart Pugh came to the conclusion that he 'still did not know how best to organise the design process'. Any text which assists the student engineer in progressing through this 'minefield' is welcome, and the use of this text during project work and in support of formal classes will make a significant contribution.

Without a thorough understanding of and exposure to engineering detail design a student cannot hope to make an impact on the nebulous area of conceptual design. The detailed information allowing component determination in the main body of this text will enable students to evaluate alternative concepts. During project work students are frequently required to answer questions, such as 'gear or a belt drive?' and 'journal or rolling element bearing?'. The information in this text, along with the sensible advice to contact component manufacturers and ask their opinions, will enable optimum decisions to be made. The comprehensive review of design literature and information sources will prove very useful in this respect.

Within the text there are many illustrations and worked examples. The worksheets, with answers, at the end of each chapter reinforce all aspects and could be used in tutor-led form. However, any capable student sufficiently motivated by design could use this text as a self learning aid before or during project work.

This book has a carefully chosen title and has clearly defined aims which it meets completely. From it the student engineer will gain an appreciation of the Total Design process and a working insight into the common detail design component selection decisions which designers are often called upon to make. Clearly the intention is not to employ the techniques and information in a mechanistic manner, rather to use the text as a supporting framework which will lead to product design success.

Dr Ken Hurst
Chairman of SEED (Sharing Experience in Engineering Design)
Senior Lecturer in Mechanical Engineering Design, The University of Hull

1 Mechanical design

The aims of this chapter are to introduce the process of mechanical design and design methodology and to outline the approach and content of this book.

1.1 Introduction

The term design is popularly used to refer to an object's aesthetic appearance. We often speak about designer clothes and well-designed cars. The word design comes from the Latin *designare*, which means to designate or mark out. Mechanical design refers to the design of devices, products and systems of a mechanical nature such as engines, machines, tools and instruments.

Design can be taken to mean all the processes of conception, invention, visualisation, calculation, refinement and specification of details which determine the form of a product. Design generally begins with either a need or requirement or, alternatively, an idea. It ends with a set of drawings or computer representations and other information which enables a product to be manufactured and utilised. Here design is defined as 'the total activity necessary to provide a product or process to meet a market need' (Pugh, 1990).

The bulk of this text is concerned with the selection and design of mechanical machinery components. However, whilst good skills in these areas are essential to the successful operation of a product utilising them, no amount of good detailed design can make up for products which are poor in conception. In other words if the proposal is a non-starter, it does not matter how well its constituent components are designed and put together. For this reason this chapter introduces the principles of the design process and methodology in Sections 1.2 and 1.3. This will impart valuable understanding and enable the context of the remainder of the book to be appreciated.

1.2 The design process

Probably from your own experience you will know that design can consist of looking at a design need and working on the problem by means of sketches, models, brainstorming, calculations as necessary, development of styling as appropriate, making sure the product fits together and can be manufactured, and calculation of the costs. The process of design can be schematically represented to levels of increasing formality and complexity as shown in Figs 1.1 and 1.2. Figure 1.1 represents the traditional approach associated with lone inventors comprising the generation of the 'bright idea', drawings and calculations giving form to the idea, judgement of the design and re-evaluation if necessary, resulting in the generation of the end product. Figure 1.2 shows a more formal description of the design process which might be associated with engineers operating within a company's management structure. The various terms used in Fig. 1.2 are described below.

- **Recognition of need.** Often design begins when an individual or company recognises a need, or identifies a potential market, for a product, device or process. Alternatively 'need' can be

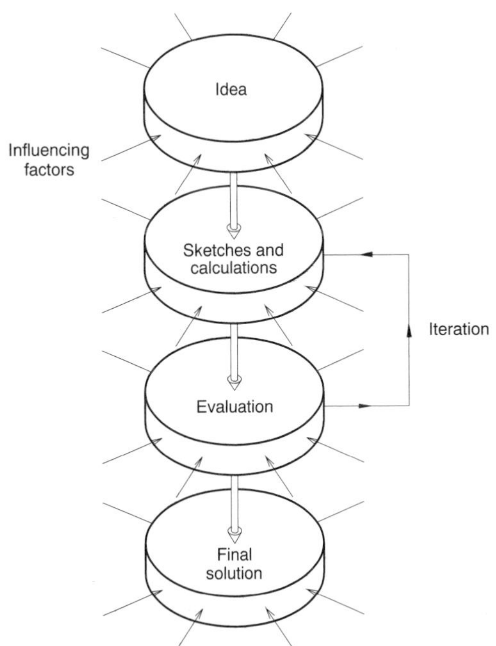

Fig. 1.1 The traditional and familiar 'inventor's' approach to design.

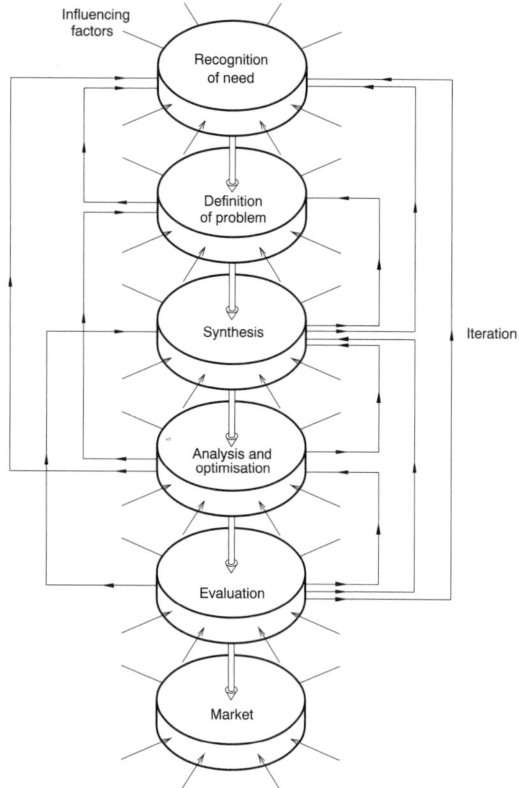

Fig. 1.2 The design process.

defined as when a company decides to re-engineer one of its existing products (for example, producing a new car model).

- **Definition of problem.** This involves all the specification of the product or process to be designed. For example, this would include inputs and outputs, characteristics, dimensions and limitations on quantities.
- **Synthesis.** This is the process of combining the ideas developed into a form or concept which offers a potential solution to the design requirement. The term synthesis may be familiar from its use in chemistry where it is used to describe the process of producing a compound by a series of reactions of other substances.
- **Analysis.** This involves the application of engineering science – subjects explored extensively in traditional engineering courses such as statics and dynamics, mechanics of materials, fluid flow and heat transfer. These engineering 'tools' and techniques can be used to examine the design to give quantitative information such as whether it is strong enough or will operate at an acceptable temperature. Analysis and synthesis invariably go together. Synthesis means putting something together and analysis means resolving something into its constituent parts or taking it to pieces. Designers have to synthesise something before it can be analysed – the famous chicken-and-egg scenario! When a product is analysed, some kind of deficiency or inadequacy may be identified requiring the synthesis of a new solution prior to reanalysis and repetition of the process until an adequate solution is obtained.
- **Optimisation.** This is the process of repetitively refining a set of often conflicting criteria to achieve the best compromise.
- **Evaluation.** This is the process of identifying whether the design satisfies the original requirements. It may involve assessment of the analysis, prototype testing and market research.

Although Figs 1.1 and 1.2 at first sight show design occurring in a sequential fashion, with one task following another, the design process may actually occur in a step-forward, step-back fashion. For instance, you may propose a solution to the design need and then perform some calculations or judgements which indicate that the proposal is inappropriate. A new solution will need to be put forward and further assessments made. This is known as the iterative process of design and forms an essential part of refining and improving the product proposal.

Note that the flow charts shown in Figs 1.1 and 1.2 do not represent a method of design but rather a description of what actually occurs within the process of design. The method of design used is often unique to the engineer or design team. Design methodology is not an exact science and there are indeed no guaranteed methods of design. Some designers work in a progressive fashion; others work on several aspects simultaneously.

1.2.1 Case study

The process identified in Fig. 1.2 can be illustrated by example. Following some initial market assessments, the board of a plant machinery company has decided to proceed with the design of a new product for transporting pallets around factories. The board have in mind a forklift truck but do not wish to constrain the design team to this concept alone. The process of the design can be viewed in terms of the labels used in Fig. 1.2 as listed below.

- **Recognition of need.** The company has identified a potential market for a new pallet moving device.

Table 1.1 Morphological chart for a pallet moving device

Feature	Means				
Support	Track	(Wheels)	Air cushion	Slides	Pedipulators
Propulsion	(Driven wheels)	Air thrust	Moving cable	Linear induction	
Power	Electric	Diesel	(Petrol)	Bottled gas	Steam
Transmission	Belts	Chains	(Gears and shafts)	Hydraulics	Flexible cable
Steering	(Turning wheels)	Air thrust	Rails		
Stopping	(Brakes)	Reverse thrust	Ratchet		
Lifting	(Hydraulic ram)	Rack and pinion	Screw	Chain or rope hoist	Linkage
Operator	Standing	Walking	(Seated at front)	Seated at rear	Remote control

Reproduced with modifications from Cross (1994).

- **Definition of problem.** A full specification of the product desired by the company should be written. This is sometimes called the brief, or specification of need. This allows the design team to identify whether their design proposals meet the original request. In this case writing of the brief may require further information from the board – what size of pallet is to be moved, what lifetime is required for the product and so on.
- **Synthesis.** This is often identified as the formative and creative stage of design. Some initial ideas must be proposed or generated in order for them to be assessed and improved. Concepts can be generated by your imagination, experience or by the use of design techniques such as morphological charts, as illustrated in Table 1.1. The use of morphological charts requires the designer to consider the function of components rather than their specific details. Our pallet moving device would probably comprise support, propulsion, power, transmission, steering, stopping and lifting systems. Within each of these criteria all plausible options should be listed, as shown in Table 1.1. The task of the designer is to look at this chart and select a justifiable option from each criteria producing the make-up of the pallet device. In this case there would not necessarily be any surprises in the final make-up of the product and selection of 'wheels', 'driven wheels', 'petrol', 'gears and shafts', 'turning wheels', 'brakes', 'hydraulic ram' and 'seated at front' would result in the traditional forklift truck illustrated in Fig. 1.3. However, the use of this method can assist in the production of alternatives as shown in Table 1.2 and Fig. 1.4, where the option for a linkage-based lifting method has been considered. Some evaluation should be made at this stage to reduce the number of concepts requiring further work. Various techniques are available for this, including merit and adequacy assessments. In this case the linkage option is selected as this offers elements of market novelty and therefore potential marketing advantage.
- **Analysis.** Once a concept has been proposed it can then be analysed to determine whether constituent components can meet the demands placed on them in terms of performance, manufacture, cost and any other specified criteria. Alternatively, analysis techniques can be used to determine what size components need to be to meet the required functions.

Fig. 1.3 Forklift truck.

Fig. 1.4 The Teletruk concept by JCB. (Teletruk photograph courtesy of J.C. Bamford Excavators Ltd.)

- **Optimisation.** Inevitably there are conflicts between requirements. In the case of the forklift truck, size, manoeuvrability, cost, aesthetic appeal, ease of use, stability and speed are not necessarily all in accordance with each other. Cost minimisation may call for compromises on material usage and manufacturing methods. These considerations form part of the optimisation of the product producing the best or most

Table 1.2 Morphological chart for a pallet moving device with alternative choices identified

Feature	Means				
Support	Track	**Wheels**	Air cushion	Slides	Pedipulators
Propulsion	**Driven wheels**	Air thrust	Moving cable	Linear induction	
Power	Electric	Diesel	**Petrol**	Bottled gas	Steam
Transmission	Belts	Chains	**Gears and shafts**	Hydraulics	Flexible cable
Steering	**Turning wheels**	Air thrust	Rails		
Stopping	**Brakes**	Reverse thrust	Ratchet		
Lifting	Hydraulic ram	Rack and pinion	Screw	Chain or rope hoist	**Linkage**
Operator	Standing	Walking	**Seated at front**	Seated at rear	Remote control

Reproduced with modifications from Cross (1994).

acceptable compromise between the desired criteria.

- **Evaluation.** So a concept has been proposed and selected and the details of component sizes, materials, manufacture, costs and performance worked out. Does it meet the original request? This is where an assessment of the overall performance analysis should be made. If the envisaged performance is acceptable, then market analysis, possibly consisting of current customer reactions to product drawings and models, can be performed. Following this, prototype testing can take place and possibly some limited trialling with customers.

1.3 Design methodology

The process of design is currently the focus of detailed study with approaches being proposed for design methodologies. Design methodology is a framework within which the designer can practise with thoroughness. One such approach called 'total design' has been proposed by the SEED (Sharing Experience in Engineering Design) programme (SEED, 1985) and Pugh (1990), and is illustrated schematically in Fig. 1.5 as a design core and in Fig. 1.6 as the total design process.

Almost any product such as an engine, gear box, turbocharger or burner requires input from people of many disciplines including engineering, law and marketing, requiring considerable coordination. In industrial terms, the integration comes about as a result of the partial design inputs from each discipline. Industry is usually concerned with total design. Total design is the systematic activity necessary from the identification of a market need to the commercialisation of the product to satisfy the market need. Total design can be regarded as having a central core of activities consisting of the market potential, product specification, conceptual design, detailed design, manufacture and marketing. The constituent elements of the design core illustrated in Figs 1.5 and 1.6 are briefly described below prior to reference in later chapters and a more detailed treatment in Chapter 9.

Several aspects of the design core given in Figs 1.5 and 1.6 are similar to those described for Fig. 1.2. The market phase refers to the assessment of sales opportunities or perceived need to update an existing product. Specification involves the formal statement of the required function features and performance of the product or process to be designed. Recommended practice from the outset of design work is to produce a product design specification which should be formulated from

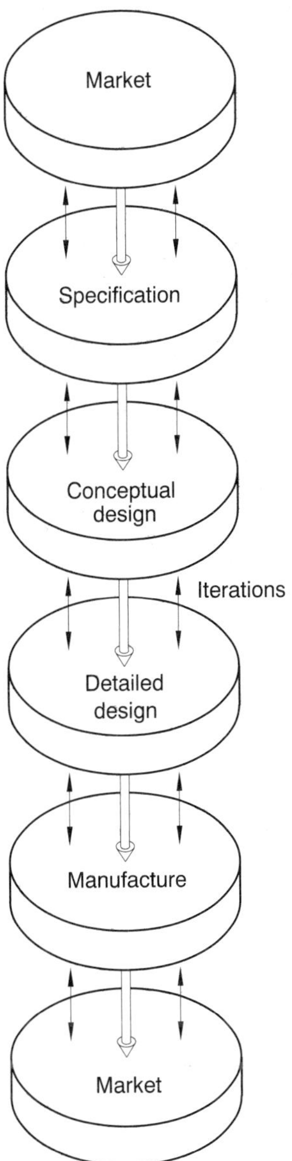

Fig. 1.5 The total design core (after Pugh, 1990).

the statement of need (or brief as it is sometimes known). The product design specification is the formal specification of the product to be designed. It acts as the control for the total design activity because it sets the boundaries for the subsequent design. The early stages of design where the major decisions are to be made is sometimes called conceptual design. During this phase a rough idea is developed as to how a product will function and what it will look like. The process of conceptual design can also be described as the

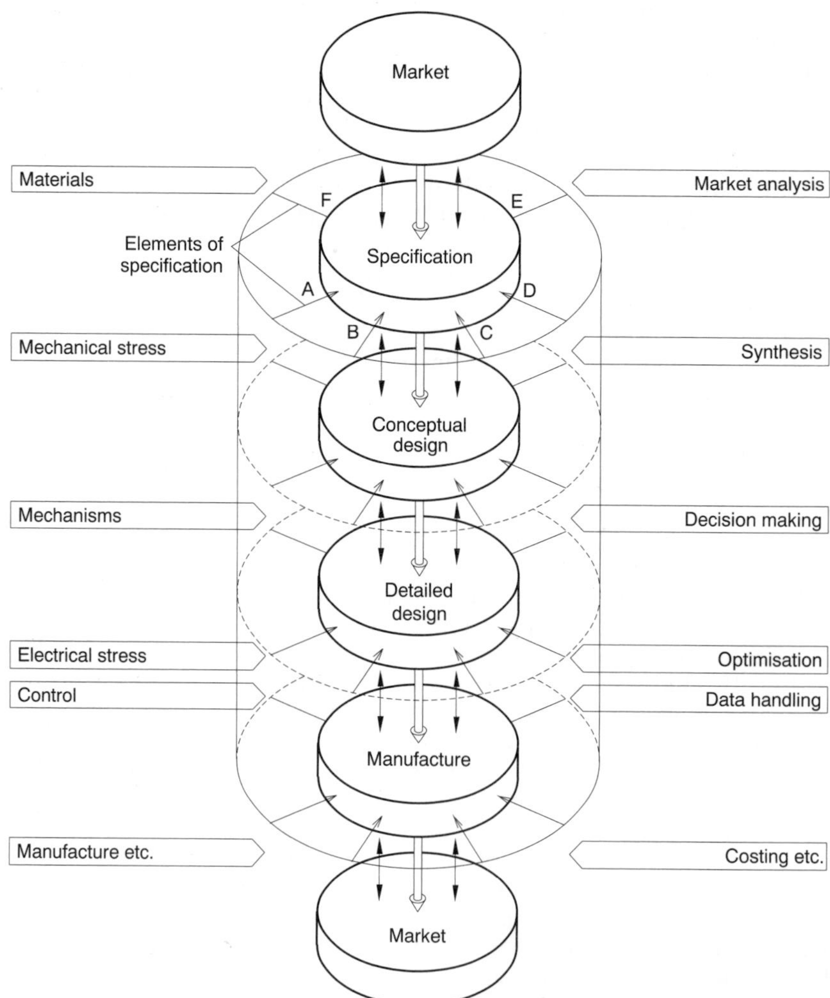

Fig. 1.6 The total design process (after Pugh, 1990).

definition of the product's morphology, how it is made up and its layout. The detailed design phase consists of the determination of the specific shape and size of individual components, what materials should be used, how they fit together and the method of manufacture. Detailed design makes use of the many skills acquired and developed by engineers in the areas of analysis. It is the detailed design phase which can take up the bulk of the time spent on a design. However, as implied earlier, it is wise to only spend time on details once a sensible concept has been selected. The manufacture phase, although identified as distinct within the structure, is typical of other phases in that it influences all the others. The design of any item must be such that it is feasible to manufacture it! The materials selected must be compatible with the manufacturing facilities and skills available and at acceptable costs to match marketing requirements. The finish of the components must match the aesthetic and safety requirements. The last phase, selling, is of course essential and should match the expectations of the initial market phase. This phase too has an impact on other phases within the design core. Information such as customer reaction to the product, any component failures or wear, damage during packaging and transportation should be fed back to influence the design of product revisions and future designs.

The double arrows shown in Figs 1.5 and 1.6 represent the flow of information and control from one activity to another, as well as the iterative nature of the process. For instance, detailed design activity may indicate that an aspect of the conceptual design is not feasible and must be reconsidered. Alternatively, conceptual work may yield features which have the potential for additional marketing opportunities. In other words, the activity on one level can and does interact dynamically with activities on other levels.

In Fig. 1.6 additional activities, such as market analysis, stressing and optimisation, have been added to the design core as inputs. The effective and efficient design of any product invariably requires the use of different techniques and skills. The disciplines indicated are the designer's toolkit and indicate the multidisciplinary nature of design. The forklift truck example mentioned in Section 1.2 will require engine management and control systems, as well as the design of mechanical components. Although this text concentrates on mechanical design, this is just one, albeit an important, interesting and necessary aspect of the holistic or total design activity.

The additional circumferential inputs shown in Fig. 1.6, labelled A to F – although the exact number depends on the actual case under consideration – represent elements of the specification listed in order of relative importance. The priority order of these specifications may alter for different phases of the design activity.

As illustrated in the design core presented in Fig. 1.6, materials are fundamental to the process. Engineers should develop a knowledge or have access to a database which indicates what materials are available and how they perform under different operating conditions. Although this text concentrates on the use of metals and to a lesser extent plastics, an open attitude to materials enables new opportunities to be exploited. The range of existing and well tried technology is extensive. Many items are available as standard components and can be purchased from specialist suppliers. In addition, standard practice and design methodology has been specified for many others. The scope of mechanical machine elements available to the designer is outlined in Table 1.3. This list is not exhaustive but does give an idea of some of the 'building blocks' available. The items covered in this book are indicated by shading. Allied to a knowledge of materials and existing technology is the general requirement for engineers to know what is technically and scientifically feasible so that 'far-fetched' ideas can be objectively ruled out.

The model illustrated in Figs 1.5 and 1.6 was originally proposed some years ago (1985). Since then the ethos of design has developed to include the design of products for total life recycling. This means that the end of a product should be considered at the design stage. When the product

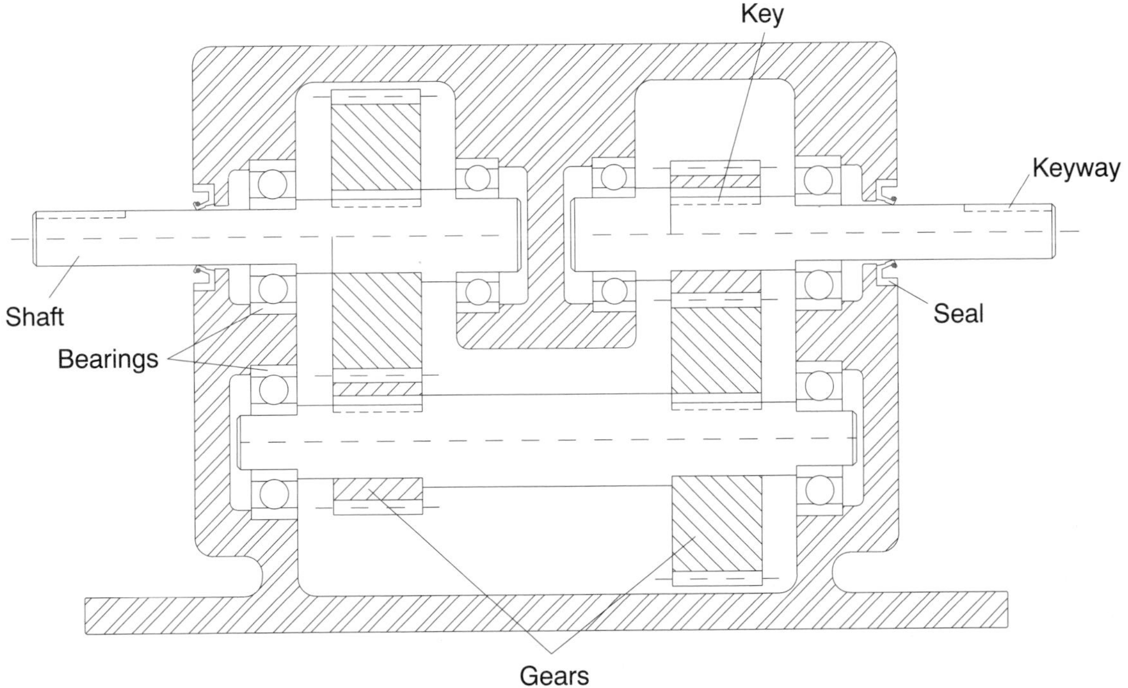

Fig. 1.7 Double reduction gearbox illustrating use of bearings, gears, shafts and seals.

Table 1.3 An overview of the scope of machine elements

Energy conversion	Energy transmission	Energy storage	Locating	Friction reduction	Switching	Sealing	Sensors	Miscellaneous mechanisms
Turbomachinery:	Gears:	Flywheel	Threaded	Rolling element	Clutches:	Dynamic seals:	Motion	Hinges, pivots
Gas turbine	Spur	Springs:	fasteners:	bearings:	Square jaw	Mechanical	Dimensional	Linkages
engines	Helical	Helical	Bolts	Deep groove	Multiple serration	face	Mass	Levers
Rotodynamic	Bevel	Leaf	Nuts and	Cylindrical roller	Sprag	Lip ring	Force	Tools:
pumps and	Worm	Belleville	lock nuts	Needle roller	Roller	Bush	Torque	Cutters
compressors	Conformal	Rubber	Grub screws	Taper roller	Disc	Labyrinth	Power	Shears
Fans	Belts:	Spiral	Studs	Angular contact	Drum	Brush	Pressure:	Drills
Propellers	Flat	Garter	Screws	Self-aligning	Cone	Ferrofluidic	Pitot tubes	Formers
Turbines	Vee	Fluid accumulator	Washers	Thrust	Magnetic	Rim seals	Static tappings	Grips
Internal combustion	Wedge	Gas spring	Nails	Recirculating ball	Synchromesh	O rings	Manometers	Guides
engines:	Round	Reservoir	Pins:	Sliding bearings:	Ratchet and pawl	Packings	Piezoelectric	Followers
Rotary	Synchronous	Pressure vessel	Cylindrical	Plain rubbing	Geneva	Piston rings	Sound	Housings:
Reciprocating	Chains:	Solid mass	Taper	Hydrodynamic	mechanisms	Static seals	Flow:	Frames
Boilers and	Roller	Chemical	Spring	Hydrostatic	Valves	Gaskets	Laser doppler	Casings
combustors	Leaf	Charge	Rivets	Hydrodynamic	Latches	O rings	Hot wire	Enclosures
Electric motors	Conveyor	Torsion bar	Tolerance rings	slideway	Triggers	Gaskets	Ultrasonic	Sprayers
Alternators and	Silent		Expanding bolts	Wheels	Bimetallic strips	Sealants	Level	Shutters
generators	Cables and ropes		Keys:	Rollers			Humidity	Hooks
Solenoids	Couplings:		Flat, round	Brushes			Temperature:	Pulleys
Pneumatic and	Rigid		Profiled				Thermocouples	Handles
hydraulic actuators	Flexible		Gib head				Resistance	Rollers and
Brakes:	Universal		Woodruff				Thermometers	drums
Disc	Cranks		Splines				Pyrometers	Centrifuges
Drum and band	Cams		Circlips				Heat flux:	Filters
Pumps	Power screws and		Snap rings				Thermopile	
Rockets	threads		Clamps				Gardon gauges	
Heat exchangers	Levers and linkages		Retaining rings				Strain and stress	
Guns	Pipes, hoses and		Shoulders				Time	
Dampers and shock	ducts		Spacers				Chemical	
absorbers	Ball screws		Grooves				Composition	
			Fits:					
			Clearance					
			Transition					
			Interference					
			Adhesives					
			Welds					

After personal communication: C. McMahon, Bristol University, 1997.

finally fails or is deliberately scheduled for withdrawal consideration should be given to its disposal or recycling. For instance, are there any environmentally threatening aspects that must be dealt with? Some automotive companies, for example, have developed recycling policies whereby approaching 100% of an old exchanged vehicle is recycled. This has the added benefit of giving the company the opportunity to sell a new product to the customer.

Although the total design model has been adopted for this text, others have been developed and widely referenced and valued. The reader is recommended to review the models proposed by Pahl and Beitz (1977), March (1976) and Cross (1994). For example in the total design model the conceptual design phase includes the development of bulk concepts as well as the process of providing form and layout to the design. The model presented by Pahl and Beitz (1977) models this process as distinct conceptual and embodiment phases. In the embodiment phase the layout and form of the design concept are quantified with consideration of the overall objectives of the specification. The subject of design has been practised for thousands of years, but understanding of the process of design is in its infancy and all these models will no doubt be developed further in the future.

1.4 Text approach

Design as illustrated in Figs 1.1, 1.2, 1.5 and 1.6 can appear somewhat abstract. As with many subjects it is best appreciated by practice. In order for case studies in mechanical design to be developed, the approach adopted here is to introduce a series of relevant mechanical components and techniques such as bearings, shafts, gears, seals, belt and chain drives, clutches and brakes and tolerancing prior to a more detailed approach to the design process. Figure 1.7 shows a double reduction gearbox and, as can be seen, the components described above feature prominently in its make-up. These subjects will be covered in Chapters 2–8, allowing experience of the design of machine elements to be developed. In Chapter 9 the design process will be further explored.

References and sources of information

Cross, N. 1994: *Engineering design methods. Strategies for product design*, 2nd edition. Wiley.

March, L.J. 1976: *The architecture of form*. Cambridge University Press.

Pahl, G. and Beitz, W. 1977, 1988: *Engineering design: a systematic approach*. Design Council.

Pugh, S. 1990: *Total design*. Addison Wesley.

SEED 1985: *Curriculum for design. Engineering undergraduate courses*. Sharing Experience in Engineering Design.

Web sites

http://www.hyster.co.uk (Forklift trucks)

http://www.nissanforklift.com/nissan-html (Forklift trucks)

http://www.strath.ac.uk.~clds 13/SEED/seed.htm (Sharing Experience in Engineering Design)

2 Bearings

The purpose of a bearing is to support a load whilst allowing relative motion between two elements of a machine. The aims of this chapter are to describe the range of bearing technology, to outline the identification of which type of bearing to use for a given application and to introduce the design of hydrodynamic bearings and the selection of standard rolling element bearings.

2.1 Introduction

The term 'bearing' typically refers to contacting surfaces through which a load is transmitted. Lubrication is often required in a bearing to reduce friction between surfaces and to remove heat. Bearings may roll or slide or do both simultaneously. The range of bearing types available is extensive, although they can be broadly split into two categories: sliding bearings, where the motion is facilitated by a thin layer of film of lubricant; and rolling element bearings, where the motion is aided by a combination of rolling motion and lubrication. Figure 2.1 illustrates two of the more commonly known bearings: a deep groove ball bearing and a journal bearing. A general classification scheme for the distinction of bearings is given in Fig. 2.2.

As can be seen from Fig. 2.2 the scope of choice for a bearing is extensive. For a given application it may be possible to use different bearing types: e.g. for a small gas turbine engine either rolling bearings or journal bearings can be used. Figure 2.3 can be used to give guidance for which kind of bearing has the maximum load capacity at a given speed and shaft size; Table 2.1 gives an indication of the performance of the various bearing types for some criteria other than load capacity.

The design of lubricant film sliding bearings is introduced in Section 2.2 and the selection and installation of rolling element bearings is introduced in Section 2.3.

2.2 Sliding bearings

The term sliding bearing refers to bearings where two surfaces move relative to each other without the benefit of rolling contact. The two surfaces slide over each other; this motion can be facilitated by means of a lubricant which gets squeezed by the motion of the components and can generate

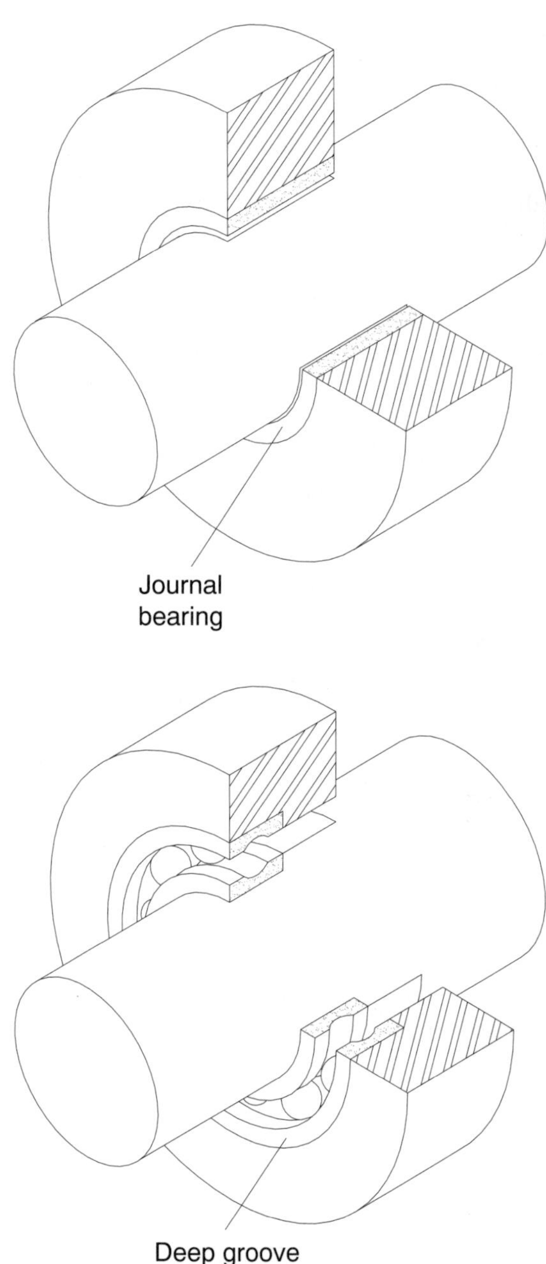

Fig 2.1 Journal and deep groove ball bearings.

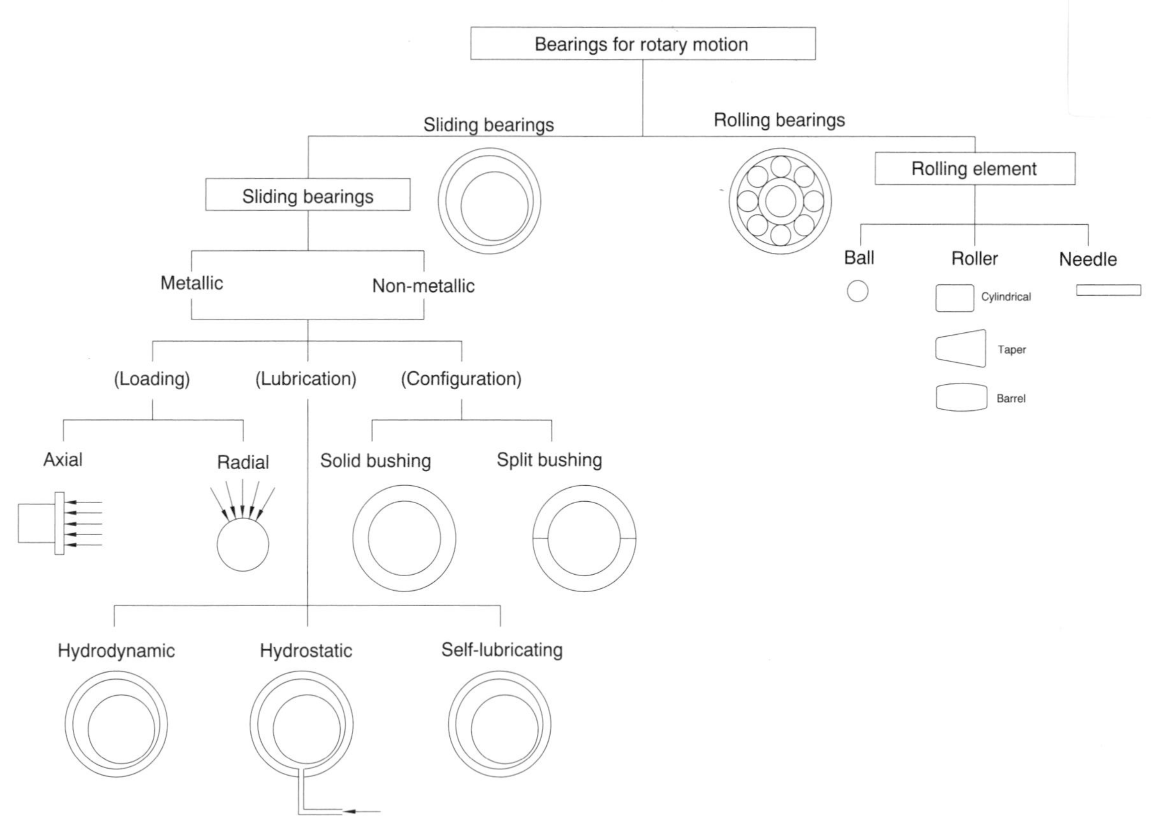

Fig. 2.2 Bearing classification. (Reproduced with adaptations from Hindhede *et al.*, 1983.)

sufficient pressure to separate them, thereby reducing frictional contact and wear.

A typical application of sliding bearings is to allow rotation of a load carrying shaft. The portion of the shaft at the bearing is referred to as the journal, and the stationary part, which supports the load, is called the bearing (*see* Fig. 2.4). For this reason sliding bearings are often collectively referred to as journal bearings, although this term ignores the existence of sliding bearings which support linear translation of components. Another common term is plain surface bearings. This section is principally concerned with bearings for rotary motion and the terms journal and sliding bearing are used interchangeably.

There are three regimes of lubrication for sliding bearings:

1. boundary lubrication
2. mixed film lubrication
3. full film lubrication.

Boundary lubrication typically occurs at low relative velocities between the journal and the bearing surfaces and is characterised by actual physical contact. The surfaces, even if ground to a low value of surface roughness, will still consist of a series of peaks and troughs, as illustrated schematically in Fig. 2.5. Although some lubricant may be present, the pressures generated within it are not significant and the relative motion of the surfaces brings the corresponding peaks periodically into contact. Mixed film lubrication occurs when the relative motion between the surfaces is sufficient to generate high enough pressures in the lubricant film which can partially separate the surfaces for periods of time. There is still contact in places between the two components. Full film lubrication occurs at higher relative velocities. Here the motion of the surface generates high pressures in the lubricant which separate the two components and journals can 'ride' on a wedge of fluid. All of these types of lubrication can be encountered in a bearing without external pressuring of the bearing. If lubricant under pressure is supplied to the bearing it is called a hydrostatic bearing.

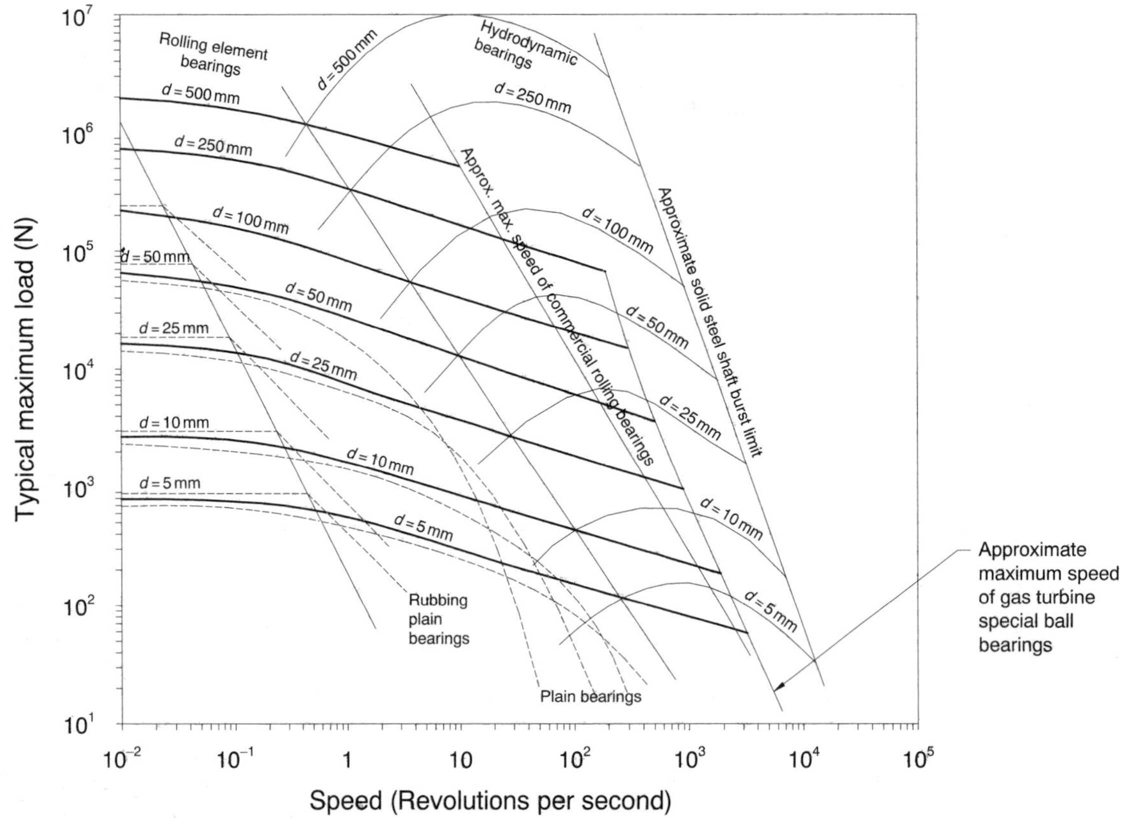

Fig. 2.3 Bearing type selection by load capacity and speed. (Reproduced courtesy of Neale, 1993.)

The performance of a sliding bearing differs markedly depending on which type of lubrication is physically occurring. This is illustrated in Fig. 2.6, which shows the variation of the coefficient of friction with a group of variables called the bearing parameter which is defined by

$$\frac{\mu N}{P} \tag{2.1}$$

where

μ = viscosity of lubricant (Pa · s)
N = speed (for this definition normally in rpm)
P = load capacity given by $P = W/LD$ (N/m^2)
W = applied load (N)
L = bearing length (m)
D = journal diameter (m)

The bearing parameter, $\mu N/P$, groups several of the bearing design variables into one number. Normally, of course, a low coefficient of friction is desirable. In general, boundary lubrication is used for slow-speed applications where the surface speed is less than approximately 1.5 m/s. Mixed film lubrication is rarely used as it is difficult to quantify the actual value of the coefficient of friction (note the steep gradient in Fig. 2.6 for this zone). The design of boundary lubricated bearings is outlined in Section 2.2.2 and full film hydrodynamic bearings in Section 2.2.3.

2.2.1 Lubricants

As can be seen from Fig. 2.6, bearing performance is dependent on the type of lubrication occurring and the viscosity of the lubricant. The viscosity is a measure of a fluid's resistance to shear. Lubricants can be solid, liquid or gaseous, although the most commonly known are oils and greases. The principal classes of liquid lubricants are mineral oils and synthetic oils. Their viscosity is highly dependent on temperature and are typically solid at −35°C, thin as paraffin at 100°C, and they burn above 240°C. Many additives are used to affect their performance. For example, EP (extreme pressure) additives add fatty acids and other compounds to the oil which attack the metal surfaces to form 'contaminant' layers, which

Table 2.1 Comparison of bearing performance for continuous rotation

Bearing type	Accurate radial location	Combined axial and radial load capability	Low starting torque capability	Silent running	Standard parts available	Lubrication simplicity
Rubbing plain bearings (non-metallic)	Poor	Some in most cases	Poor	Fair	Some	Excellent
Porous metal plain bearings, oil impregnated	Good	Some	Good	Excellent	Yes	Excellent
Fluid film hydrodynamic bearings	Fair	No; separate thrust bearing needed	Good	Excellent	Some	Usually requires a recirculation system
Hydrostatic bearings	Excellent	No; separate thrust bearing needed	Excellent	Excellent	No	Poor; special system needed
Rolling bearings	Good	Yes in most cases	Very good	Usually satisfactory	Yes	Good when grease lubricated

Reproduced courtesy of Neale (1993).

protect the surfaces and reduce friction even when the oil film is squeezed out by high contact loads. Greases are oils mixed with soaps to form a thicker lubricant which can be retained on surfaces.

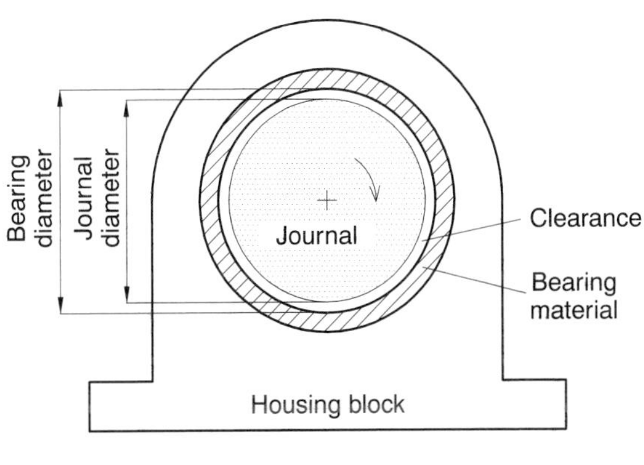

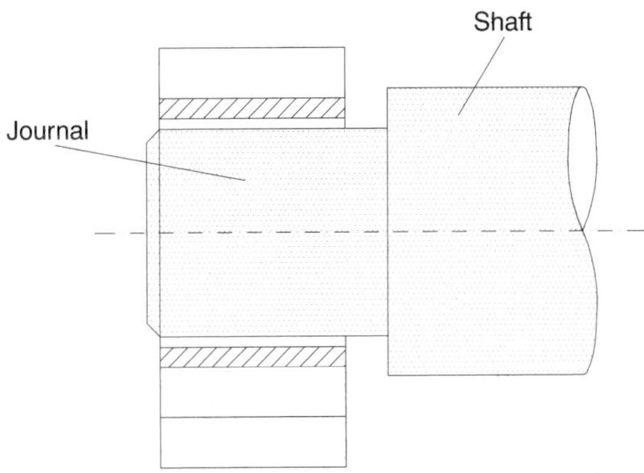

Fig. 2.4 A plain surface, sliding or journal bearing.

The viscosity variation with temperature of oils has been standardised and oils are available with a class number, for example SAE 10, SAE 20, SAE 30, SAE 40, SAE 5W, SAE 10W etc. The origin of this identification system developed by the Society of Automotive Engineers was to class oils for general purpose use and winter use, the 'W' signifying the latter. The lower the numerical value, the thinner or less viscous the oil. A multigrade oil, e.g. SAE 10W/40, is formulated to meet the viscosity requirements of two oils giving some of the benefits of the constituent parts. An equivalent identification system is also available from the International Organisation for Standardisation (ISO 3448).

2.2.2 Design of boundary lubricated bearings

As described in Section 2.2, the journal and bearing surfaces in a boundary lubricated bearing are in direct contact in places. These bearings are typically used for very low-speed applications such as bushes and linkages, where their simplicity and compact nature are advantageous. General considerations in the design of a boundary lubricated bearing are: the coefficient of friction (both static and dynamic), the load capacity, the relative velocity between the stationary and moving components, the operating temperature, wear limitations and the production capability. A useful measure in the design of boundary lubricated bearings is the PV factor (load capacity $\times$ peripheral speed) which indicates the ability of the bearing material to accommodate the frictional energy generated

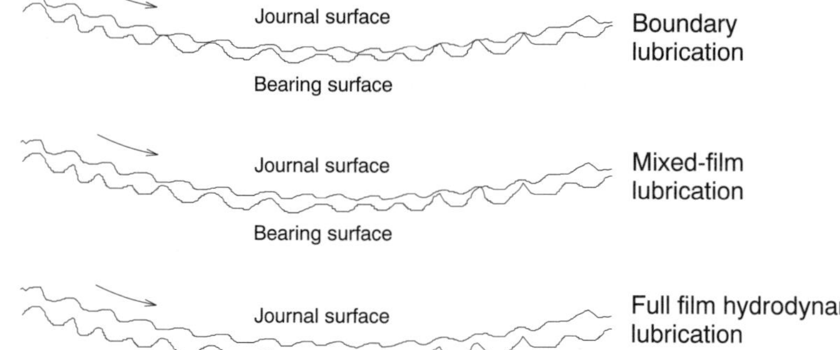

Fig. 2.5 A schematic representation of the surface roughness for sliding bearings and the relative position depending on the type of lubrication.

in the bearing. At the limiting PV value the temperature will be unstable and failure will occur rapidly. A practical value for PV is half the limiting PV value. Values for PV for various bearing materials are given in Table 2.2. The preliminary design of a boundary lubricated bearing essentially consists of setting the bearing proportions, its length and its diameter, and selecting the bearing material such that an acceptable PV value is obtained. Common practice is to set the length to diameter ratio between 0.5 and 1.5.

Example A bearing is to be designed to carry a radial load of 700 N for a shaft of diameter 25 mm running at a speed of 75 rpm (*see* Fig. 2.7). The bearing length is 25 mm. Calculate the PV value and by comparison with the available materials listed in Table 2.2 determine a suitable bearing material.

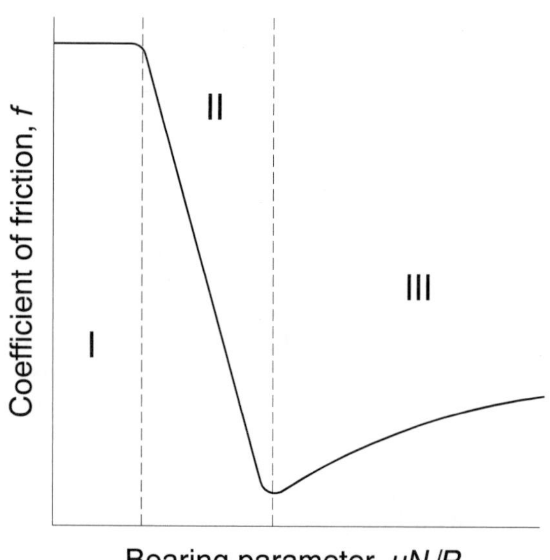

Fig. 2.6 Variation of bearing performance with lubrication. I: boundary; II: mixed film; III: full film hydrodynamic.

Solution

$$P = \frac{W}{LD} = \frac{700}{0.025 \times 0.025} = 1.12\,\text{MN/m}^2$$

$$V = \left(\frac{2\pi}{60}\right)N\left(\frac{D}{2}\right)$$

$$= 0.1047 \times 75\left(\frac{0.025}{2}\right) = 0.09816\,\text{m/s}$$

$$PV = 0.11\,(\text{MN/m}^2)(\text{m/s})$$

A material with a PV factor double or more than this, such as filled PTFE (limiting PV value up to 0.35) or PTFE with filler bonded to a steel backing (limiting PV value up to 1.75), would give acceptable performance.

2.2.3 Design of full film hydrodynamic bearings

In a full film hydrodynamic bearing, the load on the bearing is supported on a continuous film of lubricant so that no contact between the bearing and the rotating journal occurs. The motion of the journal inside the bearing creates the necessary pressure to support the load. Hydrodynamic bearings can consist of a full circumferential surface or a partial surface around the journal. The methodology and design charts presented here are applicable to a full circumferential bearing.

The design of a journal bearing includes the specification of the journal radius r, the radial clearance c, the axial length of the bearing surface L, the type of lubricant and its viscosity μ, the journal speed N and the load W. Values for the speed, the load and possibly the journal radius are usually

Table 2.2 Characteristics of rubbing materials

Material	Maximum load capacity P (MN/m^2)	Limiting PV value (MN/m^2 per second)	Maximum operating temperature (°C)	Coefficient of friction	Coefficient of expansion ($\times 10^{-6}$/°C)	Comments	Typical application
Carbon/graphite	1.4–2	0.11	350–500	0.1–0.25 dry	2.5–5.0	For continuous dry operation	Food and textile machinery
Carbon/graphite with metal	3.4	0.145	130–350	0.1–0.35 dry	4.2–5		
Graphite impregnated metal	70	0.28–0.35	350–600	0.1–0.15 dry	12–13		
Graphite/ thermosetting resin	2	0.35	250	0.13–0.5 dry	3.5–5	Suitable for sea water operation	
Reinforced thermosetting plastics	35	0.35	200	0.1–0.4 dry	25–80		Roll-neck bearings
Thermoplastic material without filler	10	0.035	100	0.1–0.45 dry	100		Bushes and thrust washers
Thermoplastic with filler or metal backed	10–14	0.035–0.11	100	0.15–0.4 dry	80–100		Bushes and thrust washers
Thermoplastic material with filler bonded to metal back	140	0.35	105	0.2–0.35 dry	27	With initial lubrication only	For conditions of intermittent operation or boundary lubrication, e.g. ball joints, suspension, steering
Filled PTFE	7	Up to 0.35	250	0.05–0.35 dry	60–80	Glass, mica, bronze, graphite	For dry operations where low friction and wear required
PTFE with filler, bonded to steel backing	140	Up to 1.75	280	0.05–0.3 dry	20	Sintered bronze bonded to steel backing impregnated with PTFE/lead	Aircraft controls, linkages, gearbox, clutch, conveyors, bridges
Woven PTFE reinforced and bonded to metal backing	420	Up to 1.6	250	0.03–0.3	–	Reinforcement may be interwoven glass fibre or rayon	Aircraft and engine controls, linkages, engine mountings, bridge bearings

Reproduced courtesy of Neale (1993).

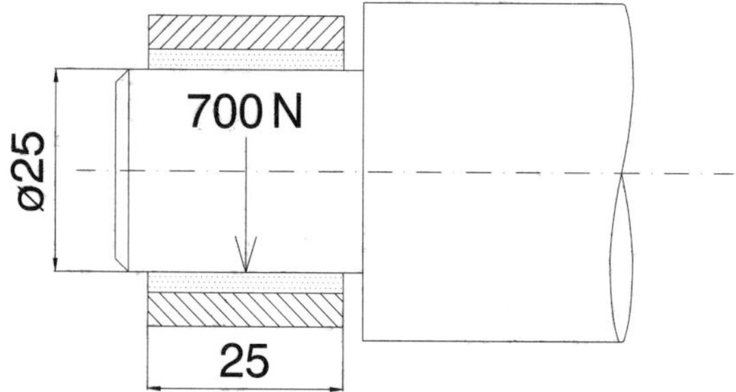

Fig. 2.7 Boundary lubricated bearing design example.

specified by the machine requirements and stress and deflection considerations. As such the journal bearing design consists of the determination of the radial clearance, the bearing length and the lubricant viscosity. The design process for a journal bearing is usually iterative. Trial values for the clearance, the length and the viscosity are chosen, various performance criteria calculated and the process repeated until a satisfactory or optimised design is achieved. Criteria for optimisation may be minimising of the frictional loss, minimising the lubricant temperature rise, minimising the lubricant supply, maximising the load capability, and minimising production costs. The optimisation process is best performed using bearing design software.

The clearance between the journal and the bearing depends on the nominal diameter of the journal, the precision of the machine, surface roughness and thermal expansion considerations. An overall guideline is for the radial clearance, c, to be in the range $0.001r < c < 0.002r$, where r is the nominal bearing radius ($0.001D < 2c < 0.002D$). Figure 2.8 shows values for the recommended diametral clearance ($2c$) as a function of the journal diameter and rotational speed for steadily loaded bearings.

For a given combination of r, c, L, μ, N and W, the performance of a journal bearing can be calculated. This requires determining the pressure distribution in the bearing, the minimum film thickness h_0, the location of the minimum film thickness $\theta_{p_{\max}}$, the coefficient of friction f, the lubricant flow Q, the maximum film pressure $p_{\max}$ and the temperature rise ΔT.

The pressure distribution in a journal bearing (*see* Fig. 2.9) can be determined by solving the relevant form of the Navier–Stokes fluid flow equations, which in the reduced form for journal bearings is called the Reynolds equation. This is given in its steady form as

$$\frac{\partial}{\partial x}\left(\frac{h^3}{\mu}\frac{\partial p}{\partial x}\right) + \frac{\partial}{\partial z}\left(\frac{h^3}{\mu}\frac{\partial p}{\partial z}\right) = 6V\frac{\partial h}{\partial x} + 6h\frac{\partial V}{\partial x} \quad (2.2)$$

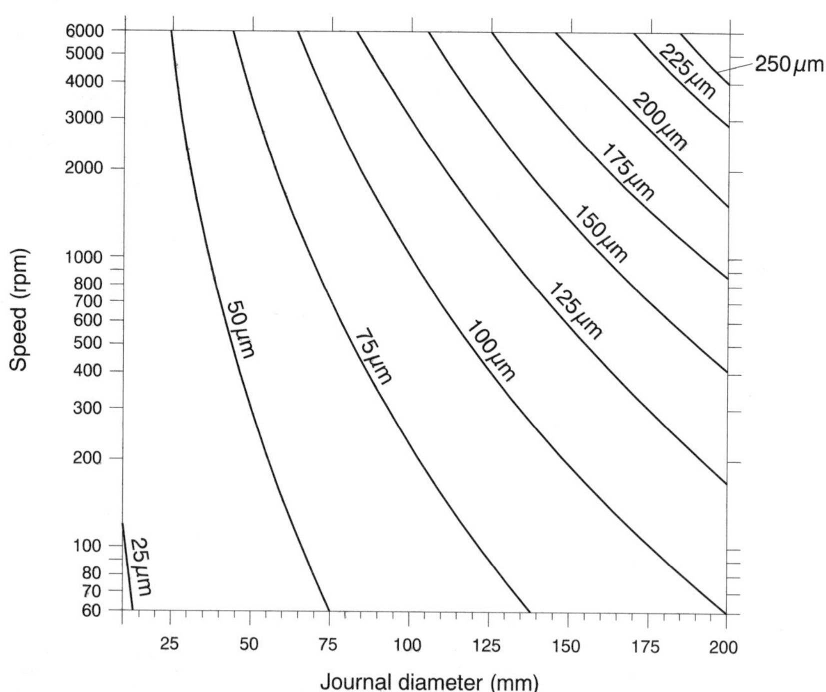

Fig. 2.8 Minimum recommended values for the diametral clearance ($2c$) for steadily loaded journal bearings. (Reproduced from Welsh, 1983.)

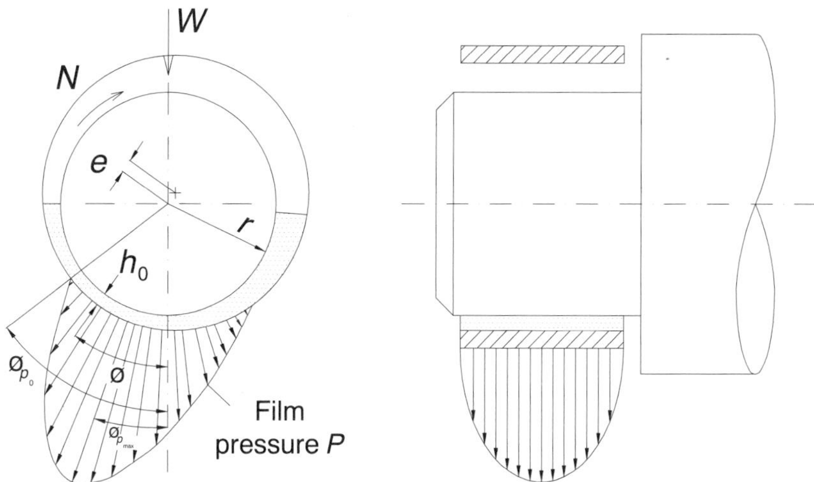

Fig. 2.9 Pressure distribution in a full film journal bearing. Beyond h_0, the minimum film thickness, the pressure terms go negative and the film is ruptured.

where h is the film thickness, μ is the dynamic viscosity, p is the fluid pressure, V is the journal velocity and (x, y, z) are rectangular coordinates. The Reynolds equation can be solved by approximate mathematical methods or numerically. Once the pressure distribution has been established the journal performance can be determined in terms of the bearing load capacity, frictional losses, lubricant flow requirements and the lubricant temperature rise.

If the designer wishes to avoid the direct solution of equation 2.2, use can be made of a series of design charts. These were originally produced by Raimondi and Boyd (1958a, b, c) who used an iterative technique to solve the Reynolds equations. These charts give the film thickness, coefficient of friction, lubricant flow, lubricant side flow ratio, minimum film thickness location, maximum pressure ratio, maximum pressure ratio position and film termination angle versus the Sommerfield number (*see* Figs 2.14–2.21).

The Sommerfield number (S), which is also known as the bearing characteristic number, is defined as

$$S = \left(\frac{r}{c}\right)^2 \frac{\mu N_s}{P} \tag{2.3}$$

where

r is the journal radius (m)
c is the radial clearance (m)
μ is the absolute viscosity (Pa · s)
N_s is the journal speed (revolutions per second, rps)
$P = W/LD$ is the load per unit of projected bearing area (N/m^2)
W is the load on the bearing (N)
D is the journal diameter (m)
L is the journal bearing length (m).

It encapsulates all the parameters usually defined by the designer. Great care needs to be taken with units in the design of journal bearings. Many design charts have been produced using English units (psi, reyn, Btu etc.). As long as a consistent set of units is maintained, use of the charts will yield sensible results. In particular note the use of revolutions per second in the definition of speed in the Sommerfield number.

Consider the journal shown in Fig. 2.10. As the journal rotates it will pump lubricant in a clockwise direction. The lubricant is pumped into a wedge-shaped space and the journal is forced over to the opposite side. The angular position where the lubricant film is at its minimum thickness h_0 is called the attitude angle. The centre of the journal is displaced from the centre of the bearing by a distance e called the eccentricity:

$$h_0 = c - e \tag{2.4}$$

The ratio of the eccentricity e to the radial clearance c is called the eccentricity ratio:

$$\varepsilon = \frac{e}{c} \tag{2.5}$$

The relationship between the film thickness, radial clearance and eccentricity ratio is therefore

$$\frac{h_0}{c} = 1 - \varepsilon \tag{2.6}$$

One of the assumptions made in the analysis of Raimondi and Boyd is that the viscosity of the

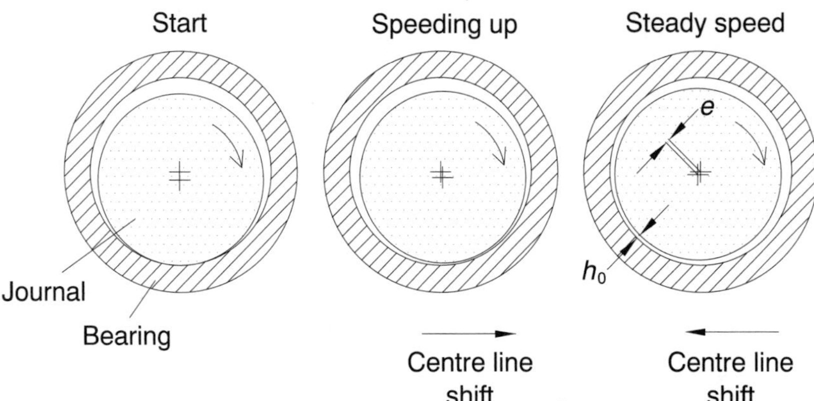

Fig. 2.10 Full film hydrodynamic bearing motion from start up.

lubricant is constant as it passes through the bearing. But work is done on the lubricant in the bearing and the temperature of the lubricant leaving the bearing zone will be higher than the entrance value.

Figure 2.11 shows the variation of viscosity with temperature for some of the SAE and ISO defined lubricants, and it can be seen that the value of the viscosity for a particular lubricant is highly depen-

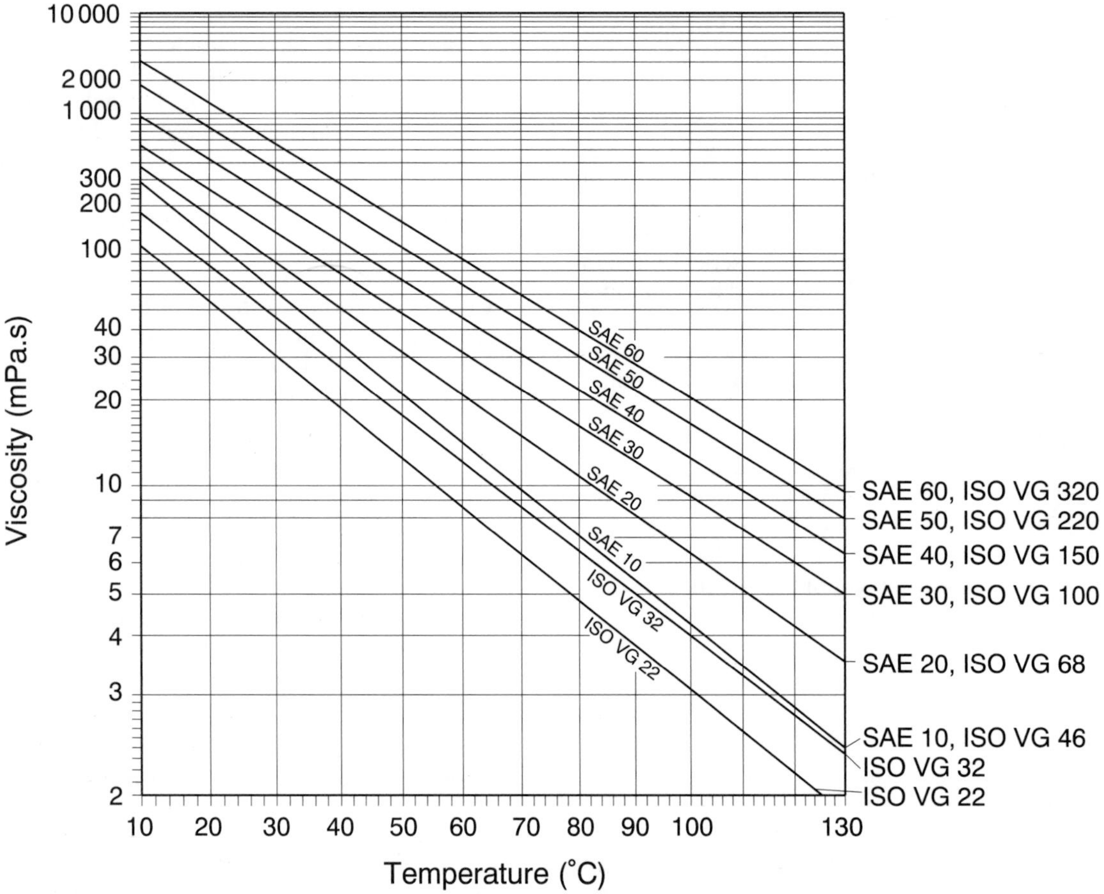

Fig. 2.11 Variation of absolute viscosity with temperature for various lubricants.

dent on the temperature. Some of the lubricant entrained into the bearing film emerges as side flow which carries away some of the heat. The remainder flows through the load-bearing zone and carries away the remainder of the heat generated. The temperature used for determining the viscosity can be taken as the average of the inlet and exit lubricant temperatures:

$$T_{av} = T_1 + \frac{T_2 - T_1}{2} = T_1 + \frac{\Delta T}{2} \qquad (2.7)$$

where T_1 is the temperature of the lubricant supply and T_2 is the temperature of the lubricant leaving the bearing.

One of the parameters that needs to be determined is the bearing lubricant exit temperature T_2. This is a trial-and-error or iterative process. A value of the temperature rise, ΔT, is guessed, the viscosity for a standard oil determined corresponding to this value and the analysis performed. If the temperature rise calculated by the analysis does not correspond closely to the guessed value the process should be repeated using an updated value for the temperature rise until the two match. If the temperature rise is unacceptable it may be necessary to try a different lubricant or modify the bearing configuration.

Given the length, the journal radius, the radial clearance, the lubricant type and its supply temperature, the steps for determining the various bearing operating parameters are listed below.

1. Calculate the load capacity $P = W/LD$. Values for P should typically be between $0.34\,\mathrm{MN/m^2}$ for light machinery and $13.4\,\mathrm{MN/m^2}$ for heavy machinery.
2. Estimate a value for the temperature rise ΔT across the bearing. The value taken for the initial estimate is relatively unimportant. As a guide, a value of $\Delta T = 10°\mathrm{C}$ is generally a good starting guess. This value can be increased for high-speed bearings and for low bearing clearances.
3. Determine the average lubricant temperature $T_{av} = T_1 + \Delta T/2$ and find the corresponding value for the viscosity for the chosen lubricant.
4. Calculate the Sommerfield number, $S = (r/c)^2 \mu N_s/P$, and the length to diameter ratio.
5. Use the charts (Figs 2.15–2.17) to determine values for the coefficient of friction variable, the total lubricant flow variable, and the ratio of the side flow to the total lubricant flow with the values for the Sommerfield number and the L/D ratio.
6. Calculate the temperature rise of the lubricant through the bearing using

   $$\Delta T = \frac{8.30 \times 10^{-6} P}{1 - \frac{1}{2}(Q_s/Q)} \times \frac{(r/c)f}{[Q/(rcN_s L)]} \qquad (2.8)$$

   If this calculated value does not match the estimated value for ΔT to within say 0.5°C, repeat the procedure from step 3 using the updated value of the temperature rise to determine the average lubricant temperature.
7. If the value for the temperature rise across the bearing has converged, values for the total lubricant flow rate Q, the side flow rate Q_s and the coefficient of friction f can be calculated. The journal bearing must be supplied with the value of the total lubricant calculated in order for it to perform as predicted by the charts.
8. The charts given in Figs 2.14–2.21 can be used to determine values for the maximum film thickness, the maximum film pressure ratio, the location of the maximum film pressure and its terminating angular location as required.
9. The torque required to overcome friction in the bearing can be calculated by

   $$\mathrm{Torque} = fWr \qquad (2.9)$$

10. The power lost in the bearing is given by

    $$\mathrm{Power} = \omega \times \mathrm{Torque}$$
    $$= 2\pi N_s \times \mathrm{Torque} \qquad (2.10)$$

    where ω is the angular velocity (rad/s).
11. Specify the surface roughness for the journal and bearing surfaces. A ground journal with an arithmetic average surface roughness of 0.4–0.8 μm is recommended for bearings of good quality. For high-precision equipment both surfaces can be lapped or polished to give a surface roughness of 0.2 to 0.4 μm. The specified roughness should be less than the minimum film thickness.
12. Design the recirculation and sealing system for the bearing lubricant.

Lubricant oil selection is a function of speed or compatibility with other lubricant requirements. Generally as the design speed rises, oils with a lower viscosity should be selected.

The chart for the minimum film thickness variable (Fig. 2.14) indicates the optimal operating region for minimal friction and maximum load capability. If the operating Sommerfield number and L/D ratio combination do not fall within this zone, then it is likely that the bearing design can be improved by altering the values for c, L, D, the lubricant type and the operating temperature, as appropriate.

Example A full journal bearing has a nominal diameter of 50.0 mm and a bearing length of 25.0 mm (*see* Fig. 2.12). The bearing supports a load of 3000 N, and the journal design speed is 3000 rpm. The radial clearance has been specified as 0.04 mm. An SAE 10 oil has been chosen and the lubricant supply temperature is 50°C.

Find the temperature rise of the lubricant, the lubricant flow rate, the minimum film thickness, the torque required to overcome friction and the heat generated in the bearing.

Solution The primary data are $D = 50.0$ mm, $L = 25$ mm, $W = 3000$ N, $N = 3000$ rpm, $c = 0.04$ mm, SAE 10, $T_1 = 50$°C.

Guess a value for the lubricant temperature rise ΔT across the bearing to be, say, $\Delta T = 20$°C:

$$T_{av} = T_1 + \frac{\Delta T}{2} = 50 + \frac{20}{2} = 60°\text{C}$$

From Fig. 2.11 for SAE 10 at 60°C, $\mu = 0.014$ Pa · s.

$N_s = 3000/60 = 50$ rps, $L/D = 25/50 = 0.5$.

$$P = \frac{W}{LD} = \frac{3000}{0.025 \times 0.05} = 2.4 \times 10^6 \text{ N/m}^2$$

$$S = \left(\frac{r}{c}\right)^2 \frac{\mu N_s}{P}$$

$$= \left(\frac{25 \times 10^{-3}}{0.04 \times 10^{-3}}\right)^2 \frac{0.014 \times 50}{2.4 \times 10^6} = 0.1139$$

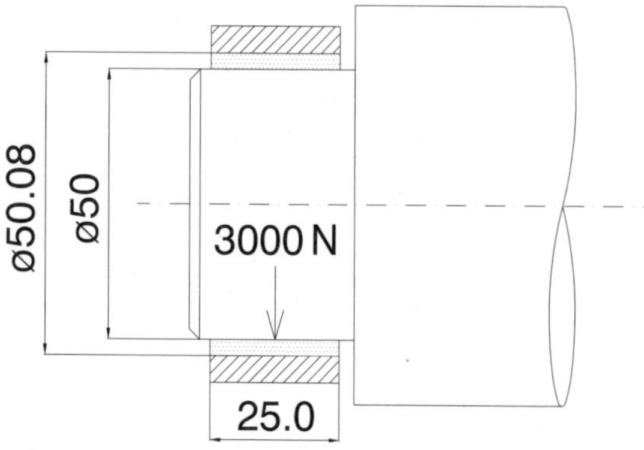

Fig. 2.12 Bearing design example.

From Fig. 2.15 with $S = 0.1139$ and $L/D = 0.5$, $(r/c)f = 3.8$.

From Fig. 2.16 with $S = 0.1139$ and $L/D = 0.5$, $Q/(rcN_sL) = 5.34$.

From Fig. 2.17 with $S = 0.1139$ and $L/D = 0.5$, $Q_s/Q = 0.852$.

The value of the temperature rise of the lubricant can now be calculated using equation 2.8:

$$\Delta T = \frac{8.3 \times 10^{-6} P}{1 - 0.5(Q_s/Q)} \times \frac{(r/c)f}{[Q/(rcN_sL)]}$$

$$= \frac{8.3 \times 10^{-6} \times 2.4 \times 10^6}{1 - (0.5 \times 0.852)} \times \frac{3.8}{5.34} = 24.70°\text{C}$$

As the value calculated for the lubricant temperature rise is significantly different to the estimated value, it is necessary to repeat the above calculation but using the new improved estimate for determining the average lubricant temperature.

Using $\Delta T = 24.70$°C to calculate T_{av} gives

$$T_{av} = 50 + \frac{24.70}{2} = 62.35°\text{C}$$

Repeating the procedure using the new value for T_{av} gives

$\mu = 0.0136$ Pa · s

$S = 0.1107$

From Fig. 2.15 with $S = 0.1107$ and $L/D = 0.5$, $(r/c)f = 3.7$.

From Fig. 2.16 with $S = 0.1107$ and $L/D = 0.5$, $Q/(rcN_sL) = 5.35$.

From Fig. 2.17 with $S = 0.1107$ and $L/D = 0.5$, $Q_s/Q = 0.856$.

$$\Delta T = \frac{8.3 \times 10^{-6} \times 2.4 \times 10^6}{1 - (0.5 \times 0.856)} \times \frac{3.7}{5.35} = 24.08°\text{C}$$

$$T_{av} = 50 + \frac{24.08}{2} = 62.04°\text{C}$$

This value for T_{av} is close to the previous calculated value, suggesting that the solution has converged. For $T_{av} = 62.04$°C, $\mu = 0.0136$ Pa · s and $S = 0.1107$.

The other parameters can now be found:

$$Q = rcN_sL \times 5.35 = 25 \times 0.04 \times 50 \times 25 \times 5.35$$

$$= 6688 \text{ mm}^3/\text{s}$$

From Fig. 2.14, $h_0/c = 0.22$. $h_0 = 0.0088$ mm.

$f = 3.7 \times (c/r) = 3.7 \times (0.04/25) = 0.00592$

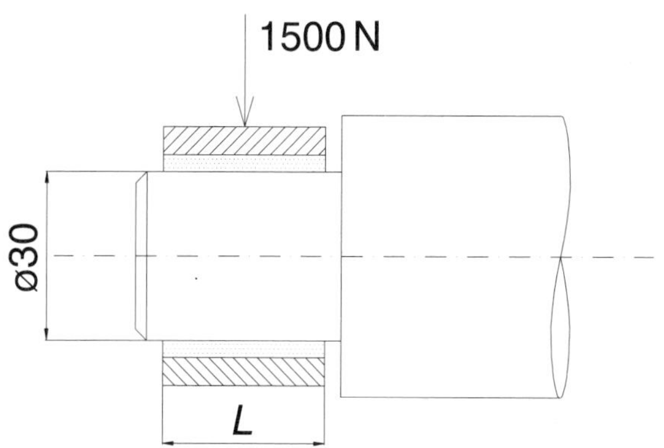

Fig. 2.13 Bearing design example.

The torque is given by

$$\text{Torque} = fWr = 0.00592 \times 3000 \times 0.025$$
$$= 0.444\,\text{N}\cdot\text{m}$$

The power dissipated in the bearing is given by

$$\text{Power} = 2\pi \times \text{Torque} \times N_s = 139.5\,\text{W}$$

Example A full journal bearing is required for a 30 mm diameter shaft rotating at 1200 rpm, supporting a load of 1500 N (*see* Fig. 2.13). An SAE 10 lubricant oil has been selected and, from previous experience, the lubrication system available is capable of delivering the lubricant to the bearing at 40°C. Select an appropriate radial clearance and length for the bearing and determine the temperature rise of the lubricant, the total lubricant flow rate required and the power absorbed in the bearing. Check that your chosen design operates within the optimum zone indicated by the dotted lines on the minimum film thickness ratio chart (Fig. 2.14).

Solution The primary design information is: $D = 30.0\,\text{mm}$, $W = 1500\,\text{N}$, $N = 1200\,\text{rpm}$, SAE 10, $T_1 = 40°\text{C}$.

From Fig. 2.8, for a speed of 1200 rpm and a nominal journal diameter of 30 mm, a suitable value for the diametral clearance is 0.05 mm. The radial clearance is therefore 0.025 mm. The next step is to set the length of the bearing. Typical values for L/D ratios are between 0.5 and 1.5. Here a value for the ratio is arbitrarily set as $L/D = 1$, i.e. $L = 30\,\text{mm}$.

The procedure for determining the average lubricant temperature is the same as for the previous example. As a first estimate the temperature rise of the lubricant is taken as 10°C and the

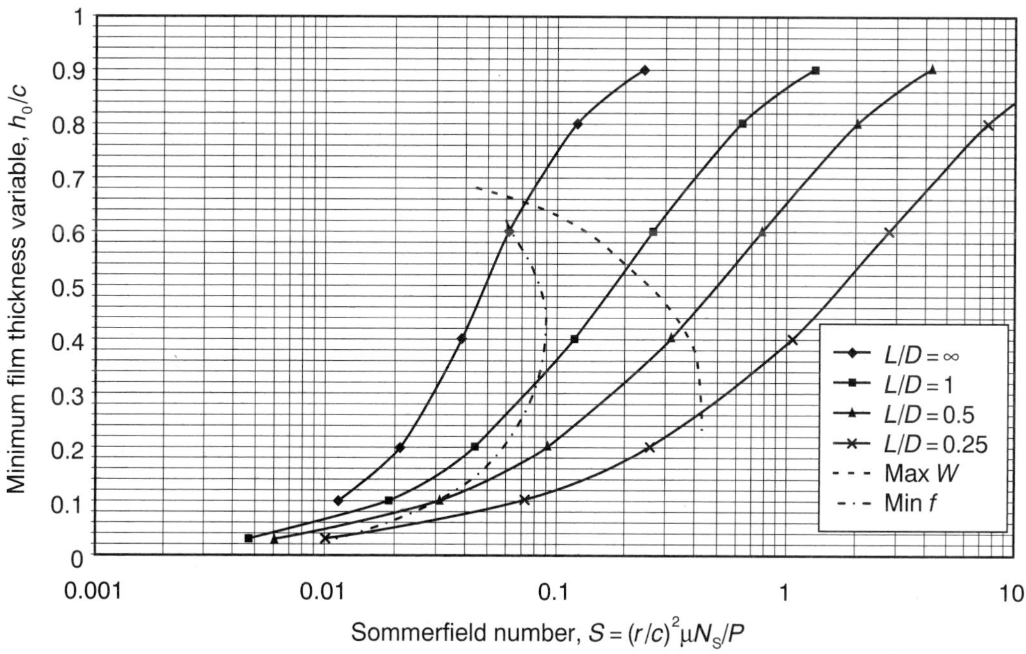

Fig. 2.14 Chart for the minimum film thickness variable (h_0/c) versus the Sommerfield number. (Data from Raimondi and Boyd, 1958c.)

results of the iterative procedure are tabulated below:

$$P = \frac{W}{LD} = \frac{1500}{0.03 \times 0.03} = 1.667 \times 10^6\ \mathrm{N/m^2}$$

$$S = \left(\frac{r}{c}\right)^2 \frac{\mu N_s}{P}$$

$$= \left(\frac{15 \times 10^{-3}}{0.025 \times 10^{-3}}\right)^2 \frac{\mu 20}{1.667 \times 10^6} = 4.319\mu$$

$$T_{av} = T_1 + \frac{\Delta T}{2} = 40 + \frac{10}{2} = 45^\circ\mathrm{C}$$

Temperature rise, ΔT (°C)	10	15.90	13.76	13.76
Average lubricant temperature, T_{av} (°C)	45	47.95	46.88	46.88
Average lubricant viscosity, μ (Pa · s)	0.028	0.023	0.024	Converged
Sommerfield number, S	0.1209	0.09934	0.1037	
Coefficient of friction variable, $(r/c)f$	3.3	2.8	2.8	
Flow variable, $Q/(rcN_sL)$	4.35	4.4	4.4	
Side flow to total flow ratio, Q_s/Q	0.68	0.72	0.72	

The total lubricant flow rate required is given by

$$Q = rcN_sL \times 4.4$$

$$= 15 \times 0.025 \times 20 \times 30 \times 4.4$$

$$= 990\ \mathrm{mm^3/s}$$

With $S = 0.1037$ and $L/D = 1$ the design selected is within the optimum operating zone for minimum friction and optimum load capacity indicated in Fig. 2.14.

$$\text{Friction factor, } f = 2.8 \times (c/r)$$

$$= 2.8 \times 0.025/15$$

$$= 0.004667$$

$$\text{Torque} = fWr = 0.004667 \times 1500 \times 0.015$$

$$= 0.105\ \mathrm{N \cdot m}$$

$$\text{Power} = 2\pi N_s \times \text{Torque} = 13.19\ \mathrm{W}$$

For values of the L/D ratio other than those shown in the charts given in Figs 2.14–2.21, values for the various parameters can be found

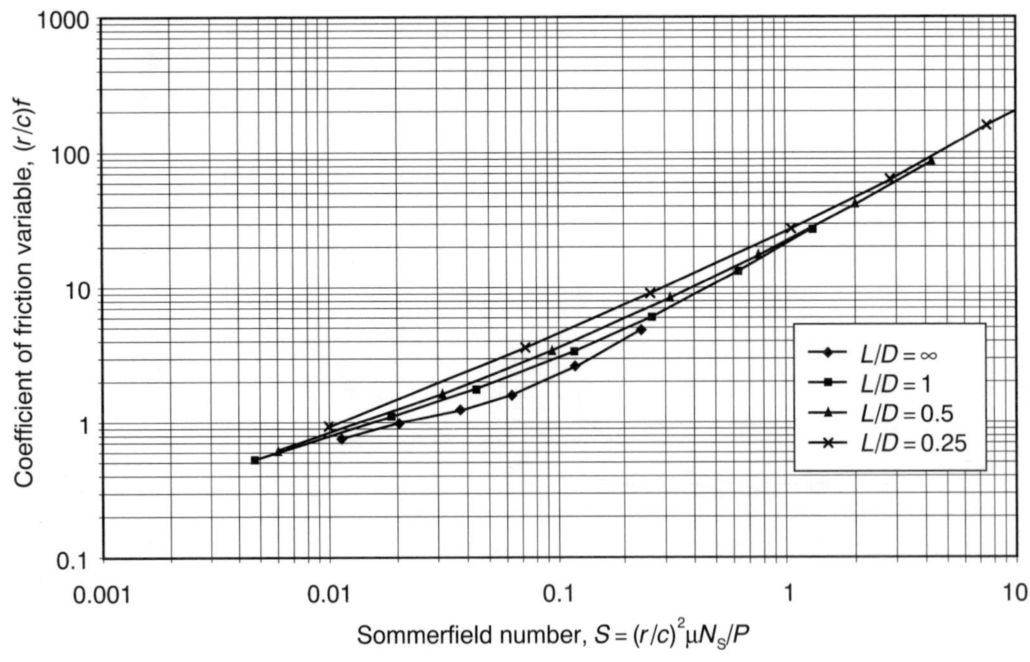

Fig. 2.15 Chart for determining the coefficient of friction variable, $(r/c)f$. (Data from Raimondi and Boyd, 1958c.)

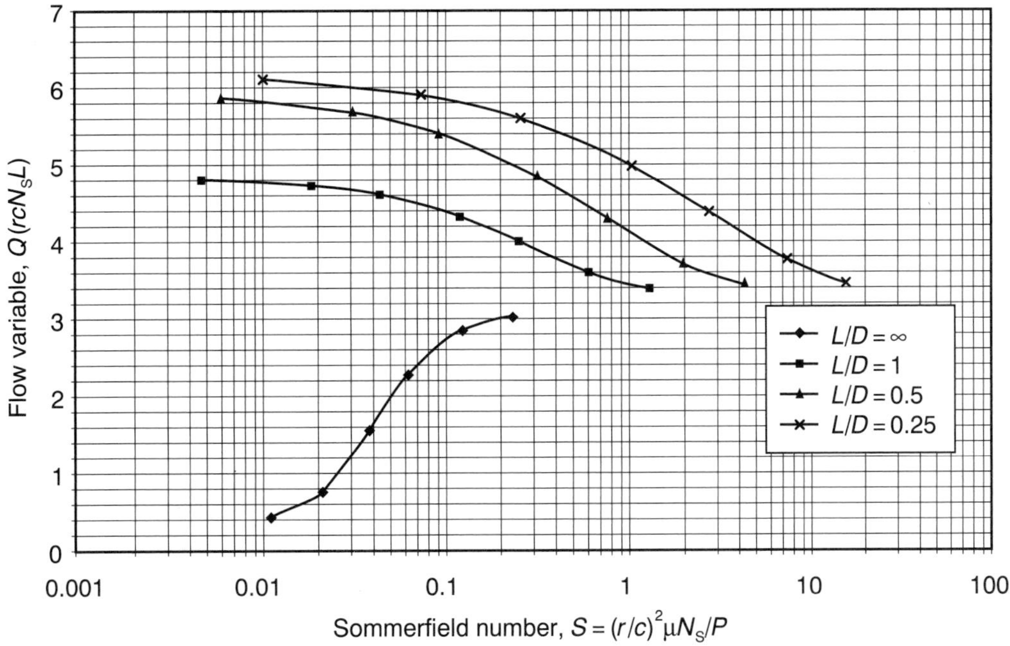

Fig. 2.16 Chart for determining the flow variable, $Q/(rcN_sL)$. (Data from Raimondi and Boyd, 1958c.)

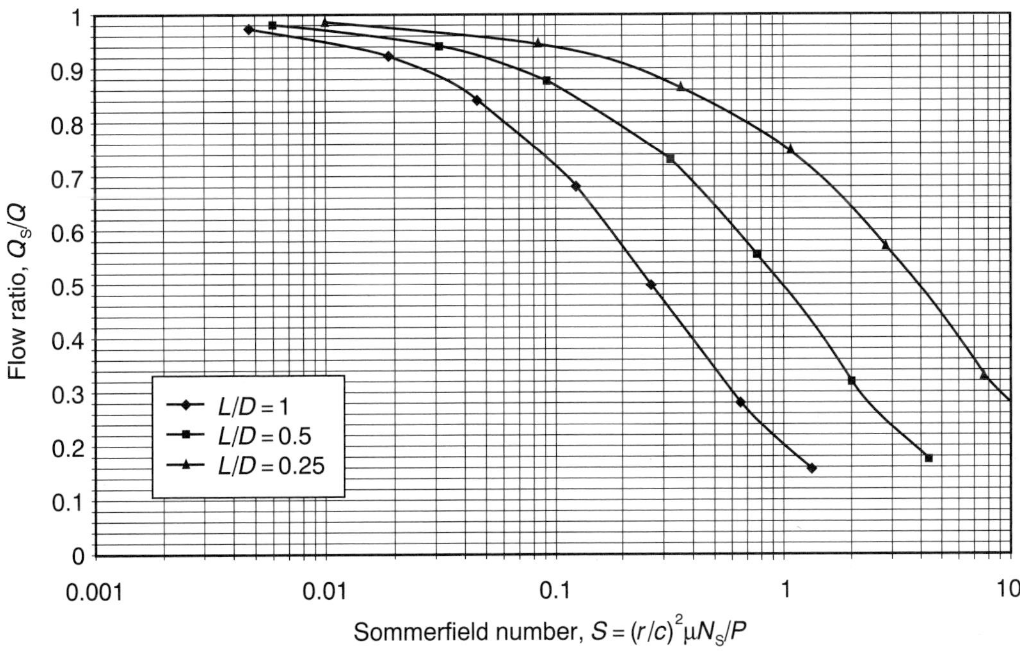

Fig. 2.17 Chart for determining the ratio of side flow, Q_s, to total flow, Q. (Data from Raimondi and Boyd, 1958c.)

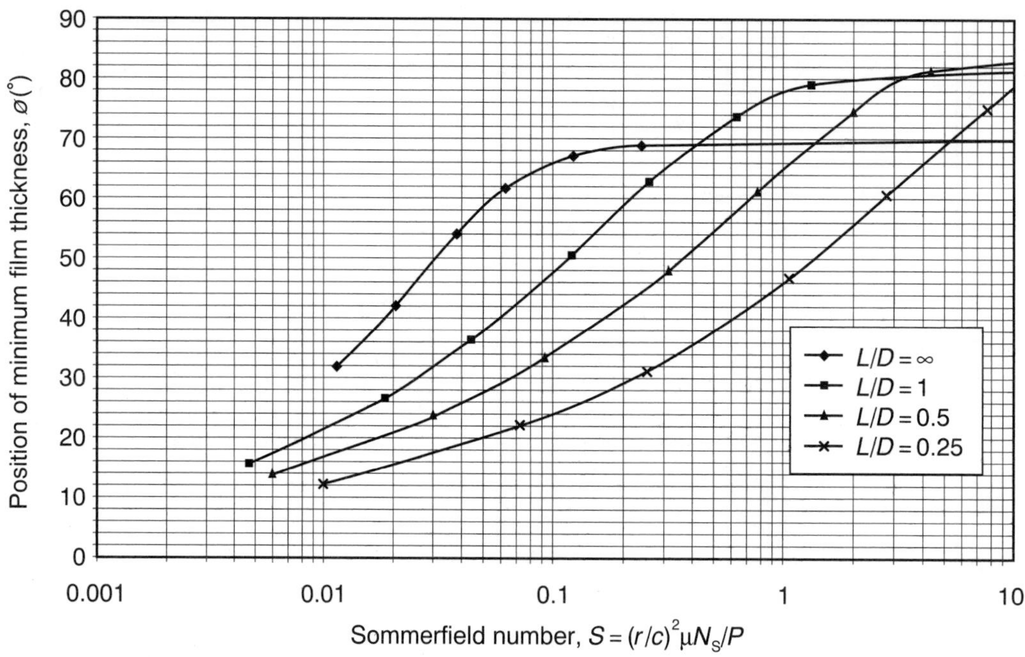

Fig. 2.18 Chart for determining the position of the minimum film thickness, ϕ. (Data from Raimondi and Boyd, 1958c.)

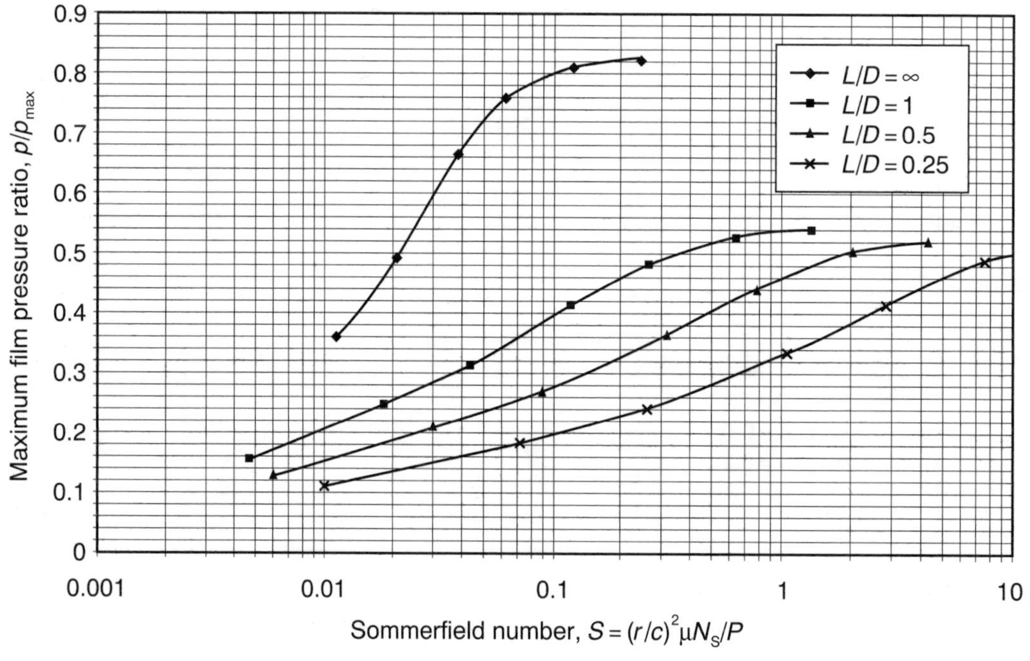

Fig. 2.19 Chart for determining the maximum film pressure ratio, p/p_{max}. (Data from Raimondi and Boyd, 1958c.)

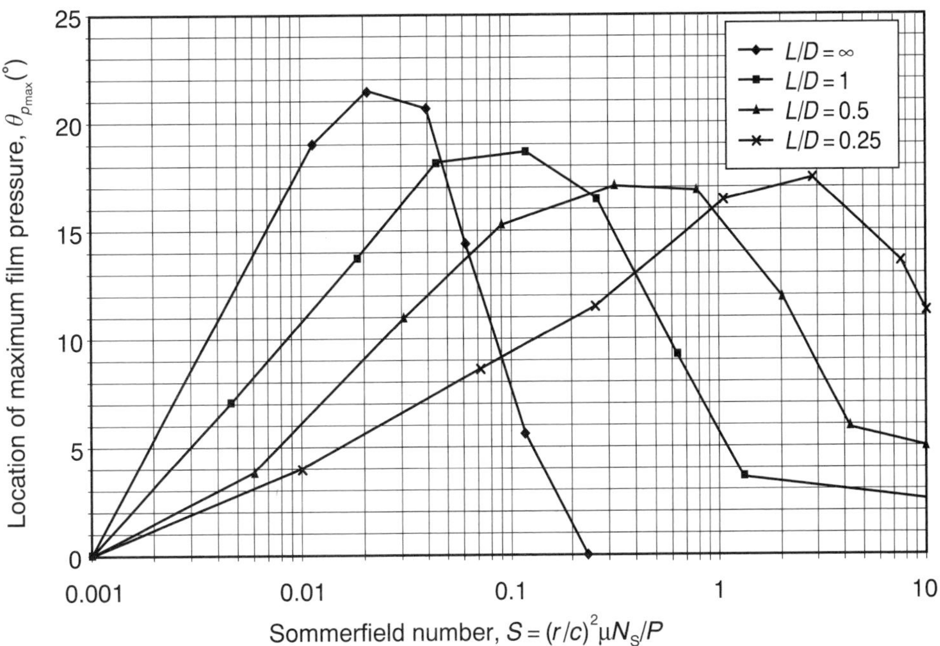

Fig. 2.20 Chart for determining the position of maximum pressure, $\theta_{p_{max}}$. (Data from Raimondi and Boyd, 1958c.)

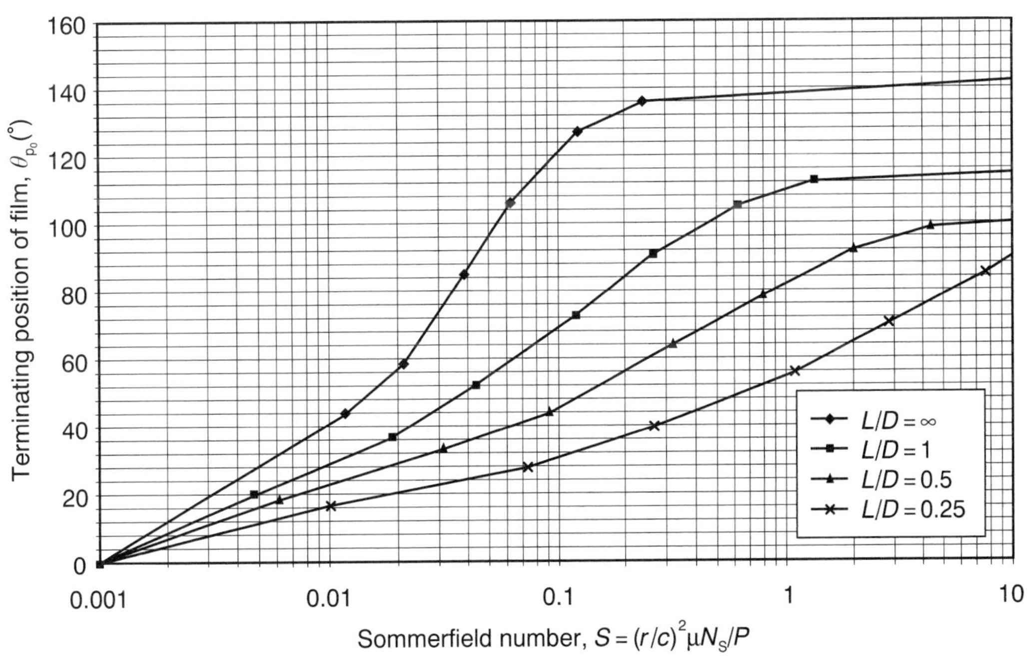

Fig. 2.21 Chart for determining the film termination angle, θ_{p_0}. (Data from Raimondi and Boyd, 1958c.)

by interpolation using (Raimondi and Boyd, 1958b):

$$y = \frac{1}{(L/D)^3} \times \left[-\frac{1}{8}\left(1-\frac{L}{D}\right)\left(1-2\frac{L}{D}\right)\left(1-4\frac{L}{D}\right)y_\infty + \frac{1}{3}\left(1-2\frac{L}{D}\right)\left(1-4\frac{L}{D}\right)y_1 - \frac{1}{4}\left(1-\frac{L}{D}\right)\left(1-4\frac{L}{D}\right)y_{1/2} + \frac{1}{24}\left(1-\frac{L}{D}\right)\left(1-2\frac{L}{D}y_{1/4}\right)\right] \quad (2.11)$$

Here y is the desired variable and y_∞, y_1, $y_{1/2}$ and $y_{1/4}$ are the variable at $L/D = \infty$, 1, $\frac{1}{2}$ and $\frac{1}{4}$ respectively.

2.2.4 *Alternative method for the design of full film hydrodynamic bearings*

An alternative method to using the design charts of Raimondi and Boyd is to use an approximate method developed by Reason and Narang (1982). They defined an approximation for the pressure variation in a journal bearing given by

$$p = \frac{\dfrac{3\mu V}{rc^2}\left(\dfrac{L^2}{4} - z^2\right)\left(\dfrac{\varepsilon \sin\theta}{(1+\varepsilon\cos\theta)^3}\right)}{1 + \dfrac{2+\varepsilon^2}{2r^2}\left(\dfrac{L^2}{4} - z^2\right)\left(\dfrac{1}{(1+\varepsilon\cos\theta)(2+\varepsilon\cos\theta)}\right)} \quad (2.12)$$

where

μ = viscosity
V = surface velocity
r = journal radius
c = radial clearance
L = bearing length
z = axial coordinate
ε = eccentricity variable
θ = circumferential coordinate.

This equation was integrated using an approximate method and the solution given in terms of two integrals I_s and I_c. Both I_s and I_c are functions of the L/D ratio and the eccentricity ratio ε. The values of these integrals are given in Table 2.3 along with the corresponding Sommerfield number calculated using

$$S = \frac{1}{6\pi\sqrt{I_c^2 + I_s^2}} \quad (2.13)$$

Knowing values for the integrals I_s and I_c enables the bearing parameters to be calculated as shown below.

- Given W, N_s, L, D, c, the type of lubricant and T_1, guess a value for the temperature rise of the lubricant ΔT and calculate T_{av}. Determine the viscosity at T_{av} from Fig. 2.11.
- Calculate the bearing loading pressure:

$$P = \frac{W}{LD}$$

- Calculate the Sommerfield number:

$$S = \left(\frac{r}{c}\right)^2 \frac{\mu N_s}{P}$$

- Knowing the length to diameter ratio and the Sommerfield number, search the correct L/D column in Table 2.3 for a matching Sommerfield number. This number will give the eccentricity ratio reading across the table to the left. Interpolation may be necessary for values of S between the values listed in the table. Once the correct eccentricity has been found, determine the values for the integrals I_s and I_c to correspond with the value of the Sommerfield number.

Recall that for a linear function, $y = f(x)$, the value for y between the bounding points (x_1, y_1) and (x_2, y_2) for a given value of x can be found by linear interpolation:

$$y = \left(\frac{x - x_1}{x_2 - x_1}\right)(y_2 - y_1) + y_1 \quad (2.14)$$

- Calculate the minimum film thickness $h_0 = c(1-\varepsilon)$.
- Calculate the attitude angle ϕ (rad/s):

$$\phi = \tan^{-1}\left(-\frac{I_s}{I_c}\right) \quad (2.15)$$

- Calculate the flow variable $Q/(rcN_sL)$, where Q is the lubricant supply:

$$\frac{Q}{rcN_sL} = \pi(1+\varepsilon) - \pi\varepsilon\eta_1 \times \left[1 - \eta_1\left(\frac{D}{L}\right)^2 \frac{1}{\sqrt{\left(\frac{D}{L}\right)^2\frac{\eta_1}{2} + \frac{1}{4}}} \times \tanh^{-1}\left(\frac{1}{2\sqrt{\left(\frac{D}{L}\right)^2\frac{\eta_1}{2} + \frac{1}{4}}}\right)\right] \quad (2.16)$$

Table 2.3 Values of the integrals I_s and I_c, the Sommerfield number S, as functions of the L/D ratio and the eccentricity ratio ε

	I_s, I_c and S for $L/D =$						
ε	0.25	0.5	0.75	1.0	1.5	2	∞
0.10	$I_s = 0.0032$	0.0120	0.0244	0.0380	0.0636	0.0839	0.1570
	$I_c = -0.0004$	−0.0014	−0.0028	−0.0041	−0.0063	−0.0076	−0.0100
	$S = 16.4506$	4.3912	2.1601	1.3880	0.8301	0.6297	0.3372
0.20	$I_s = 0.0067$	0.0251	0.0505	0.0783	0.1300	0.1705	0.3143
	$I_c = -0.0017$	−0.0062	−0.0118	−0.0174	−0.0259	−0.0312	−0.0408
	$S = 7.6750$	2.0519	1.0230	0.6614	0.4002	0.3061	0.1674
0.30	$I_s = 0.0109$	0.0404	0.0804	0.1236	0.2023	0.2628	0.4727
	$I_c = -0.0043$	−0.0153	−0.0289	−0.0419	−0.0615	−0.0733	−0.0946
	$S = 4.5276$	1.2280	0.6209	0.4065	0.2509	0.1944	0.1100
0.40	$I_s = 0.0164$	0.0597	0.1172	0.1776	0.2847	0.3649	0.6347
	$I_c = -0.0089$	−0.0312	−0.0579	−0.0825	−0.1183	−0.1391	−0.1763
	$S = 2.8432$	0.7876	0.4058	0.2709	0.1721	0.1359	0.0850
0.50	$I_s = 0.0241$	0.0862	0.1656	0.2462	0.3835	0.4831	0.8061
	$I_c = -0.0174$	−0.0591	−0.1065	−0.1484	−0.2065	−0.2391	−0.2962
	$S = 1.7848$	0.5076	0.2694	0.1845	0.1218	0.0984	0.0618
0.60	$I_s = 0.0363$	0.1259	0.2345	0.3396	0.5102	0.6291	0.9983
	$I_c = -0.0338$	−0.1105	−0.1917	−0.2590	−0.3474	−0.3949	−0.4766
	$S = 1.0696$	0.3167	0.1752	0.1242	0.0859	0.0714	0.0480
0.70	$I_s = 0.0582$	0.1927	0.3430	0.4793	0.6878	0.8266	1.2366
	$I_c = -0.0703$	−0.2161	−0.3549	−0.4612	−0.5916	−0.6586	−0.7717
	$S = 0.5813$	0.1832	0.1075	0.0798	0.0585	0.0502	0.0364
0.80	$I_s = 0.1071$	0.3264	0.5425	0.7220	0.9771	1.1380	1.5866
	$I_c = -0.1732$	−0.4797	−0.7283	−0.8997	−1.0941	−1.1891	−1.3467
	$S = 0.2605$	0.0914	0.0584	0.0460	0.0362	0.0322	0.0255
0.90	$I_s = 0.2761$	0.7079	1.0499	1.3002	1.6235	1.8137	2.3083
	$I_c = -0.6644$	−1.4990	−2.0172	−2.3269	−2.6461	−2.7932	−3.0339
	$S = 0.0737$	0.0320	0.0233	0.0199	0.0171	0.0159	0.0139
0.95	$I_s = 0.6429$	1.3712	1.8467	2.1632	2.5455	2.7600	3.2913
	$I_c = -2.1625$	−3.9787	−4.8773	−5.3621	−5.8315	−6.0396	−6.3776
	$S = 0.0235$	0.0126	0.0102	0.0092	0.0083	0.0080	0.0074
0.99	$I_s = 3.3140$	4.9224	5.6905	6.1373	6.6295	6.8881	8.7210
	$I_c = -22.0703$	−28.5960	−30.8608	−31.9219	−32.8642	−33.2602	−33.5520
	$S = 0.0024$	0.0018	0.0017	0.0016	0.0016	0.0016	0.0015

Source: Reason and Narang (1982).

where

$$\eta_1 = \frac{(1+\varepsilon)(2+\varepsilon)}{2+\varepsilon^2}$$

- Calculate the flow through the minimum film thickness Q_π:

$$\frac{Q_\pi}{rcN_sL} = \pi(1-\varepsilon) + \pi\varepsilon\eta_2$$
$$\times \left[1 - \eta_2\left(\frac{D}{L}\right)^2 \frac{1}{\sqrt{\left(\frac{D}{L}\right)^2 \frac{\eta_2}{2} + \frac{1}{4}}} \times \tanh^{-1}\left(\frac{1}{2\sqrt{\left(\frac{D}{L}\right)^2 \frac{\eta_2}{2} + \frac{1}{4}}}\right)\right] \quad (2.17)$$

where

$$\eta_2 = \frac{(1-\varepsilon)(2-\varepsilon)}{2+\varepsilon^2}$$

Note that

$$\tanh^{-1} x = \frac{1}{2}\ln\left(\frac{1+x}{1-x}\right)$$

- The flow ratio can be calculated using

$$\frac{Q_s}{Q} = 1 - \frac{Q_\pi}{Q} \quad (2.18)$$

- Calculate the friction variable $(r/c)f$:

$$f\left(\frac{r}{c}\right) = 6\pi S\left(\frac{\pi}{3\sqrt{(1-\varepsilon^2)}} + \frac{\varepsilon}{2}I_s\right) \quad (2.19)$$

- Calculate the temperature rise:

$$\Delta T = \frac{8.30 \times 10^{-6} P}{1 - \frac{1}{2}(Q_s/Q)} \times \frac{(r/c)f}{Q/(rcN_sL)} \quad (2.20)$$

Although these equations may appear complex, the advantage of the method developed by Reason and Narang is that it is suitable for conventional programming or utilization within a spreadsheet.

Example A full journal bearing has a nominal diameter of 20.0 mm and a bearing length of 20.0 mm (*see* Fig. 2.22). The bearing supports a load of 1000 N, and the journal design speed is 6000 rpm. The radial clearance has been specified as 0.02 mm. An SAE 20 oil has been chosen to be compatible with requirements elsewhere in the machine and the lubricant supply temperature is 60°C. Using the approximate method of Reason and Narang, find the temperature rise of the lubricant, the total lubricant flow rate, the minimum film thickness, the torque required to overcome friction and the heat generated in the bearing.

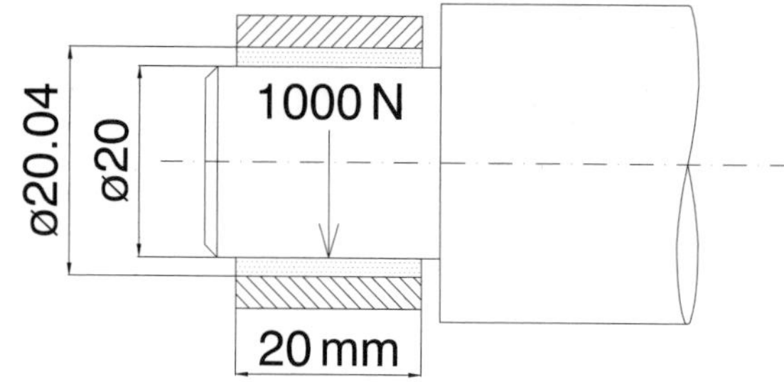

Fig. 2.22 Bearing design example.

Solution The primary design data are $D = 20$ mm, $L = 20$ mm, $W = 1000$ N, $N = 6000$ rpm, $c = 0.02$ mm, SAE 20, $T_1 = 60$°C.

Guess $\Delta T = 15$°C. Hence:

$$T_{av} = T_1 + \frac{\Delta T}{2} = 60 + \frac{15}{2} = 67.5\text{°C}$$

For SAE 20, at 67.5°C, from Fig. 2.11, $\mu = 0.016$ Pa · s.

$$P = \frac{W}{LD} = \frac{1000}{0.02 \times 0.02} = 2.5 \text{ MPa}$$

$$S = \left(\frac{r}{c}\right)^2 \frac{\mu N_s}{P}$$

$$= \left(\frac{0.01}{0.00002}\right)^2 \frac{0.016 \times 100}{2.5 \times 10^6} = 0.16$$

From Table 2.3 for $L/D = 1$, the nearest bounding values of the Sommerfield number to $S = 0.16$ are for $\varepsilon = 0.5$ ($S = 0.1845$) and $\varepsilon = 0.6$ ($S = 0.1242$).

Interpolating to find the values of ε, I_s and I_c which correspond to $S = 0.16$:

$$\varepsilon = \frac{0.16 - 0.1845}{0.1242 - 0.1845}(0.6 - 0.5) + 0.5 = 0.5406$$

$$I_s = \frac{0.16 - 0.1845}{0.1242 - 0.1845}(0.3396 - 0.2462) + 0.2462$$
$$= 0.2841$$

$$I_c = \frac{0.16 - 0.1845}{0.1242 - 0.1845}(-0.2590 - -0.1484)$$
$$+ -0.1484 = -0.1933$$

From equation 2.16 with $\eta_1 = 1.708$,

$$\frac{Q}{rcN_sL} = 4.84 - 2.901 \times [1 - 1.708 \times 0.9517 \tanh^{-1}(0.4759)] = 4.379$$

From equation 2.17 with $\eta_2 = 0.2925$,

$$\frac{Q_\pi}{rcN_sL} = 1.443 + 0.4968 \times [1 - 0.2925 \times 1.589 \tanh^{-1}(0.7943)] = 1.690$$

From equation 2.18,

$$Q_s/Q = 0.6141$$

From equation 2.19,

$$f\left(\frac{r}{c}\right) = 3.016(1.245 + 0.07679) = 3.987$$

Thus

$$\Delta T = \left(\frac{8.3 \times 10^{-6} P}{1 - 0.5(Q_s/Q)}\right) \frac{(r/c)f}{Q/(rcN_sL)} = 27.26°\text{C}$$

Repeating the procedure with this value for ΔT gives

$T_{av} = 73.6°\text{C}$, hence $\mu = 0.013\ \text{Pa}\cdot\text{s}$

$S = 0.13$, hence $\varepsilon = 0.5904$

$Q/(rcN_sL) = 4.492$

$Q_\pi/(rcN_sL) = 1.531$

$(r/c)f = 3.418$

Using these values gives $\Delta T = 23.56°\text{C}$.

Repeating the procedure with this value for ΔT gives

$T_{av} = 71.78°\text{C}$, hence $\mu = 0.014\ \text{Pa}\cdot\text{s}$

$S = 0.14$, hence $\varepsilon = 0.5738$

$Q/(rcN_sL) = 4.454$

$Q_\pi/(rcN_sL) = 1.585$

$(r/c)f = 3.613$

Using these values gives $\Delta T = 24.83°\text{C}$. Hence $T_{av} = 72.4°\text{C}$. This is close enough to the previous value, indicating that the correct value for T_{av} has been determined.

The values for the total lubricant supply, minimum film thickness, coefficient of friction, torque and the heat generated in the bearing can now be determined:

$Q = 1782\ \text{mm}^3/\text{s}$

$h_0 = c(1 - \varepsilon) = 0.0085\ \text{mm}$

$f = 0.007226$

$\text{Torque} = 0.072\ \text{N}\cdot\text{m}$

$\text{Power} = 45.4\ \text{W}$

2.3 Rolling contact bearings

The term rolling contact bearings encompasses the wide variety of bearings that use spherical balls or some type of roller between the stationary and moving elements, as illustrated in Fig. 2.23. The most common type of bearing supports a rotating shaft resisting a combination of radial and axial (or thrust) loads. Some bearings are designed to carry only radial or only thrust loads.

Selection of the type of bearing to be used for a given application can be aided by the many comparison charts, as given for example in Table 2.4. Several bearing manufacturers produce excellent catalogues (e.g. NSK/RHP, SKF, FAG, INA), including design guides. The reader is commended to gain access to this information which is available in hard copy (just telephone and ask the manufacturer for a brochure), via the internet and in the form of electronic handbooks (e.g. *see* SKF, 1995). The comparative ratings shown in Table 2.4 can be justified for say radial load carrying capacity; roller bearings are better than ball bearings because of their shape and the area over which the load is spread. Thrust load capacity varies dramatically with design. The grooves in the races of the deep groove ball bearing permit the transfer of moderate thrust in combination with radial load. But the angular contact bearing (Fig. 2.23f) is better than the single row ball bearing because the races are higher on one side, providing a more favourable load path. Cylindrical and needle bearings should not generally be subjected to any thrust load.

2.3.1 Bearing life and selection

The load on a rolling contact bearing is exerted on a very small area, as illustrated in Fig. 2.24. The resulting contact stresses are very high and of the order of 2000 MPa. Despite very strong steels (e.g. BS 970 534 A99, AISI 52100) all bearings have a finite life and will eventually fail due to fatigue. For two groups of apparently identical bearings tested under loads F_1 and F_2,

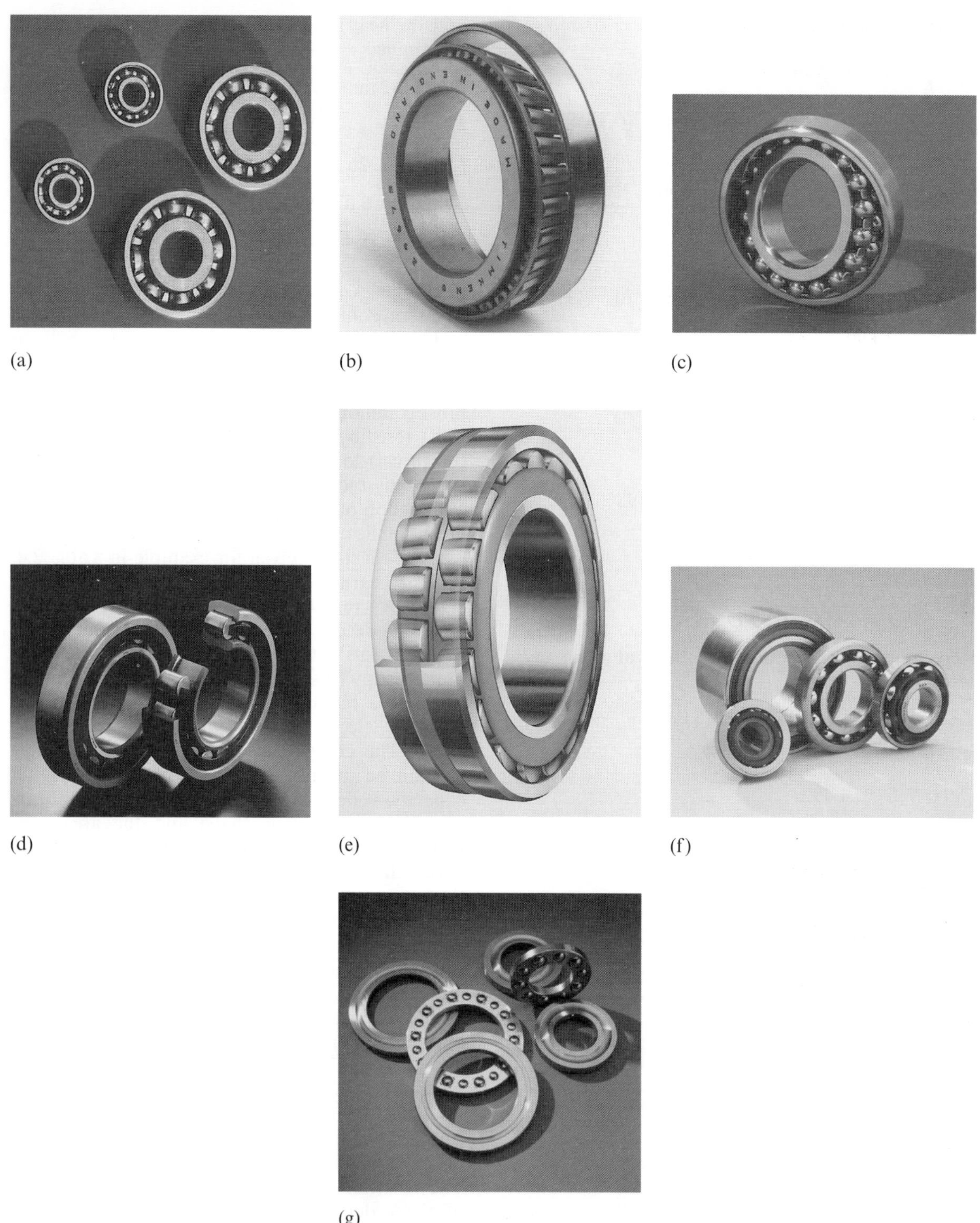

Fig. 2.23 A selection of the various types of rolling element bearings. (a) Deep groove ball bearing; (b) taper roller bearing; (c) self-aligning ball bearing; (d) cylindrical roller bearings; (e) spherical roller bearing; (f) angular contact ball bearings; (g) thrust ball bearings. (Photographs (a) and (c)–(f) courtesy of NSK RHP Bearings.)

Table 2.4 Merits of different rolling contact bearings

Bearing type	Radial load capacity	Axial or thrust load capacity	Misalignment capability
Single row	Good	Fair	Fair
Double row deep groove ball	Excellent	Good	Fair
Angular contact	Good	Excellent	Poor
Cylindrical roller	Excellent	Poor	Fair
Needle roller	Excellent	Poor	Poor
Spherical roller	Excellent	Fair/good	Excellent
Tapered roller	Excellent	Excellent	Poor

the respective lives L_1 and L_2 are related by

$$\frac{L_1}{L_2} = \left(\frac{F_2}{F_1}\right)^k \tag{2.21}$$

where $k = 3$ for ball bearings and $k = 3.33$ for roller bearings.

Rolling element bearings are generally standard items and can be purchased from specialist manufacturers. The selection of a bearing from a manufacturer's catalogue involves consideration of the bearing load carrying capacity and the bearing geometry. For a given bearing the load carrying capacity is given in terms of the basic dynamic load rating and the basic static load rating. The various commonly used definitions for rolling element bearing life specification are outlined and illustrated by example below.

The **basic dynamic load rating**, C, is the constant radial load which a bearing can endure for 1×10^6 revolutions.

The **life** of a ball bearing, L, is the number of revolutions (or hours at some constant speed) which the bearing runs before the development of fatigue in any of the bearing components.

Fatigue occurs over a large number of cycles of loading. For a bearing this would mean a large number of revolutions. Fatigue is a statistical phenomena with considerable spread of the actual life of a group of bearings of a given design. The **rated life** is the standard means of reporting the results of many tests of bearings. It represents the life that 90% of the bearings would achieve successfully at a rated load. The rated life is referred to as the L_{10} life at the rated load. The rated life, L_{10}, of a group of apparently identical bearings is defined as the number of revolutions (or hours at some constant speed) that 90% of the group of bearings will complete before the first evidence of fatigue develops.

The life of a bearing with basic dynamic load rating C with a load P is given by

$$L = \left(\frac{C}{P}\right)^k \text{ million revolutions} \tag{2.22}$$

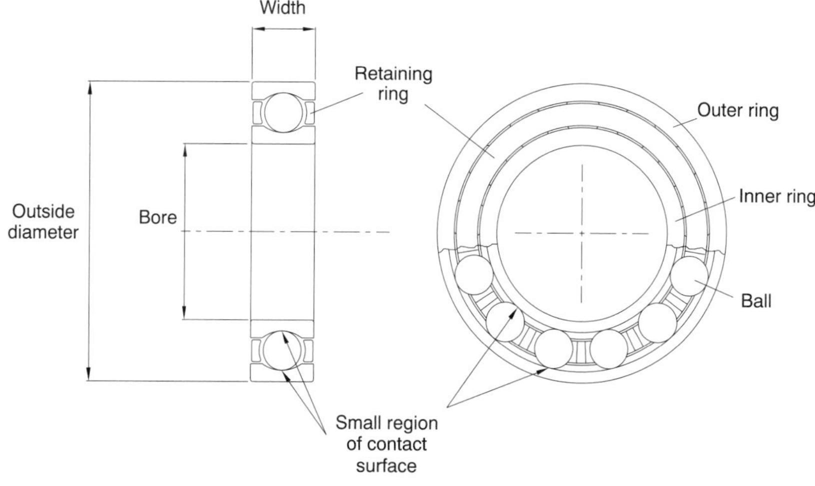

Fig. 2.24 Contact area for a ball bearing.

The required basic dynamic load rating C for a given load P and life L is given by

$$C = P\left(\frac{L}{10^6}\right)^{(1/k)} \qquad (2.23)$$

The **basic static load rating** is the load the bearing can withstand without any permanent deformation of any component. If this load is exceeded it is likely the bearing races will be indented by the rolling elements (called Brinelling). Subsequently the operation of the bearing would be noisy and impact loads on the indented area would produce rapid wear and progressive failure of the bearing would ensue.

Many bearing manufacturers publish ratings for bearings corresponding to a particular number of hours life at a specified rotational speed. The designer's task is to determine which value of catalogue rating to use given a set of particular values for P_d, L_d and n_d:

$$C_{cat} = P_d\left(\frac{L_d n_d}{L_{cat} n_{cat}}\right)^{1/k} \qquad (2.24)$$

where

C_{cat} = catalogue radial rating ((N), (kN))
P_d = required radial design load ((N), (kN))
L_d = required design life (revolutions or hours)
L_{cat} = catalogue rated life (revolutions or hours)
n_d = required design speed (rpm)
n_{cat} = catalogue rated speed (rpm).

Loads on bearings often vary with time and may not be entirely radial. The **equivalent load**, P, is defined as the constant radial load, which if applied to a bearing would give the same life as that which the bearing would attain under the actual conditions of load and rotation. When both radial and thrust loads are exerted on a bearing, the equivalent load is the constant radial load that would produce the same rated life for the bearing as the combined loading.

Normally,

$$P = VXR + YT \qquad (2.25)$$

where

P = equivalent load (N)
$V = 1.2$ if mounting rotates, $V = 1.0$ if shaft rotates
X = radial factor (given in bearing catalogues)
R = applied radial load (N)
Y = thrust factor (given in bearing catalogues)
T = applied thrust load (N).

Example A straight cylindrical roller bearing operates with a load of 7.5 kN. The required life is 8760 hours at 1000 rpm. What load rating should be used for selection from the catalogue?

Solution Using equation 2.23,

$$C = P_d(L_d/10^6)^{1/k}$$

$$= 7500\left(\frac{8760 \times 1000 \times 60}{10^6}\right)^{1/3.33} = 49.2\ \text{kN}$$

Example A catalogue lists the basic dynamic load rating for a ball bearing to be 33 800 N for a rated life of 1 million revolutions. What would be the expected L_{10} life of the bearing if it were subjected to 15 000 N and determine the life in hours that this corresponds to if the speed of rotation is 2000 rpm.

Solution $C_{cat} = 33\,800$ N, $P_d = 15\,000$ N, $L_{cat} = 10^6$ (L_{10} life at load C), $k = 3$ (ball). Using equation 2.22, the life is given by

$$L = 10^6\left(\frac{33\,800}{15\,000}\right)^3$$

$$= 11.44 \times 10^6 \text{ revolutions } (=L_{10} \text{ life at } 15\,000\ \text{N})$$

If the rotational speed is 2000 rpm, $L = 11.44 \times 10^6/(2000 \times 60) = 95$ hours operation. This is not very long and illustrates the need to use a bearing with a high basic dynamic load rating.

Example A ball bearing for an industrial grinder is chosen to withstand a radial load of 1300 N and have an L_{10} life of 3600 hours at 3000 rpm. The manufacturer's catalogue rating is based on an L_{10} life of 3800 hours at 1000 rpm. What load should be used to enter into the catalogue for selection?

Solution Using equation 2.24:

$$C_{cat} = P_d\left(\frac{L_d n_d}{L_{cat} n_{cat}}\right)^{1/k}$$

$$= 1300\left(\frac{3600 \times 3000}{3800 \times 1000}\right)^{1/3} = 1841\ \text{N}$$

Tables 2.5–2.8 give an overview of the information typically available in bearing manufacturers' catalogues and the example given below illustrates their basic use.

Table 2.5 Selected example of single-row deep groove ball bearing ratings

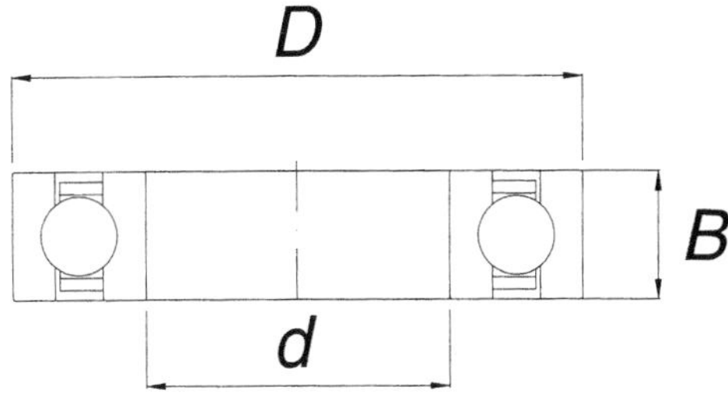

d (mm)	D (mm)	B (mm)	Basic dynamic load rating, C (N)	Basic static load rating, C_0(N)	Speed limit for grease lubrication (rpm)	Speed limit for oil lubrication (rpm)	Code
15	24	5	1 570	800	28 000	34 000	61802
	32	8	5 600	2 850	22 000	28 000	16002
	32	9	5 600	2 850	22 000	28 000	6002
	35	11	7 850	3 750	19 000	24 000	6202
	42	13	11 500	5 400	17 000	20 000	6302
17	26	5	1 690	930	24 000	30 000	61803
	35	8	6 060	3 250	19 000	24 000	16003
	35	10	6 060	3 250	19 000	24 000	6003
	40	12	9 550	4 750	17 000	20 000	6203
	47	14	13 600	6 550	16 000	19 000	6303
	62	17	23 000	10 800	12 000	15 000	6403
20	32	7	2 750	1 500	19 000	24 000	61804
	42	8	6 900	4 050	17 000	20 000	16004
	42	12	9 400	5 000	17 000	20 000	6004
	47	14	12 800	6 550	15 000	18 000	6204
	52	15	16 000	7 800	13 000	16 000	6304
	72	19	30 800	15 000	10 000	13 000	6404
25	37	7	4 400	2 600	17 000	20 000	61805
	47	8	7 600	4 750	14 000	17 000	16005
	47	12	11 300	6 550	15 000	18 000	6005
	52	15	14 050	7 800	12 000	15 000	6205
	62	17	22 600	11 600	11 000	14 000	6305
	80	21	36 000	19 300	9 000	11 000	6405
30	42	7	4 500	2 900	15 000	18 000	61806
	55	9	11 300	7 350	12 000	15 000	16006
	55	13	13 400	8 300	12 000	15 000	6006
	62	16	19 600	11 200	10 000	13 000	6206
	72	19	28 200	16 000	9 000	11 000	6306
	90	23	43 700	23 600	8 500	10 000	6406

Example A bearing is required to carry a radial load of 2.8 kN and provide axial location for a shaft of 30 mm diameter rotating at 1500 rpm. An L_{10} life of 10 000 hours is required. Select and specify an appropriate bearing.

Solution Axial shaft location is required, so a deep groove ball bearing which provides axial location capability in both directions would be suitable.

The total number of revolutions in life is

$10\,000 \times 1500 \times 60 = 900$ million

so $L = 900$. The load is purely radial, so $P = 2800$ N. The required dynamic loading is given by

$C = PL^{1/3} = 2800 \times 900^{1/3} = 27\,033$ N

Reference to the deep groove bearing chart (Table 2.5) shows a suitable bearing could be:

- ISO designation 6306

Table 2.6 Selected example of angular contact ball bearing ratings

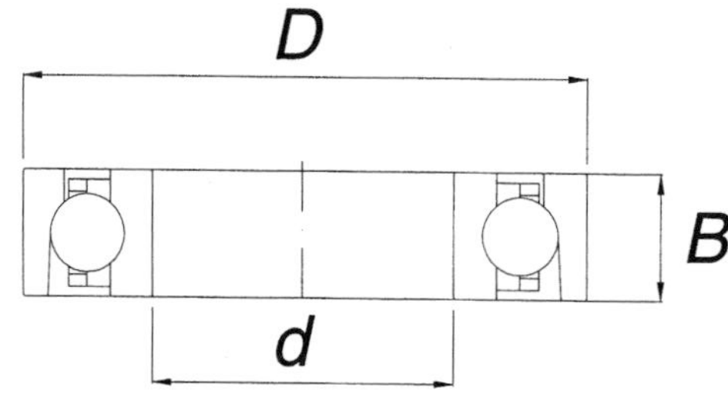

d (mm)	D (mm)	B (mm)	Basic dynamic load rating, C (N)	Basic static load rating, C_0 (N)	Speed limit for grease lubrication (rpm)	Speed limit for oil lubrication (rpm)	Code
15	35	11	8 850	4 800	17 000	24 000	7202B
	42	13	13 050	6 700	15 000	20 000	7302B
17	40	12	11 200	6 100	15 000	20 000	7203B
	47	14	16 000	8 300	13 000	18 000	7303B
20	47	14	14 050	8 300	12 000	17 000	7204B
	52	15	19 050	10 400	11 000	16 000	7304B
25	52	15	15 500	10 200	10 000	15 000	7205B
	62	17	26 100	15 600	9 000	13 000	7305B
30	62	16	23 900	15 600	8 500	12 000	7206B
	72	19	34 600	21 200	8 000	11 000	7306B
40	80	18	36 500	26 000	7 000	9 500	7208B
	90	23	49 500	33 500	6 700	9 000	7308B
50	90	20	39 050	30 500	6 000	8 000	7210B
	110	27	74 000	51 000	5 300	7 000	7310B
55	100	21	48 850	38 000	5 600	7 500	7211B
	120	29	85 300	60 000	4 800	6 300	7311B
60	110	22	57 250	45 500	5 000	6 700	7212B
	130	31	95 700	69 500	4 500	6 000	7312B
65	120	23	66 400	54 000	4 500	6 000	7213B
	140	33	109 000	80 000	4 300	5 600	7313B
70	125	24	71 600	60 000	4 300	5 600	7214B

- bore diameter 30 mm, outer diameter 72 mm
- width 19 mm
- $C = 18\,200$ N, $C_0 = 16\,000$ N
- speed limit (using grease) 9000 rpm
- speed limit (using oil) 11 000 rpm.

The equations and examples developed so far have been for steady load or steady running conditions. Bearings are, however, often subjected to cyclic or unsteady loading as in a start, load, unload, stop, restart cycle. An equivalent load under conditions of varying loads can be defined by the constant cubic mean load, or the mean effective load, F_m, which gives the same life as the variable loads.

If the loads are constant for periods of time then the mean effective load is given by

$$F_m = \left(\frac{\sum F_i^3 N_i}{L_n}\right)^{1/3} = \left(\frac{F_1^3 N_1 + F_2^3 N_2 + F_3^3 N_3 + \cdots}{L_n}\right)^{1/3} \qquad (2.26)$$

where

F_m is the mean cubic load (N)
F_i is the force acting for N_i revolutions
L_n is the total number of revolutions.

Table 2.7 Selected example of cylindrical roller bearing ratings

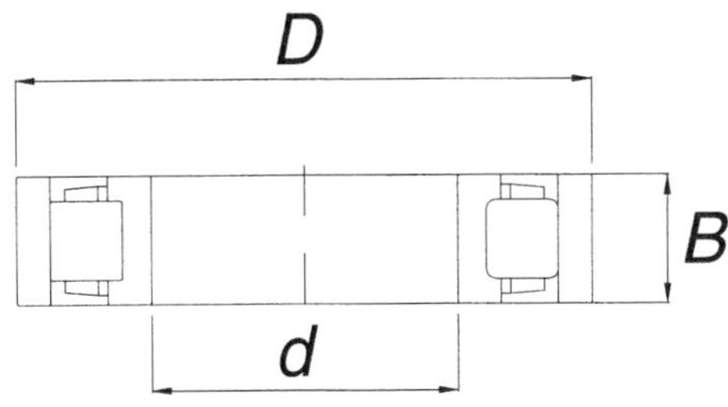

d (mm)	D (mm)	B (mm)	Basic dynamic load rating, C (N)	Basic static load rating, C_0 (N)	Speed limit for grease lubrication (rpm)	Speed limit for oil lubrication (rpm)	Code
15	35	11	12 600	10 200	18 000	22 000	NU202E
	42	13	19 500	15 300	16 000	19 000	NU302E
25	52	15	28 700	27 000	11 000	14 000	NU205E
	62	17	40 300	36 500	9 500	12 000	NU305E
30	62	16	38 100	36 500	9 500	12 000	NU206E
	72	19	51 300	48 000	9 000	11 000	NU306E
50	90	20	64 500	69 500	6 300	7 500	NU210E
	110	27	111 000	112 000	5 000	6 000	NU310E
	130	31	131 000	127 000	5 000	6 000	NU410
100	180	34	252 000	305 000	3 200	3 800	NU220E
	250	58	430 000	475 000	2 400	3 000	NU420
200	360	58	766 000	1 060 000	1 500	1 800	NU240E
	420	80	990 000	1 320 000	1 300	1 600	NU340
600	870	118	2 750 000	510 000	600	700	NU10/600

Table 2.8 Selected example of precision angular contact ball bearing ratings

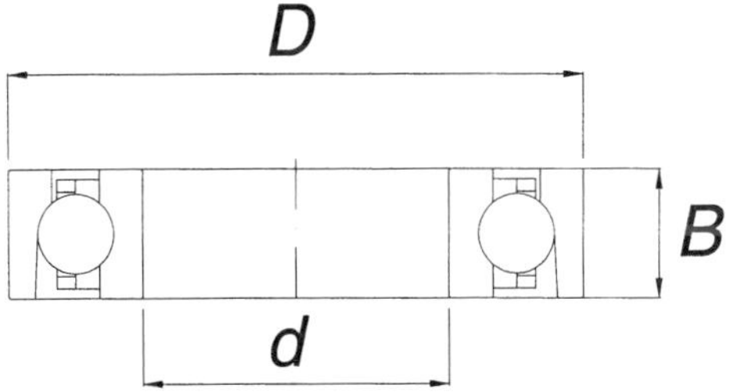

d (mm)	D (mm)	B (mm)	Basic dynamic load rating, C (N)	Basic static load rating, C_0 (N)	Speed limit for grease lubrication (rpm)	Speed limit for oil lubrication (rpm)	Code
15	35	11	7 420	3 350	48 000	70 000	7202CD
17	40	12	9 200	4 150	43 000	63 000	7203CD
20	47	14	12 000	5 850	36 000	53 000	7204CD
25	52	15	13 600	7 200	30 000	45 000	7205CD
30	62	16	24 300	16 000	24 000	38 000	7206CD
40	80	18	41 100	28 000	18 000	30 000	7208CD
50	90	20	45 000	34 000	16 000	26 000	7210CD
60	110	22	67 700	53 000	13 000	20 000	7212CD
70	125	24	79 400	64 000	11 000	18 000	7214CD
80	125	22	65 500	61 000	10 000	17 000	7016CD

If the loads are variable then the mean effective load can be found by integration:

$$F_m = \left(\frac{\int_0^{L_n} F^3 \, dN}{L_n} \right)^{1/3} \tag{2.27}$$

If the speed of rotation is constant but the load varies with time, then

$$F_m = \left(\frac{\sum F^3 t}{T} \right)^{1/3} \tag{2.28}$$

where F is the force at an instant of time t and T is the time for one cycle of the load variation.

Example A radial load $F_1 = 3.2\,\text{kN}$ acts for 2 hours on a rolling bearing and then reduces to $F_2 = 2.9\,\text{kN}$ for 1 hour. The cycle repeats itself. The shaft rotates at 430 rpm. Calculate the mean cubic load F_m, which should be used in rating the bearing for 9000 hours life.

Solution Using equation 2.26,

$$F_m = \left(\frac{F_1^3 N_1 + F_2^3 N_2}{L_n} \right)^{1/3}$$

$$= \left(\frac{3200^3(6000 \times 60 \times 430) + 2900^3(3000 \times 60 \times 430)}{9000 \times 60 \times 430} \right)^{1/3}$$

$$= 3106\,\text{N}$$

A knowledge of the reliability of a bearing or bearing combination is critical to the design of a product. An idea of the impact of using bearings with a reliability only a few points less than 100 per cent can be gained by considering the example of a double reduction gear box with six bearings. If the reliability of each bearing is 90 per cent and if the probability of failure of any one bearing is independent, the overall reliability of the bearing combination is $0.9^6 = 0.5314$ or 53 per cent. This is a very poor level of reliability and indicates the need to use bearings with a high reliability. The distribution of bearing failures at constant load can be approximated by the Weibull distribution which for bearings can be approximated by (Mischke, 1990):

$$R = \exp\left[-\left(\frac{L/L_{10} - 0.02}{4.439} \right)^{1.483} \right] \tag{2.29}$$

Rearranging equation 2.29 in terms of the desired life L and the desired reliability R gives

$$L_{10} = \frac{L}{0.02 + 4.439[\ln(1/R)]^{1/1.483}} \tag{2.30}$$

Example A small fan application requires a bearing to last for 2100 hours with a reliability of 95%. What should the rated life of the bearing be?

Solution

$$L_{10} = \frac{L}{0.02 + 4.439[\ln(1/R)]^{1/1.483}}$$

$$= \frac{2100}{0.02 + 4.439[\ln(1/0.95)]^{1/1.483}} = 3392\,\text{hours}$$

2.3.2 Bearing installation

The practical use of rolling element bearings requires proper consideration of their installation as well as correct selection. Bearing installation considerations include the bearing combination, the mounting of the bearings and the provision of lubrication.

A typical application of rolling element bearings is the support of a rotating shaft. If the operating temperature of the machine varies, the shaft length can grow relative to the casing or mounting

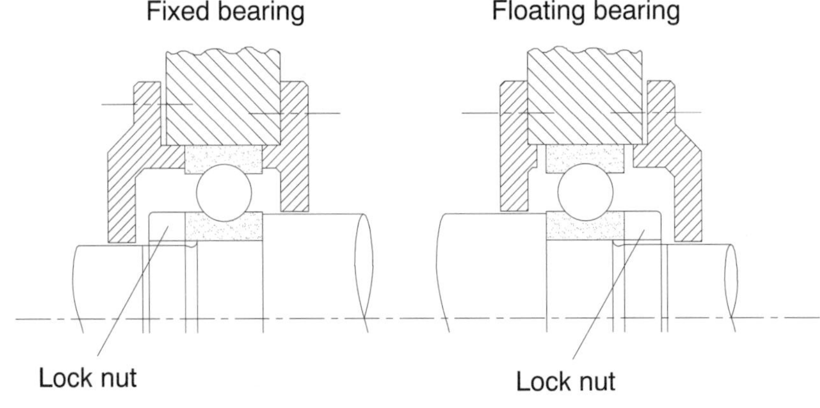

Fig. 2.25 Basic bearing mounting using two deep groove ball bearings for a rotating horizontal shaft for moderate radial and axial loading.

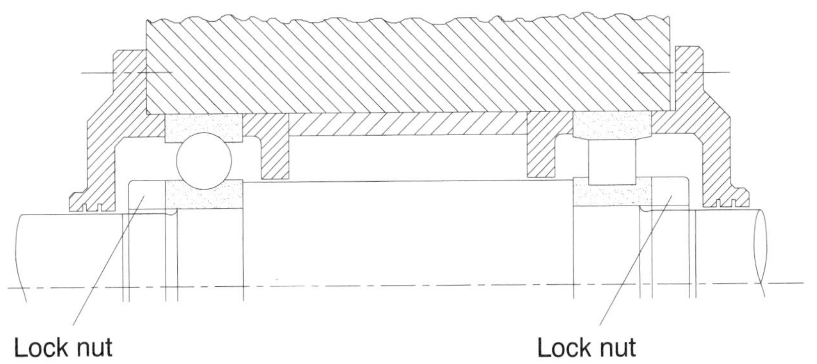

Fig. 2.26 Basic bearing mounting using a deep groove and a cylindrical roller bearing for moderate radial loads at the 'locating deep groove bearing' and high radial load capacity at the cylindrical roller bearing.

arrangement. An idea of the magnitude of the axial shaft growth can be estimated by

$$\Delta L = L\alpha\Delta T \qquad (2.31)$$

where

ΔL is the change in length (m)
L is the original length (m)
α is the coefficient of linear thermal expansion ($^\circ C^{-1}$)
ΔT is the temperature rise ($^\circ$C).

For a gas turbine engine the difference in temperature between the casing and the shaft can be 50°C. If the original length of the steel shaft was 1.0 m the growth of the shaft would be

$$\Delta L = L\alpha\Delta T$$
$$= 1.0 \times 11 \times 10^{-6} \times 50 = 5.5 \times 10^{-4}\,\text{m}$$
$$= 0.55\,\text{mm}$$

This is a considerable axial movement within a machine and must be allowed for if significant loadings and resultant stresses are to be avoided. A typical solution to this kind of situation is to allow for a limited axial movement on one bearing, as illustrated in Fig. 2.25. Here one bearing has the location of its inner and outer races 'fixed' relative to the shaft and housing by means of shoulders and locking rings. The location of the other bearing is fixed only for the inner race. If the shaft expands, the axial movement can be accommodated by limited sliding motion of the outer race of the right hand bearing within the housing bore. For this kind of arrangement the bearings are referred to as fixed and floating bearings. Similar movement is also possible for the arrangement shown in Fig. 2.26. Here the right hand bearing is a cylindrical roller bearing, and the axial location of the roller is not fixed and can move or float axially to the limited extents of the race to take up any axial movement or expansion of the shaft.

Different bearing combinations serve different purposes. The combination of two lip locating cylindrical roller bearings (*see* Fig. 2.27) provides good radial load capacity but poor axial capability. Angular contact bearings can be used in a variety of configurations, giving both axial and radial load capability. The combination shown in Fig. 2.28 gives good axial and moderate radial load capacity. Bearing manufacturers' catalogues

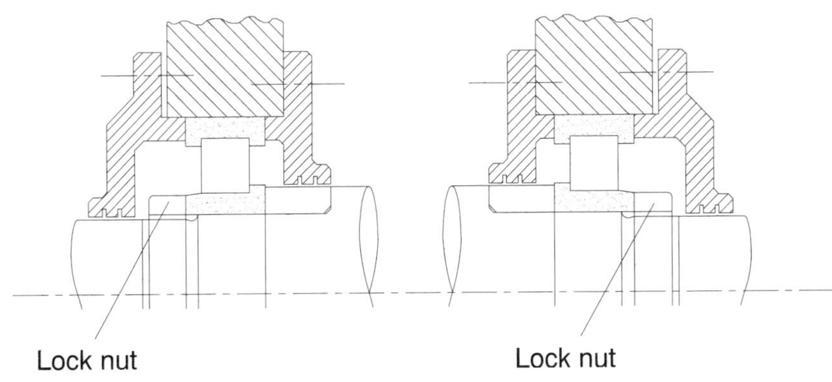

Fig. 2.27 Two lip locating roller bearings providing good radial load capacity but poor axial load capacity.

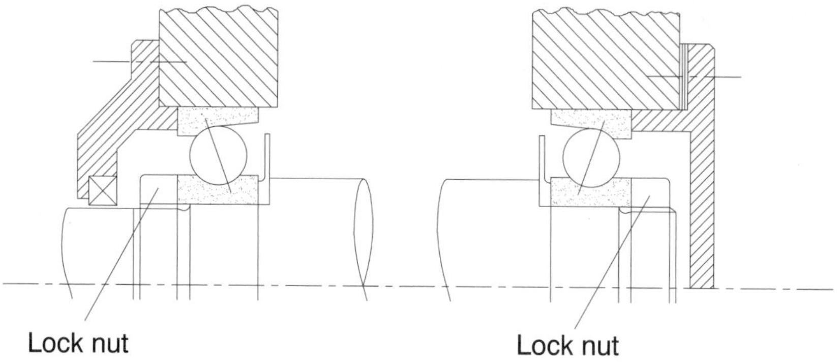

Fig. 2.28 Two angular contact bearings providing moderate radial load capacity and good axial load capacity.

normally illustrate a wide range of bearing combinations and their respective attributes.

Having chosen the bearing combination to be used the mounting of the bearings on the shaft and in the housing is critical. The bearings are designed to operate under specific clearances. For this reason, tolerances and dimensions specified by the manufacturers must be adhered to. These will include the tolerances for the shaft and the housing and the radius for mating shoulders.

Correct lubrication is essential in order to ensure the calculated life of the bearing is achieved. Too much lubrication can result in increased levels of viscous dissipation of energy in the bearings and resultant overheating of the lubricant and the bearing. Too little lubrication results in excessive wear and early failure. The form of lubrication depends on that required by the bearing application. Grease lubrication is normally the easiest and requires injection of a specific quantity of grease into the bearing and some method of dirt and dust exclusion. Bearings can be purchased which are grease filled and sealed for life. Oil lubrication can be supplied by means of partial submersion as shown in Fig. 2.29, a recirculation circuit as shown in Fig. 2.30, or oil spot lubrication. A recirculation system typically consists of a cooler, an oil pump, a filter, oil jets, scavenge or collector holes and a reservoir. Oil spot lubrication involves the application of very small quantities of lubricant directly to the bearing by means of compressed air. Once again the bearing manufacturers' handbooks are a good source of information for lubricant supply rates and systems.

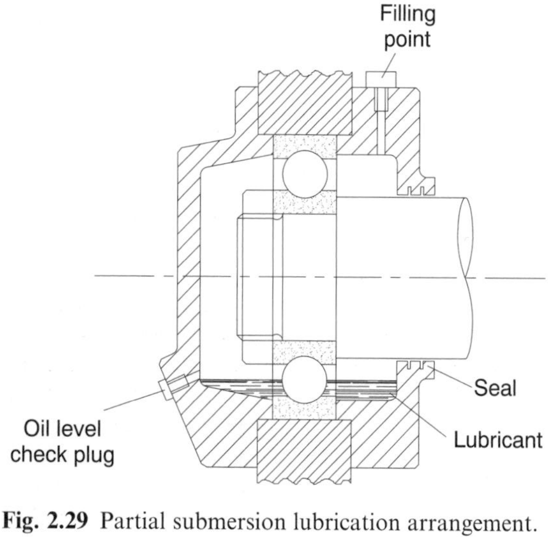

Fig. 2.29 Partial submersion lubrication arrangement.

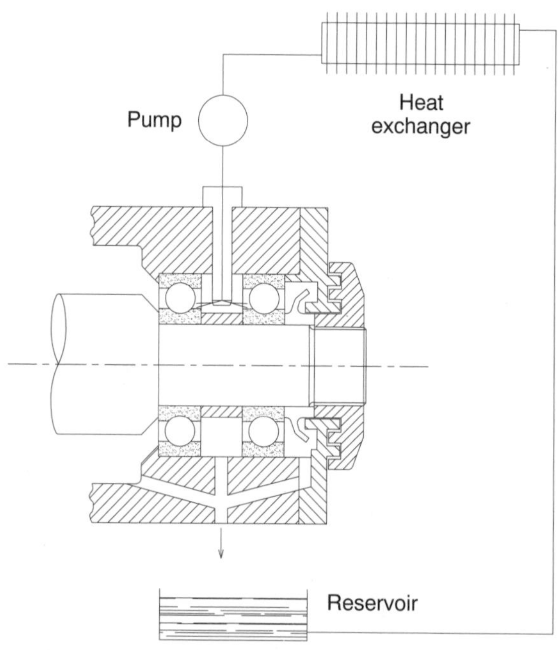

Fig. 2.30 Recirculation lubrication system.

Learning objectives checklist

Are you able to distinguish what sort of bearing to use for a given application? ☐

Can you specify when to use a boundary lubricated bearing and select an appropriate material to use for given conditions? ☐

Can you use both the Raimondi and Boyd charts to design a hydrodynamic bearing? ☐

Can you select a rolling element bearing from a manufacturer's catalogue? ☐

Can you specify the layout for a rolling bearing arrangement, sealing and lubrication system? ☐

References and sources of information

Books and papers

Cameron, A. 1981: *Basic lubrication theory*, 3rd edition. Ellis Horwood.

Engineering Sciences Data Unit 1978: ESDU 78026. Equilibrium temperatures in self contained bearing assemblies. Part I: outline of method of estimation.

Engineering Sciences Data Unit 1979: ESDU 78027. Equilibrium temperatures in self contained bearing assemblies. Part II: first approximation to temperature rise.

Engineering Sciences Data Unit 1981: ESDU 81005. Designing with rolling bearings. Part 1: design considerations in rolling bearing selection with particular reference to single row radial and cylindrical roller bearings.

Engineering Sciences Data Unit 1981: ESDU 81037. Designing with rolling bearings. Part 2: selection of single row angular contact, ball, tapered roller and spherical roller bearings.

Engineering Sciences Data Unit 1984: ESDU 84031. Calculation methods for steadily loaded axial groove hydrodynamic journal bearings.

Engineering Sciences Data Unit 1987: ESDU 87007. Design and material selection for dry rubbing bearings.

Engineering Sciences Data Unit 1990: ESDU 90027. Calculation methods for steadily loaded central circumferential groove hydrodynamic journal bearings.

Hamrock, B.J. 1994: *Fundamentals of fluid film lubrication*. McGraw Hill.

Hindhede, U., Zimmerman, J.R., Hopkins, R.B., Erisman, R.J., Hull, W.C. and Lang, J.D. 1983: *Machine design fundamentals. A practical approach*. Wiley.

IMechE 1994: Tribological design data. Part 1: Bearings. Institution of Mechanical Engineers.

Mischke, C.R. 1990: Rolling contact bearings. In Shigley, J.E. and Mischke, C.R. (eds), *Bearings and lubrication. A mechanical designers' workbook*. McGraw Hill.

Neale, M.J. 1993: *Bearings. A tribology handbook*. Butterworth Heinemann.

Palmgren, A. 1946: *Ball and roller bearing engineering*, 2nd edition. Burbank and Co.

Raimondi, A.A. and Boyd, J. 1958a: A solution for the finite journal bearing and its application to analysis and design: I. *ASLE Transactions* **1**, 159–74.

Raimondi, A.A. and Boyd, J. 1958b: A solution for the finite journal bearing and its application to analysis and design: II. *ASLE Transactions* **1**, 175–93.

Raimondi, A.A. and Boyd, J. 1958c: A solution for the finite journal bearing and its application to analysis and design: III. *ASLE Transactions* **1**, 194–209.

Reason, B.R. and Narang, I.P. 1982: Rapid design and performance evaluation of steady state journal bearings – a technique amenable to programmable hand calculators. *ASLE Transactions*, **25**(4), 429–44.

SKF 1995: *SKF electronic handbook*. (*See* 'Web sites'.)

Welsh, R.J. 1983: *Plain bearing design handbook*. Butterworth.

Standards

BS 292: Part 1: 1982. Rolling bearings: ball bearings, cylindrical and spherical roller bearings. Specification for dimensions of ball bearings, cylindrical and spherical roller bearings (metric series). BSI.

BS 6413: Part 0: 1983. Lubricants, industrial oils and related products (class L). Classification (general). BSI.

BS 6560: 1984. Glossary of terms for rolling bearings. BSI.

BS 5512: 1991. Method of calculating dynamic load rating and rating life of rolling bearings. BSI.

BS 4231: 1992. Classification for viscosity grades of industrial liquid lubricants. BSI.

BS 4480: Part 1: 1992. Plain bearings metric series. Sintered bushes. Dimensions and tolerances. BSI.

BS ISO 12128: 1995. Plain bearings. Lubrication holes, grooves and pockets. Dimensions, types, designation and their application to bearing bushes. BSI.

BS ISO 3096: 1996. Rolling bearings. Needle roller. Dimensions and tolerances. BSI.

ISO 3448 (1992). Industrial liquid lubricants – ISO viscosity classification. ISO.

SAE J300 (1989). SAE standard engine oil viscosity classification. Society of Automotive Engineers.

Web sites

http://www.skf.se/products/index.html

http://www.nsk-ltd.co.jp/eng/4-1.html

http://www.shef.ac.uk/~mpe/tribology/tribo.html (Tribology information page)

Bearings worksheet

1. A plain surface bearing has been partially specified to support a load of 2000 N at a rotational speed of 20 rpm. The nominal diameter of the journal is 50 mm and the length is 50 mm. Select an appropriate material for the bearing surface.
2. The paper feed for a photocopier is controlled by two rollers which are sprung together with a

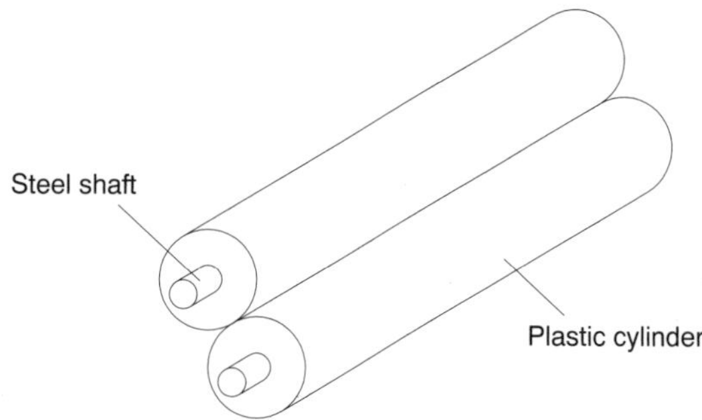

Fig. 2.31 Copier rollers. (Reproduced from IMechE, 1994.)

force of approximately 20 N. The rollers each consist of a 20 mm outer diameter plastic cylinder pressed onto a 10 mm diameter steel shaft (*see* Fig. 2.31). The maximum feed rate for the copier is 30 pages per minute. Select bearings to support the rollers. (After IMechE, 1994.)

3. A full journal bearing has a nominal diameter of 50.0 mm, a bearing length of 25.0 mm and supports a load of 2500 N when running at the design speed of 3000 rpm. The radial clearance has been specified as 0.04 mm. An SAE 10 oil has been chosen and the lubricant supply system is capable of delivering the lubricant to the bearing at 45°C. Using the Raimondi and Boyd design charts, find the temperature rise of the lubricant, the total lubricant flow rate required, the torque required to overcome friction and the heat generated in the bearing.
4. A full journal bearing has a nominal diameter of 25.0 mm and bearing length 25.0 mm. The bearing supports a load of 1450 N and the journal design speed is 2500 rpm. Two radial clearances are proposed: (a) 0.025 mm, (b) 0.015 mm. An SAE 10 oil has been chosen and the lubricant supply temperature is 50°C. Using the standard Raimondi and Boyd design charts, find the temperature rise of the lubricant, the lubricant flow rate, the torque required to overcome friction and the heat generated in the bearing for both of the proposed clearances. Determine which clearance, if any, should be selected for the bearing.
5. A full journal bearing is required for a shaft with a nominal diameter of 40 mm rotating at 1000 rpm. The load on the bearing is 2200 N. The lubricant supply available uses SAE 20 oil at an inlet temperature of 60°C. Select appropriate values for the radial clearance and bearing length, and using the Raimondi and Boyd charts determine the overall temperature rise in the bearing. Establish that the selected design lies within the optimum zone for minimum friction and maximum load on the minimum film thickness ratio chart (Fig. 2.14).
6. The proposed design for a full journal bearing, supporting a load of 1600 N and running at 5000 rpm, has a nominal diameter of 40 mm and bearing length of 20 mm. As part of the initial design, two radial clearances of (a) 50 μm and (b) 70 μm are proposed, running with SAE 10 oil with a lubricant supply temperature of 60°C. Using the standard Raimondi and Boyd design charts, determine the overall lubricant temperature rise, the total lubricant flow, the friction coefficient and the minimum film thickness for both of the proposed radial clearances, and give justification for the selection of one of the clearances for use in the bearing. As a first guess for the lubricant temperature rise, ΔT can be taken as 10°C.
7. A full journal bearing has a nominal diameter of 30.0 mm, bearing length 15.0 mm. The bearing supports a load of 1000 N, and the journal design speed is 6000 rpm. The radial clearance has been specified as 0.02 mm. An SAE 10 oil has been chosen and the lubricant supply temperature is 50°C. Using the approximate method of Reason and Narang, find the temperature rise of the lubricant, the lubricant flow rate, the minimum film thickness, the torque required to overcome friction and the heat generated in the bearing.
8. A full journal bearing has a nominal diameter of 80.0 mm, bearing length 60.0 mm. The bearing supports a load of 7500 N, and the journal design speed is 800 rpm. The radial clearance has been specified as 0.05 mm. An SAE 30 oil has been chosen and the lubricant supply temperature is 50°C. Using the approximate method of Reason and Narang for determining the design and performance of full film hydrodynamic bearings, find the temperature rise of the lubricant, the lubricant flow rate, the torque required to overcome friction, the heat generated in the bearing and the minimum film thickness.
9. The drive shaft for a combine harvester supports two pulleys: one transmitting power from the engine and the other driving the cutter. The loads on the pulleys both act in

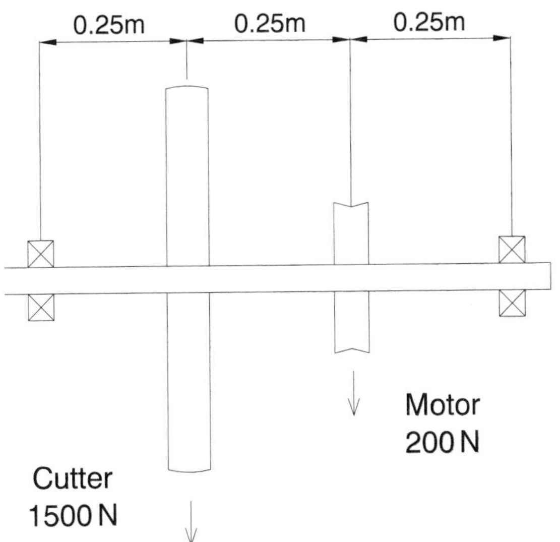

Fig. 2.32 Agricultural machinery drive shaft. (Reproduced from IMechE, 1994.)

the vertical plane, as shown in Fig. 2.32. If the diameter of the shaft is 50 mm and the maximum rotational speed is 300 rpm, design suitable hydrodynamic bearings for the shaft. (After IMechE, 1994.)

10. A straight cylindrical roller bearing operates with a load of 14.2 kN. The required life is 3800 hours at 925 rpm. What load rating should be used for selection from the catalogue.
11. A bearing is required for the floating end of a heavy duty lathe to carry a radial load of up to 9 kN. The shaft diameter is 50 mm and rotates at 3000 rpm. A life of 7500 hours for the bearings is desired. Select and specify an appropriate bearing.
12. The angular contact bearings of a dedicated lathe spindle are subjected to a cyclic load during a manufacturing process in an automated plant. During the cycle a bearing experiences a radial load of 95 N for 30 s at 2500 rpm, 75 N for 45 s at 3500 rpm, and 115 N at 3100 rpm for 15 s. Calculate the mean cubic load which should be used for 4000 hours life.

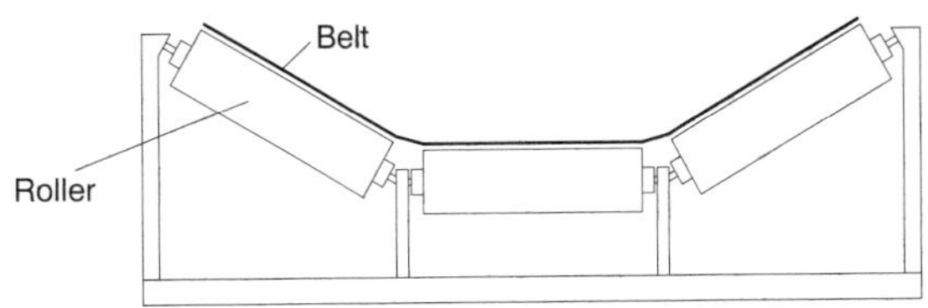

Fig. 2.33 Conveyor belt idler station. (Reproduced from IMechE, 1994.)

13. The design for a 4 km long conveyor system consists of idler stations located 1 m apart. Each idler station comprises three rollers which rotate about stationary axles fixed to supports, as illustrated in Fig. 2.33. The outer diameter of the rollers is 100 mm and each roller has a mass of 6 kg. The two end rollers are angled at 30° to the horizontal. A belt weighing 40 kg/m runs over the rollers at 3 m/s. The expected maximum loading is 3000 tonnes of aggregate per hour and experience suggests that the central idler supports approximately 60 per cent of the load. Select appropriate bearings for the rollers. (After IMechE, 1994.)

Answers

Worked solutions to the above problems are given in Appendix B.

1. $PV = 0.04188$ MN/m per second. Carbon graphite.
2. $PV = 0.0297$ MN/m per second. Thermoplastic.
3. 24.5°C, 6525 mm^3/s, 0.45 N · m, 141.4 W.
4. (a) 15.4°C, 1471 mm^3/s, 0.08 N · m, 21 W. (b) 25.9°C, 822 mm^3/s, 0.085 N · m, 22 W.
5. No unique solution.
6. (a) 17.1°C, 9200 mm^3/s, 0.00675, 0.008 mm. (b) 9.9°C, 13 160 mm^3/s, 0.0063, 0.00798 mm.
7. 39.3°C, 2264 mm^3/s, 6.9 μm, 0.13 N · m, 83.5 W.
8. 19.2°C, 7544 mm^3/s, 1.7 N · m, 143 W, 20 μm.
9. No unique solution.
10. 70.7 kN.
11. $C = 78.2$ kN. No unique solution.
12. 89.7 N.
13. Deep groove ball bearing. No unique answer.

3 Shafts

The objective of this chapter is to introduce the concepts of shaft design. An overall shaft design procedure is presented, including consideration of bearing and component mounting and shaft dynamics. At the end of this chapter the reader should be able to scheme out a general shaft arrangement, determine deflections and critical speeds, and specify shaft dimensions for strength and fluctuating load integrity.

3.1 Introduction

The term shaft usually refers to a component of circular cross-section that rotates and transmits power from a driving device, such as a motor or engine, through a machine. Shafts can carry gears, pulleys and sprockets to transmit rotary motion and power via mating gears, belts and chains. Alternatively, a shaft may simply connect to another via a coupling. A shaft can be stationary and support a rotating member such as the short shafts which support the non-driven wheels of automobiles, often referred to as spindles. Some common shaft arrangements are illustrated in Fig. 3.1.

Shaft design considerations include:

1. size and spacing of components (as on a general assembly drawing), tolerances
2. material selection and material treatments
3. deflection and rigidity
   (a) bending deflection
   (b) torsional deflection
   (c) slope at bearings
   (d) shear deflection
4. stress and strength
   (a) static strength
   (b) fatigue
   (c) reliability
5. frequency response
6. manufacturing constraints.

Shafts typically consist of a series of stepped diameters, accommodating bearing mounts and providing shoulders for locating devices such as gears, sprockets and pulleys to butt up against; keys are often used to prevent rotation, relative to the shaft, of these 'added' components. A typical arrangement illustrating the use of constant diameter sections and shoulders is shown in Fig. 3.2 for a transmission shaft supporting a gear and pulley wheel.

Shafts must be designed so that deflections are within acceptable levels. Too much deflection can, for example, degrade gear performance and cause noise and vibration. The maximum allowable deflection of a shaft is usually determined by limitations set on the critical speed, minimum deflections required for gear operation and bearing

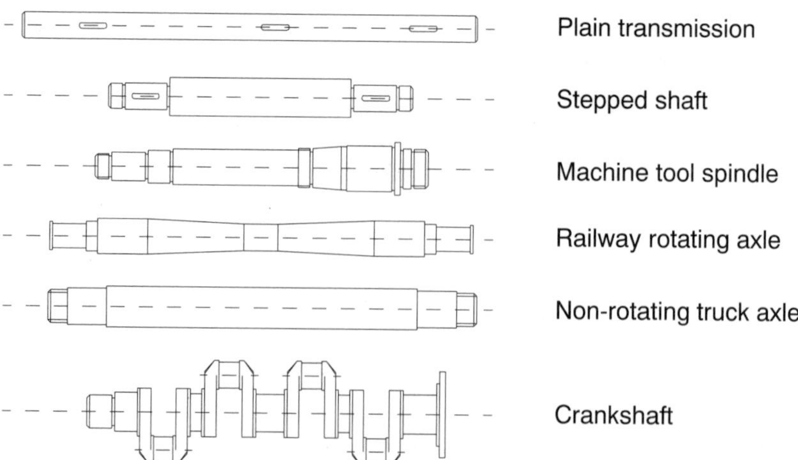

Fig. 3.1 Typical shaft arrangements. (Adapted from Reshetov, 1978.)

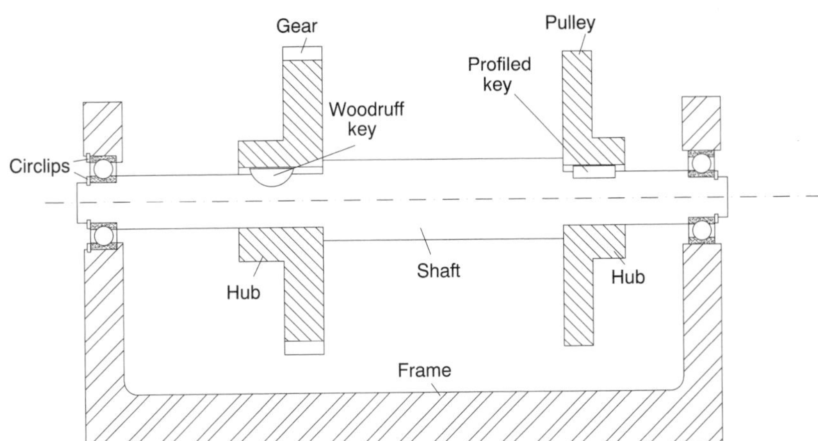

Fig. 3.2 Typical shaft arrangement incorporating constant diameter sections and shoulders for locating added components.

requirements. In general, deflections should not cause mating gear teeth to separate more than about 0.13 mm and the slope of the gear axes should not exceed about 0.03°. The deflection of the journal section of a shaft across a plain bearing should be small in comparison with the oil film thickness. Torsional and lateral deflection both contribute to lower critical speed. In addition, shaft angular deflection at rolling element bearings should not exceed 0.04°, with the exception being self-aligning rolling element bearings.

Shafts can be subjected to a variety of combinations of axial, bending and torsional loads (*see* Fig. 3.3), which may fluctuate or vary with time. Typically, a rotating shaft transmitting power is subjected to a constant torque together with a completely reversed bending load, producing a mean torsional stress and an alternating bending stress respectively.

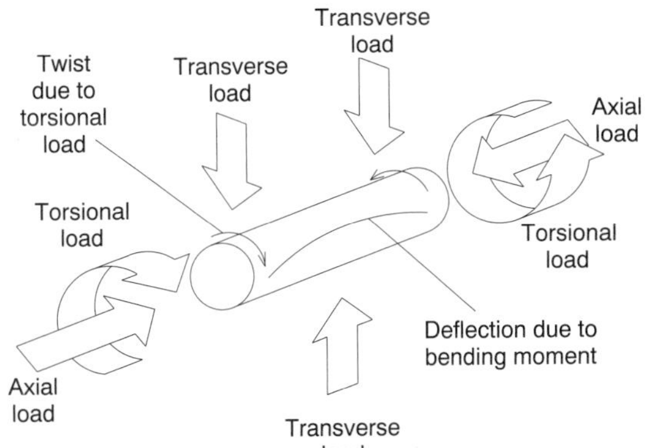

Fig. 3.3 Typical shaft loading and deflection. (Reproduced from Beswarick, 1994.)

Shafts should be designed to avoid operation at, or near, critical speeds. This is usually achieved by the provision of sufficient lateral rigidity so that the lowest critical speed is significantly above the range of operation. If torsional fluctuations are present (e.g. engine crankshafts, camshafts, compressors), the torsional natural frequencies of the shaft must be significantly different to the torsional input frequency. This can be achieved by providing sufficient torsional stiffness so that the shaft's lowest natural frequency is much higher than the highest torsional input frequency.

Rotating shafts must generally be supported by bearings. For simplicity of manufacture it is desirable to use just two sets of bearings. If more bearings are required, precise alignment of the bearings is necessary. Provision for thrust load capability and axial location of the shaft is normally supplied by just one thrust bearing taking thrust in each direction (an exception being crankshafts). It is important that the structural members supporting the shaft bearings are sufficiently strong and rigid.

The list below outlines a shaft design procedure for a shaft experiencing constant loading. The flow charts given in Figs 3.4 and 3.5 can be used to guide and facilitate design for shaft strength and rigidity and fluctuating load capability.

1. Determine the shaft rotational speed.
2. Determine the power or torque to be transmitted by the shaft.
3. Determine the dimensions of the power transmitting devices and other components mounted on the shaft and specify locations for each device.
4. Specify the locations of the bearings to support the shaft.

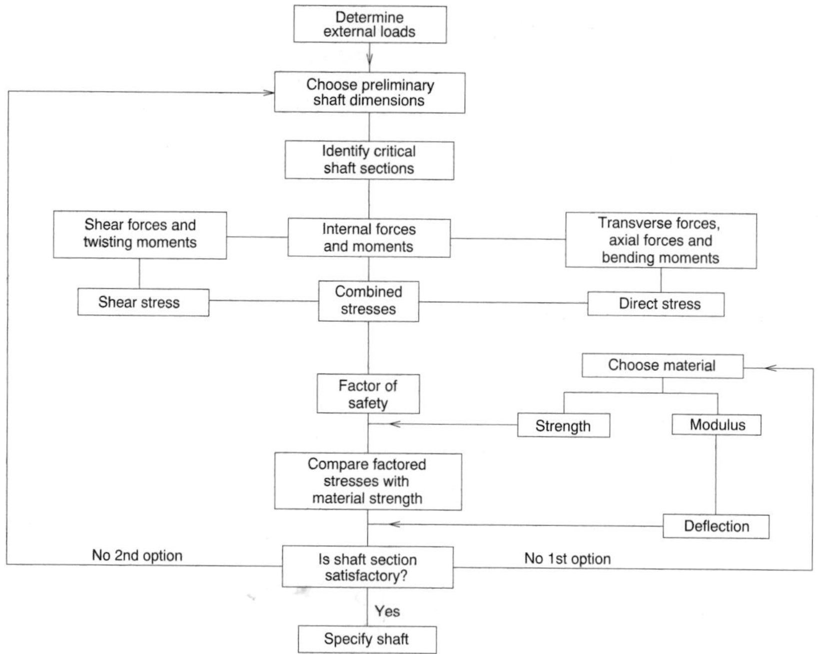

Fig. 3.4 Design procedure flow chart for shaft strength and rigidity. (Reproduced from Beswarick, 1994.)

5. Propose a general form or scheme for the shaft geometry, considering how each component will be located axially and how power transmission will take place.
6. Determine the magnitude of the torques throughout the shaft.
7. Determine the forces exerted on the shaft.
8. Produce shearing force and bending moment diagrams so that the distribution of bending moments in the shaft can be determined.
9. Select a material for the shaft and specify any heat treatments etc.

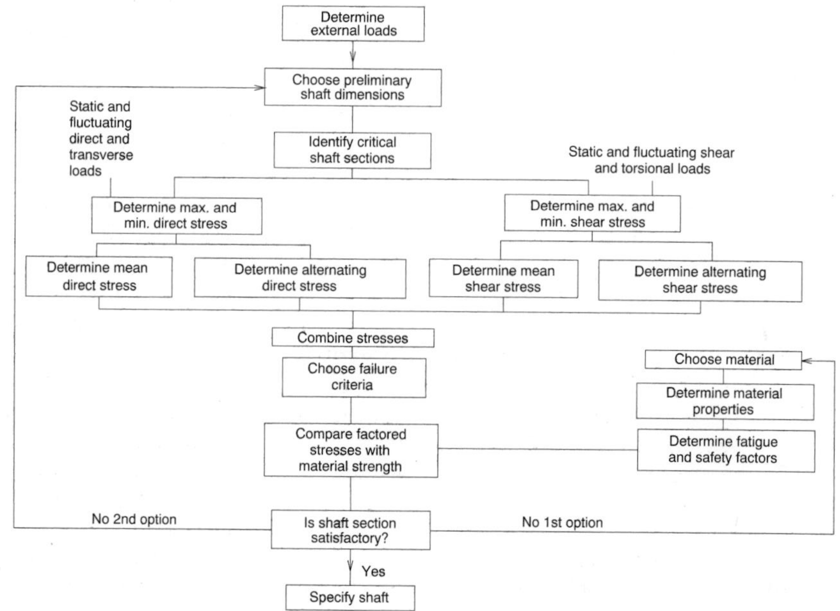

Fig. 3.5 Design procedure flow chart for a shaft with fluctuating loading. (Reproduced from Beswarick, 1994.)

Table 3.1 Merits of various shaft–hub connections

	Pin	Grub screw	Clamp	Press fit	Shrink fit	Spline	Key	Taper bush
High torque capacity	×	×	✓	×	✓	✓	✓	✓
Large axial loads	✓	×	✓	×	✓	×	×	✓
Axially compact	×	×	×	✓	✓	✓	✓	✓
Axial location provision	✓	✓	✓	✓	✓	×	×	✓
Easy hub replacement	×	✓	✓	×	×	✓	✓	✓
Fatigue	×	×	✓	✓	✓	×	×	✓
Accurate angular positioning	✓	×	×	×	×	✓	✓	(✓)
Easy position adjustment	×	✓	✓	×	×	×	×	✓

After Hurst (1994).

10. Determine an appropriate design stress, taking into account the type of loading (whether smooth, shock, repeated, reversed).
11. Analyse all the critical points on the shaft and determine the minimum acceptable diameter at each point to ensure safe design.
12. Specify the final dimensions of the shaft. This is best achieved using a detailed manufacturing drawing to a recognised standard (BSI, 1984), and the drawing should include all the information required to ensure the desired quality. Typically this will include material specifications, dimensions and tolerances (bilateral, runout, datums etc.; *see* Chapter 8), surface finishes, material treatments and inspection procedures.

The following general principles should be observed in shaft design.

- Keep shafts as short as possible with the bearings close to applied loads. This will reduce shaft deflection and bending moments and increase critical speeds.
- If possible, locate stress raisers away from highly stressed regions of the shaft. Use generous fillet radii and smooth surface finishes, and consider using local surface strengthening processes such as shot peening and cold rolling.
- If weight is critical use hollow shafts.

An overview of shaft–hub connection methods is given in Section 3.2, shaft-to-shaft connection methods in Section 3.3 and the determination of critical speeds in Section 3.4. In Section 3.5, the ASME equation for the design of transmission shafts is introduced.

3.2 Shaft–hub connection

Power-transmitting components such as gears, pulleys and sprockets need to be mounted on shafts securely and located axially with respect to mating components. In addition, a method of transmitting torque between the shaft and the component must be supplied. The portion of the component in contact with the shaft is called the hub and can be attached to, or driven by, the shaft by keys, pins, setscrews, press and shrink fits, splines and taper bushes. Table 3.1 identifies the merits of various connection methods. Alternatively, the component can be formed as an integral part of a shaft as, for example, the cam on an automotive camshaft.

Figure 3.6 illustrates the practical implementation of several shaft–hub connection methods. Gears, for example, can be gripped axially between a shoulder on a shaft and a spacer with the torque transmitted through a key. Various configurations of keys exist, including square, flat and round keys, as shown in Fig. 3.7. The grooves in the shaft and hub into which the key fits are called keyways or keyseats. A simpler and less expensive method for transmitting light loads is to use pins and various pin types are illustrated in Fig. 3.8. An inexpensive method of providing axial location of hubs and bearings on shafts is to use circlips, as shown in Figs 3.2 and 3.9. One of the simplest hub–shaft attachments is to use an interference fit, where the hub bore is slightly smaller than the shaft diameter. Assembly is achieved by press fitting, or thermal expansion of the outer ring by heating and thermal contraction of the inner ring by use of liquid nitrogen. The design of interference fits is covered in greater detail in Section 8.2.2. Mating splines, as shown in Fig. 3.10, comprise teeth cut into both the shaft and the hub, and provide one of the strongest methods of transmitting torque. Both splines and keys can be designed to allow axial sliding along the shaft.

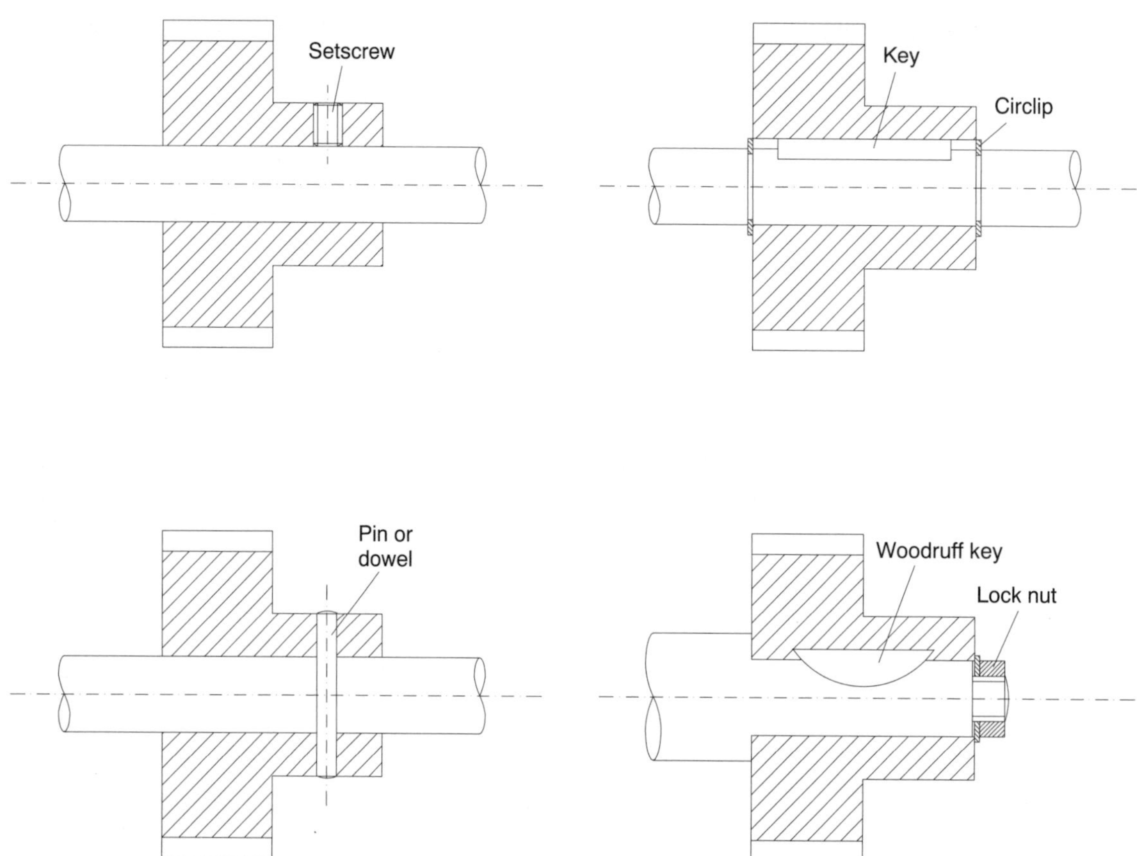

Fig. 3.6 Alternative methods of shaft–hub connection.

3.3 Shaft–shaft connections – couplings

In order to transmit power from one shaft to another a coupling or clutch can be used (for clutches *see* Chapter 7). There are two general types of coupling: rigid and flexible. Rigid couplings are designed to connect two shafts together so that no relative motion occurs between them (*see* Fig. 3.11). Rigid couplings are suitable when precise alignment of two shafts is required. If significant radial or axial misalignment occurs, high stresses may result which can lead to early failure. Flexible couplings (*see* Fig. 3.12) are designed to transmit torque whilst permitting some axial, radial and angular misalignment. Many forms of flexible coupling are available (e.g. *see* manufacturers' catalogues such as Turboflex and Fenner). Each coupling is designed to transmit a given limiting torque. Generally, flexible couplings are able to tolerate up to $\pm 3^\circ$ of angular misalignment and up to 0.75 mm parallel misalignment, depending on their design. If more misalignment is required a universal joint can be used (*see* Fig. 3.13).

3.4 Critical speeds and shaft deflection

The centre of mass of a rotating system (for example, a mid-mounted disc on a shaft supported by bearings at each end) will never coincide with the centre of rotation, due to manufacturing and operational constraints. As the shaft rotational speed is increased, the centrifugal force acting at the centre of mass tends to bow the shaft. The more the shaft bows, the greater the eccentricity and the greater the centrifugal force. Below the lowest critical speed of rotation, the centrifugal and elastic forces balance at a finite value of shaft deflection. At the critical speed, equilibrium theoretically requires infinite deflection of the centre of mass, although realistically bearing damping, internal hysteresis and windage causes equilibrium to occur at a finite displacement. However, this displacement

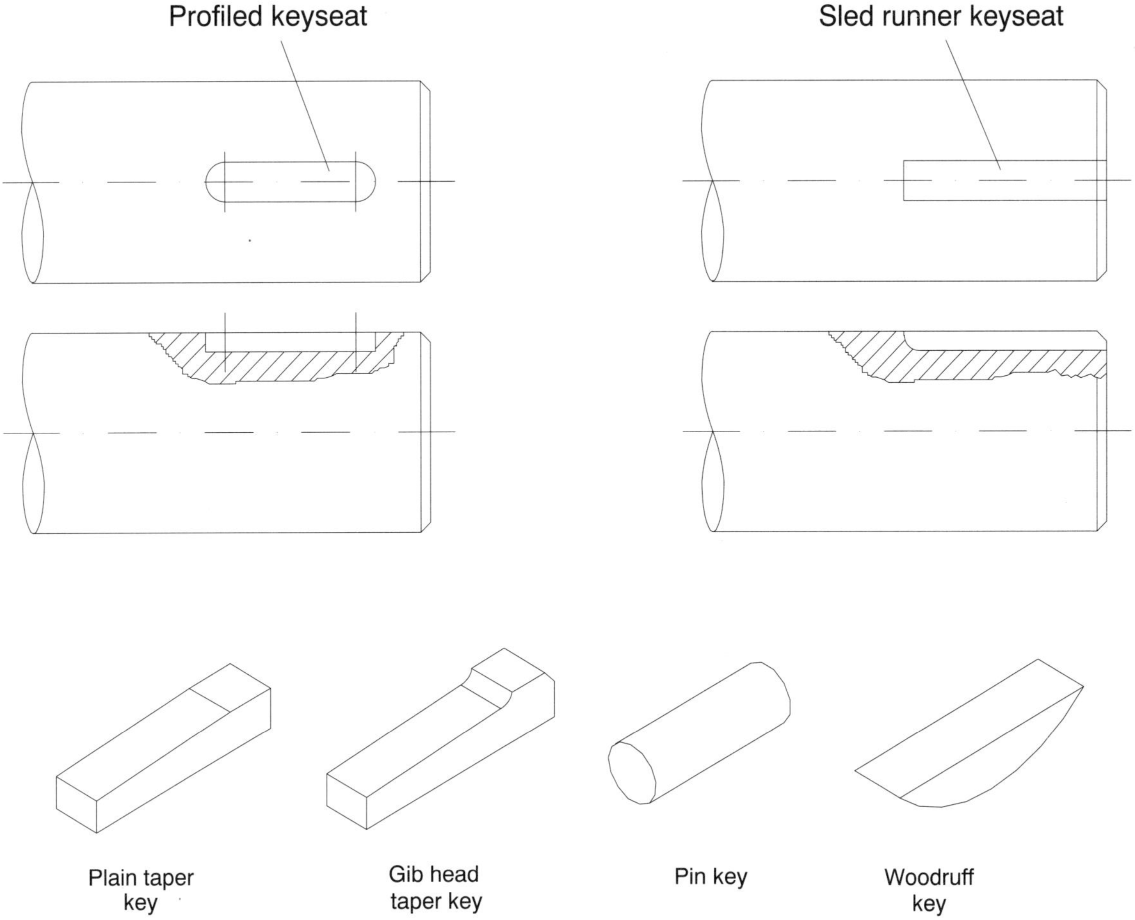

Profiled keyseat
Sled runner keyseat
Plain taper key
Gib head taper key
Pin key
Woodruff key

Fig. 3.7 Keys for torque transmission and component location.

can be large enough to break the shaft, damage bearings and cause destructive machine vibration.

The critical speed of rotation is the same as the lateral frequency of vibration, which is induced when rotation is stopped and the centre displaced laterally and suddenly released (i.e. the stationary shaft would vibrate at this frequency if struck with a hammer). For all shafts, except for the single concentrated mass shaft, critical speeds also occur at higher frequencies.

At the first critical speed the shaft will bend into the simplest possible shape, and at the second critical speed it will bend into the next simplest shape. For example, the shapes the shaft will bend into at the first two critical speeds (or

Hub
Shaft
Straight round pin
Tapered round pin
Split tubular spring pin

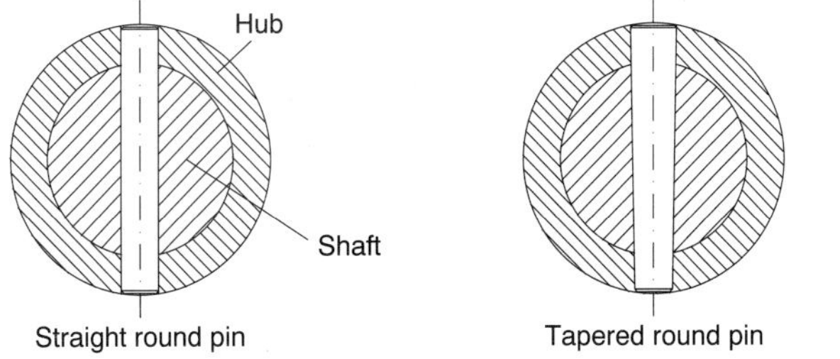

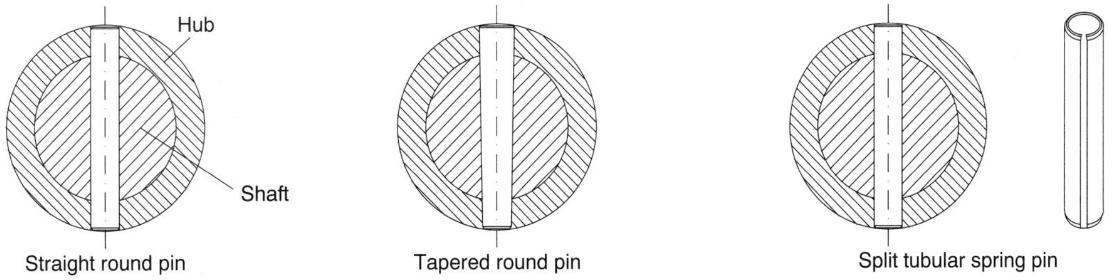

Fig. 3.8 Pins for torque transmission and component location.

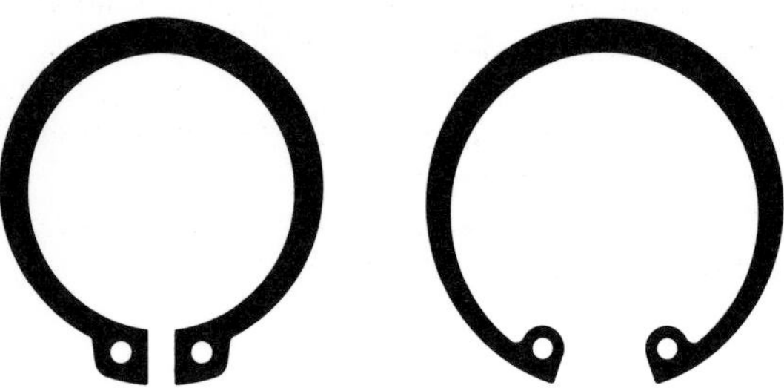
Fig. 3.9 Circlips or snap rings.

modes) for an end-supported shaft with two masses are illustrated in Fig. 3.14.

In certain circumstances the fundamental frequency of a shaft system cannot be made higher than the shaft design speed. If the shaft can be accelerated rapidly through and beyond the first resonant critical frequency, before the vibrations have a chance to build up in amplitude, then the system can be run at speeds higher than the natural frequency. This is the case with steam and gas turbines, where the size of the turbomachinery and generators give a low natural frequency but must be run at high speed due to efficiency considerations. As a general design principle, maintaining the operating speed of a shaft below half the shaft whirl critical frequency is normally acceptable.

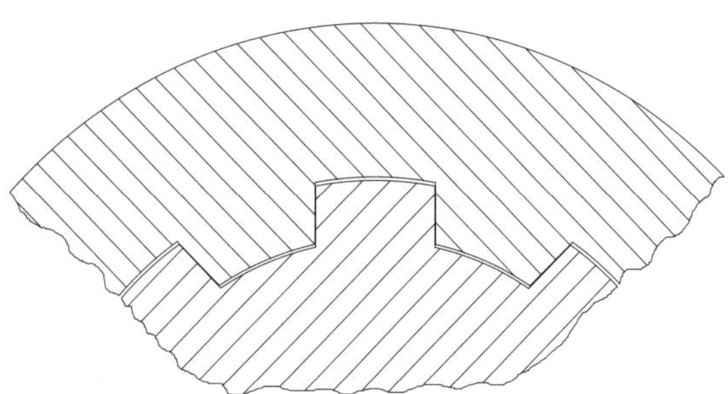

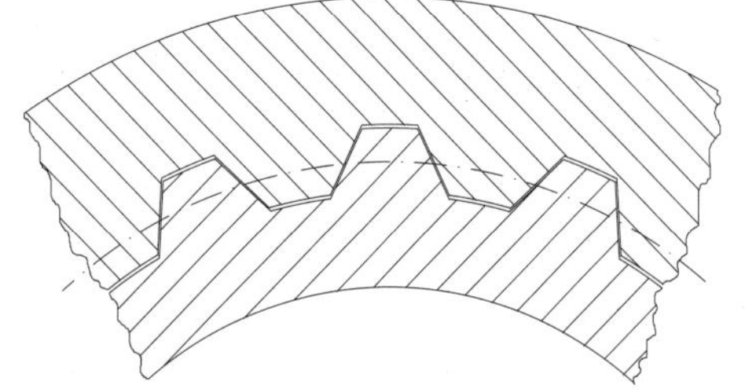

Fig. 3.10 Splines.

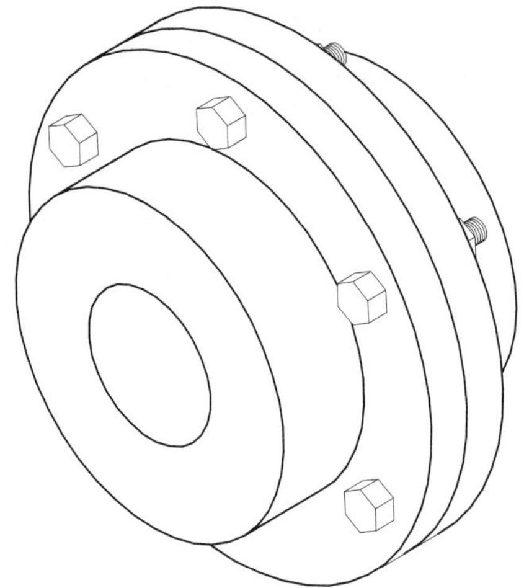
Fig. 3.11 Rigid coupling.

A complete analysis of the natural frequencies of a shaft can be performed using a finite element analysis package, such as ANSYS, called a 'nodal analysis'. This can give a large number of natural frequencies in three dimensions from the fundamental upwards. This is the sensible and easiest approach for complex systems, but a quick estimate for a simplified system can be undertaken for design purposes, as outlined in this section.

The critical speed of a shaft with a single mass attached can be approximated by

$$\omega_c = \sqrt{g/y} \qquad (3.1)$$

Fig. 3.12 Flexible couplings. (Photograph courtesy of Cross and Morse.)

Fig. 3.13 Universal joints. (Photograph courtesy of HPC Drives Ltd.)

where y is the static deflection at the location of the mass.

The first critical speed of a shaft carrying several concentrated masses is approximated by the Rayleigh–Ritz equation:

$$\omega_c = \sqrt{\frac{g\sum_{i=1}^{n} W_i y_i}{\sum_{i=1}^{n} W_i y_i^2}} \tag{3.2}$$

where ω_c is the first critical frequency, W_i is the mass or weight of node i and y_i is the static deflection of W_i.

The dynamic deflections of a shaft are generally unknown. Rayleigh showed that an estimate of the deflection curve is suitable, provided it represents the maximum deflection and the boundary conditions. The static deflection curve due to the shaft's own weight and the weight of any attached components gives a suitable estimate. Note that external loads are not considered in this analysis, only those due to gravitation. The resulting calculation gives a value higher than the actual natural frequency by a few per cent.

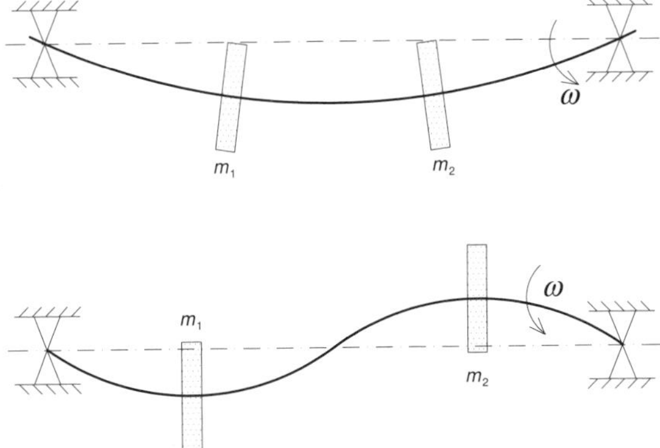

Fig. 3.14 Shaft shapes for a simply supported shaft with two masses at the first and second critical speeds (masses large in comparison with shaft mass).

Alternatively, the Dunkerley equation can be used to estimate the critical frequency:

$$\frac{1}{\omega_c^2} = \frac{1}{\omega_1^2} + \frac{1}{\omega_2^2} + \frac{1}{\omega_3^2} + \cdots \tag{3.3}$$

where ω_1 is the critical speed if only mass number 1 is present etc.

Both the Rayleigh–Ritz and the Dunkerley equations are approximations to the first natural frequency of vibration which is assumed nearly equal to the critical speed. The Dunkerley equation tends to underestimate and the Rayleigh–Ritz equation to overestimate the critical frequency.

Often we need to know the second critical speed. For a two mass system, approximate values for the first two critical speeds can be found by solving the following frequency equation:

$$\frac{1}{\omega^4} - (a_{11}m_1 + a_{22}m_2)\frac{1}{\omega^2} + (a_{11}a_{22} - a_{12}a_{21})m_1 m_2 = 0 \tag{3.4}$$

For a multimass system the frequency equation can be obtained by equating the following determinate to zero:

$$\begin{vmatrix} \left(a_{11}m_1 - \frac{1}{\omega^2}\right) & a_{12}m_2 & a_{13}m_3 & \cdots \\ a_{21}m_1 & \left(a_{22}m_2 - \frac{1}{\omega^2}\right) & a_{23}m_3 & \cdots \\ a_{31}m_1 & a_{32}m_2 & \left(a_{33}m_3 - \frac{1}{\omega^2}\right) & \cdots \\ \cdots & \cdots & \cdots & \cdots \end{vmatrix} = 0 \tag{3.5}$$

where a_{11}, a_{12} etc. are the influence coefficients. These correspond to the deflection of the shaft at the locations of the loads as a result of 1 N loads. The first subscript refers to location of the deflection, the second to the location of the 1 N force. For example, the influence coefficients for a simply supported shaft with two loads, as illustrated in Fig. 3.15, are listed below.

- a_{11} is the deflection at the location of mass 1 that would be caused by a 1 N weight at the location of mass 1.
- a_{21} is the deflection at the location of mass 2 that would be caused by a 1 N weight at the location of mass 1.

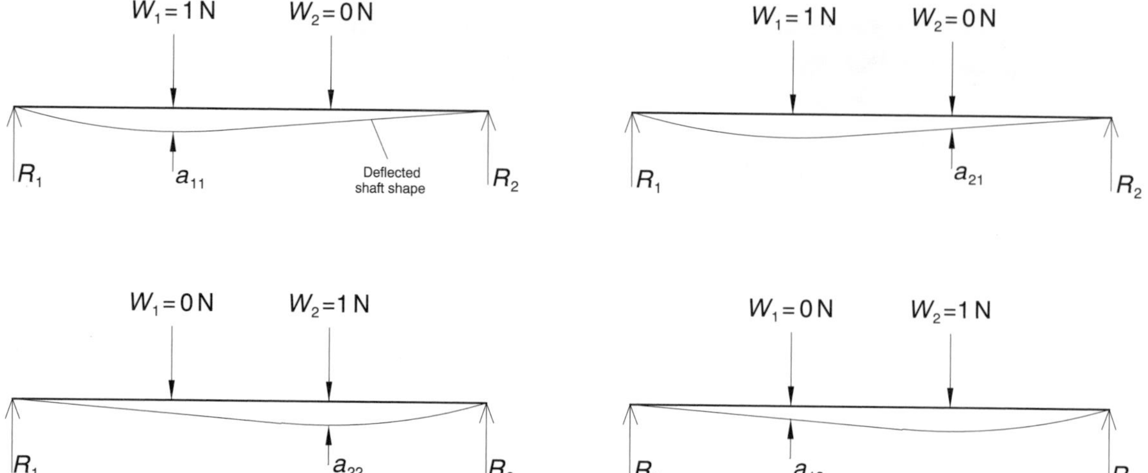

Fig. 3.15 Example of influence coefficient definition for a simply supported shaft with two concentrated loads.

- a_{22} is the deflection at the location of mass 2 that would be caused by a 1 N weight at the location of mass 2.
- a_{12} is the deflection at the location of mass 1 that would be caused by a 1 N weight at the location of mass 2.

Note that values for a_{np} and a_{pn} are equal by the principle of reciprocity (e.g. $a_{12} = a_{21}$).

It should be noted that lateral vibration requires an external source of energy. For example, vibrations can be transferred from another part of a machine and the shaft will vibrate in one or more lateral planes, regardless of whether the shaft is rotating. Shaft whirl is a self-excited vibration caused by the shaft's rotation acting on an eccentric mass.

These analysis techniques for calculating the critical frequency require the determination of the shaft deflection. Section 3.4.1 introduces a method suitable for calculating the deflection of a constant diameter shaft and Section 3.4.2 introduces methods for more complex shafts with stepped diameters.

3.4.1 Macaulay's method for calculating the deflection of beams

Macaulay's method can be used to determine the deflection of a constant cross-section shaft. The general rules for this method are given below. Having calculated the deflections of a shaft, this information can then be used to determine critical frequencies.

1. Take an origin at the left hand side of the beam.
2. Express the bending moment at a suitable section XX in the beam to include the effect of all the loads.
3. Uniformly distributed loads must be made to extend to the right hand end of the beam. Use negative loads to compensate.
4. Put in square brackets all functions of length other than those involving single powers of x.
5. Integrate as a whole any term in square brackets.
6. When evaluating the moment, slope or deflection, neglect the square brackets terms when they become negative.
7. In the moment equation, express concentrated moments in the form $M_1[x - a]^0$, where M_1 is the concentrated moment and $x - a$ is its point of application relative to the section XX.

Example As part of the preliminary design of a machine shaft, a check is undertaken to determine that the critical speed is significantly higher than the design speed of 7000 rpm. The components can be represented by three point masses, as shown in Fig. 3.16. Assume the bearings are stiff and act as simple supports. The shaft diameter is 40 mm and the material is steel, with a Young's modulus of 200×10^9 N/m^2.

Solution Macaulay's method is used to determine the shaft deflections. Resolving vertically:

$$R_1 + R_2 = W_1 + W_2 + W_3$$

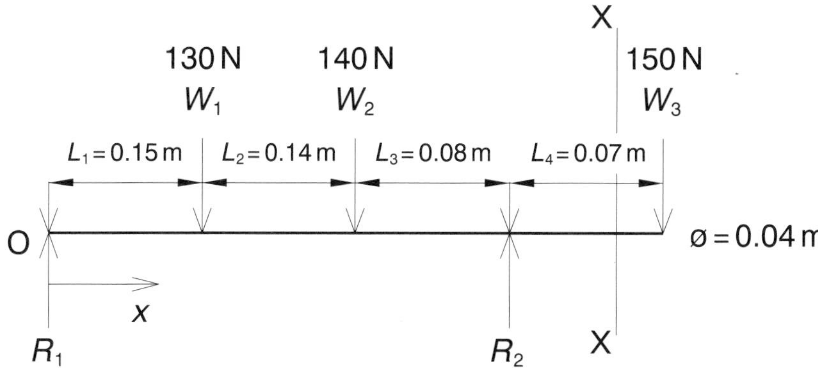

Fig. 3.16 Machine shaft example.

Clockwise moments about O:

$$W_1 L_1 + W_2(L_1 + L_2) - R_2(L_1 + L_2 + L_3) + W_3(L_1 + L_2 + L_3 + L_4) = 0$$

Hence

$$R_2 = \frac{W_1 L_1 + W_2(L_1 + L_2) + W_3(L_1 + L_2 + L_3 + L_4)}{L_1 + L_2 + L_3}$$

Calculating the moment at XX:

$$M_{XX} = -R_1 x + W_1[x - L_1] + W_2[x - (L_1 + L_2)] - R_2[x - (L_1 + L_2 + L_3)] \tag{3.6}$$

The relationship for the deflection y of a beam subjected to a bending moment M is given by

$$EI\frac{d^2y}{dx^2} = M \tag{3.7}$$

where I is the second moment of area, E is the modulus of elasticity or Young's modulus, and x is the distance from the end of the beam to the location at which the deflection is to be determined. This can be integrated once to find the slope dy/dx and twice to find the deflection y.

Integrating equation 3.6 finds the slope:

$$EI\frac{dy}{dx} = -R_1\frac{x^2}{2} + \frac{W_1}{2}[x - L_1]^2 + \frac{W_2}{2}[x - (L_1 + L_2)]^2 - \frac{R_2}{2}[x - (L_1 + L_2 + L_3)]^2 + A$$

Integrating again to find the deflection equation gives

$$EIy = -R_1\frac{x^3}{6} + \frac{W_1}{6}[x - L_1]^3 + \frac{W_2}{6}[x - (L_1 + L_2)]^3 - \frac{R_2}{6}[x - (L_1 + L_2 + L_3)]^3 + Ax + B$$

Note that Macaulay's method requires that terms within square brackets be ignored when the sign of the bracket goes negative.

It is now necessary to substitute boundary conditions to find the constants of integration. At $x = 0$, $y = 0$, hence $B = 0$. At $x = L_1 + L_2 + L_3$, $y = 0$:

$$0 = -\frac{R_1}{6}(L_1 + L_2 + L_3)^3 + \frac{W_1}{6}(L_2 + L_3)^3 + \frac{W_2}{6}(L_3)^3 + A(L_1 + L_2 + L_3)$$

Hence

$$A = \frac{\frac{R_1}{6}(L_1 + L_2 + L_3)^3 - \frac{W_1}{6}(L_2 + L_3)^3 - \frac{W_2}{6}(L_3)^3}{L_1 + L_2 + L_3}$$

$$I = \frac{\pi d^4}{64} = 1.2566 \times 10^{-7}\ \text{m}^4$$

With $W_1 = 130$ N, $W_2 = 140$ N, $W_3 = 150$ N, $\phi = 0.04$ m, and $E = 200 \times 10^9$ N/m^2, substitution of these values into the above equations gives

$R_1 = 79.19$ N

$R_2 = 340.8$ N

$A = 1.151$

At $x = 0.15$ m, $y = 5.097 \times 10^{-6}$ m.
At $x = 0.29$ m, $y = 2.839 \times 10^{-6}$ m.
At $x = 0.44$ m, $y = -1.199 \times 10^{-6}$ m.

Use *absolute* values for the displacement in the Rayleigh–Ritz equation:

$$\omega_c = \sqrt{\frac{9.81[(130 \times 5.097 \times 10^{-6}) + (140 \times 2.839 \times 10^{-6}) + (150 \times |-1.199 \times 10^{-6}|)]}{[130 \times (5.097 \times 10^{-6})^2] + [140 \times (2.839 \times 10^{-6})^2] + [150 \times (|-1.199 \times 10^{-6}|)^2]}}$$

$$= \sqrt{\frac{9.81 \times 1.240 \times 10^{-3}}{4.722 \times 10^{-9}}} = 1605\,\text{rad/s} \quad \Rightarrow$$

$$\omega_c = 15\,330\,\text{rpm}$$

Example This example explores the influence on the critical frequency of adding an overhang load to a beam.

1. Determine the critical frequency for a 0.4 m long shaft running on bearings with a single point load, as illustrated in Fig. 3.17. Assume that the bearings are rigid and act as simple supports. Ignore the self-weight of the shaft. Take Young's modulus of elasticity as $200 \times 10^9\,\text{N/m}^2$.
2. Calculate the critical frequency if an overhang load of 150 N is added at a distance of 0.15 m from the right hand bearing, as shown in Fig. 3.18.

Solution

1. Resolving vertically: $R_1 + R_2 = W_1$.
   Clockwise moments about O: $375 \times 0.25 - R_2 0.4 = 0$.
   Hence $R_2 = 234.375\,\text{N}$, $R_1 = 140.625\,\text{N}$.

$$EI\frac{d^2y}{dx^2} = -R_1 x + W_1[x - L_1]$$

$$EI\frac{dy}{dx} = -R_1\frac{x^2}{2} + \frac{W_1}{2}[x - L_1]^2 + A$$

$$EIy = -R_1\frac{x^3}{6} + \frac{W_1}{6}[x - L_1]^3 + Ax + B$$

Boundary conditions:
At $x = 0$, $y = 0$. Hence $B = 0$.
At $x = L$, $y = 0$. Hence

$$A = \frac{1}{L}\left(R_1\frac{L^3}{6} - \frac{W_1}{6}(L - L_1)^3\right)$$

$$= \frac{1}{0.4}(1.5 - 0.2109) = 3.22266$$

$$I = 1.2566 \times 10^{-7}\,\text{m}^4, \quad E = 200 \times 10^9\,\text{N/m}^2$$

$$EIy = -140.625 \times \frac{0.25^3}{6} + (3.22266 \times 0.25)$$

$$y = 1.7485 \times 10^{-5}\,\text{m}$$

Application of Macaulay's method to the geometry given in Fig. 3.17 yields $y_1 = 1.74853 \times 10^{-5}\,\text{m}$. So from equation 3.1:

$$\omega_c = \sqrt{\frac{g}{y_1}} = 749.03\,\text{rad/s} = 7152\,\text{rpm}$$

2. $$EI\frac{d^2y}{dx^2} = -R_1 x + W_1[x - L_1] - R_2[x - (L_1 + L_2)]$$

$$EI\frac{dy}{dx} = -R_1\frac{x^2}{2} + \frac{W_1}{2}[x - L_1]^2 - \frac{R_2}{2}[x - (L_1 + L_2)]^2 + A$$

$$EIy = -R_1\frac{x^3}{6} + \frac{W_1}{6}[x - L_1]^3 - \frac{R_2}{6}[x - (L_1 + L_2)]^3 + Ax + B$$

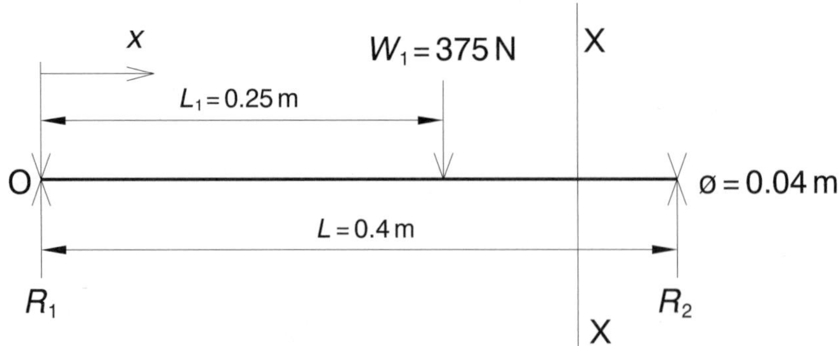

Fig. 3.17 Simple shaft with single concentrated mass.

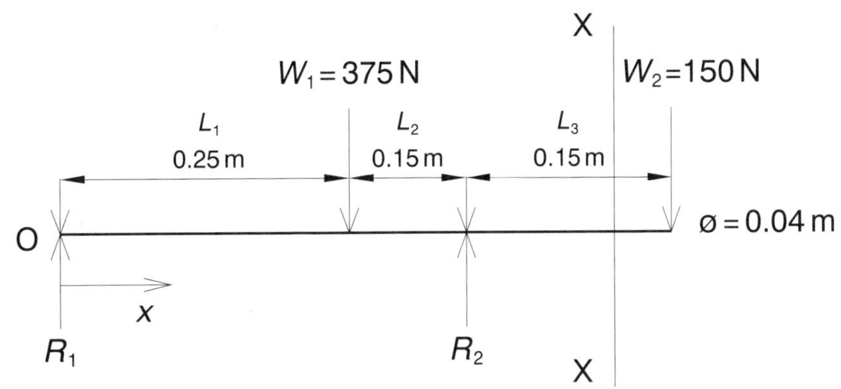

Fig. 3.18 Simple shaft with a single concentrated mass between bearing and overhang mass.

Boundary conditions:
At $x = 0$, $y = 0$. Hence $B = 0$.
At $x = L_1 + L_2$, $y = 0$. Hence

$$A = \frac{1}{L_1 + L_2}\left(\frac{R_1}{6}(L_1 + L_2)^3 - \frac{W_1}{6}L_2^3\right)$$

Resolving vertically: $R_1 + R_2 = W_1 + W_2$.
Moments about O:

$$W_1 L_1 + W_2(L_1 + L_2 + L_3) - R_2(L_1 + L_2) = 0$$

Hence

$$R_2 = \frac{1}{L_1 + L_2}(W_1 L_1 + W_2(L_1 + L_2 + L_3))$$

$$R_1 = W_1 + W_2 - R_2$$

$$R_1 = 84.375\,\text{N}$$

$$R_2 = 440.625\,\text{N}$$

Substitution for A gives $A = 1.722656$.
Evaluating y at $x = L_1$ gives

$$y = 8.3929381 \times 10^{-6}\,\text{m}$$

Evaluating y at $x = L_1 + L_2 + L_3$ gives $y = 1.8884145 \times 10^{-6}$ m.

Hence

$$\omega_c = \sqrt{\frac{9.81(3.147 \times 10^{-3} + 2.8326 \times 10^{-4})}{2.64155 \times 10^{-8} + 5.34061 \times 10^{-10}}}$$

$$= \sqrt{\frac{9.81 \times 3.4306 \times 10^{-3}}{2.695 \times 10^{-8}}} = 1117.5\,\text{rad/s}$$

$$\omega_c = 10\,671\,\text{rpm}$$

Comparison of the results from parts 1 and 2 shows that the addition of an overhang weight can have a significant effect in raising the critical frequency from ~7150 rpm to ~10 670 rpm in this case. This technique is frequently used in plant machinery.

Example

1. Calculate the first two critical speeds for the loadings on a steel shaft, indicated in Fig. 3.19. Assume the mass of the shaft can be ignored for the purpose of this calculation and that the bearings can be considered to be stiff simple supports. The internal and external diameters of the shaft are 0.03 and 0.05 m respectively. Take Young's modulus as 200 GN/m^2.

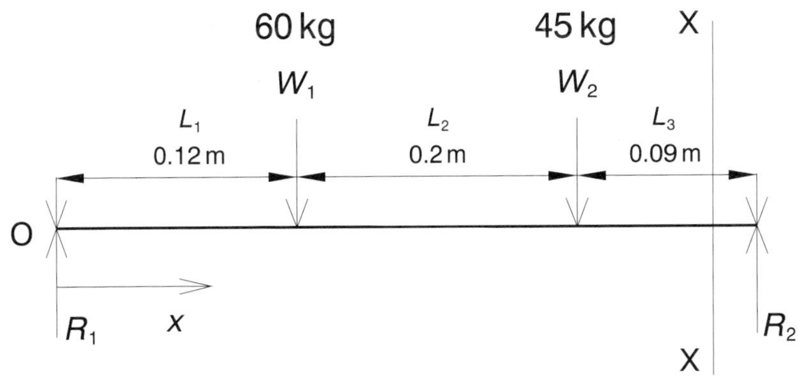

Fig. 3.19 Simple shaft.

Table 3.2 Bearing data for deflection versus load

Load on bearing (N)	Deflection at bearing (mm)
50	0.003
100	0.009
200	0.010
300	0.0115
400	0.0126
500	0.0136
1000	0.0179
2000	0.0245

2. The design speed for the shaft is nominally 5000 rpm; what would be the implication of the bearing data given in Table 3.2 on the shaft operation, assuming the bearings allow flexibility only in directions perpendicular to the shaft axis?

Solution

1. Apply Macaulay's method to find the influence coefficients: first calculate the deflections at L_1 and L_2 with $W_1 = 1\,\text{N}$ and $W_2 = 0$ to give a_{11} and a_{21}; then calculate the deflections at L_1 and L_2 with $W_1 = 0\,\text{N}$ and $W_2 = 1\,\text{N}$ to give a_{12} and a_{22}:

$$EIy = -R_1\frac{x^3}{6} + \frac{W_1}{6}[x - L_1]^3 + \frac{W_2}{6}[x - (L_1 + L_2)]^3 + Ax + B$$

With $W_1 = 1\,\text{N}$, $W_2 = 0\,\text{N}$: $R_1 = 0.707317$, $R_2 = 0.292683$, $A = 9.90244 \times 10^{-3} (B = 0)$. Hence

$$a_{11} = 1.84355 \times 10^{-8},\ a_{21} = 1.19688 \times 10^{-8}$$

With $W_1 = 0\,\text{N}$, $W_2 = 1\,\text{N}$: $R_1 = 0.2195122$, $R_2 = 0.7804878$, $A = 5.853659 \times 10^{-3}$. Hence

$$a_{22} = 1.26264 \times 10^{-8},\ a_{12} = 1.19688 \times 10^{-8}$$

Solving

$$\frac{1}{\omega^4} - (a_{11}m_1 + a_{22}m_2)\frac{1}{\omega^2} + (a_{11}a_{22} - a_{12}a_{21})m_1m_2 = 0$$

by multiplying through by ω^4 and solving as a quadratic:

$$\omega^2 = \frac{-b \pm \sqrt{b^2 - 4ac}}{2a}$$

with

$$a = (a_{11}a_{22} - a_{12}a_{21})m_1m_2$$

$$b = -(a_{11}m_1 + a_{22}m_2)$$

$$c = 1$$

gives $\omega_{c1} = 812.5\,\text{rad/s}$ (7759 rpm) and $\omega_{c2} = 2503.4\,\text{rad/s}$ (23 905 rpm).

2. The deflection at the bearings should also be taken into account in determining the critical frequency of the shaft. The reaction at each of the bearings is approximately 500 N ($R_1 = 513.2\,\text{N}$ and $R_2 = 516.8\,\text{N}$). The deflection at each bearing can be assumed roughly equal (so from Table 3.2, $y_{\text{bearing}} = 0.0136\,\text{mm}$). This deflection can be simply added to the static deflection of the shaft due to bending by superposition. If the deflections were not equal, similar triangles could be used to calculate the deflection at the mass locations.

The influence coefficients can also be used to determine the deflection under the actual loads applied, $W_1 = 60 \times 9.81 = 588.6\,\text{N}$, $W_2 = 45 \times 9.81 = 441.45\,\text{N}$:

$$y_{1\,\text{Load}} = W_1a_{11} + W_2a_{12} = 1.613 \times 10^{-5}\,\text{m}$$

$$y_{2\,\text{Load}} = W_1a_{21} + W_2a_{22} = 1.26 \times 10^{-5}\,\text{m}$$

$$y_{1\,\text{Total}} = 1.61 \times 10^{-5} + 1.36 \times 10^{-5} = 2.97 \times 10^{-5}\,\text{m}$$

$$y_{2\,\text{Total}} = 1.26 \times 10^{-5} + 1.36 \times 10^{-5} = 2.62 \times 10^{-5}\,\text{m}$$

These deflections give (using the Rayleigh–Ritz equation) a first critical frequency of

$$\omega_c = \sqrt{\frac{9.81 \times 2.96 \times 10^{-3}}{8.38 \times 10^{-8}}} = 588.6\,\text{rad/s} = 5620\,\text{rpm}$$

This is very close to the design speed. Different bearings or a stiffer shaft would be advisable.

Macaulay's method for calculating the deflection of beams and shafts is useful in that it is simple to use and easily programmed.

3.4.2 *Castigliano's theorem for calculating shaft deflections*

The strain energy, U, in a straight beam subjected to a bending moment, M, is

$$U = \int \frac{M^2\,\text{d}x}{2EI} \tag{3.8}$$

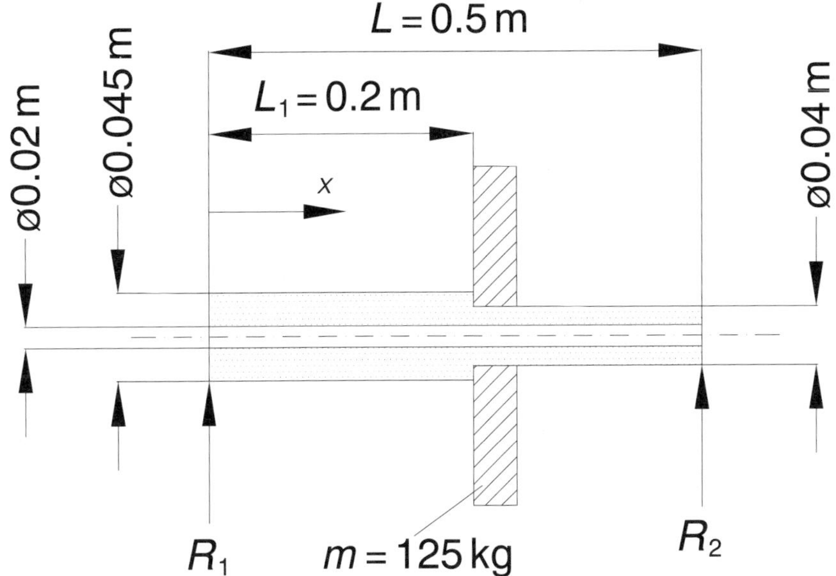

Fig. 3.20 Stepped shaft with single load.

Castigliano's strain energy equation can be applied to problems involving shafts of non-constant section to calculate deflections, and hence critical speeds.

Example Determine the deflection of the stepped shaft illustrated in Fig. 3.20 under a single concentrated load.

Solution The total strain energy due to bending is

$$U = \int \frac{M^2\,\mathrm{d}x}{2EI} = U_1 + U_2$$

where

U_1 is the energy from $x = 0$ to L_1
U_2 is the energy from $x = L_1$ to L.

In general, consider the strain energy for each section of a beam, i.e. between loads and between any change of shaft section.

Resolving vertical loads:

$$R_1 + R_2 = W$$

Taking moments about the left hand bearing:

$$R_2 = \frac{WL_1}{L}$$

Hence

$$R_1 = W\left(1 - \frac{L_1}{L}\right)$$

Splitting the beam between $x = 0$ and L_1, the moment can be expressed as $R_1 x$:

$$U_1 = \int_0^{L_1} \frac{(R_1 x)^2}{2EI_1}\,\mathrm{d}x = \left[\frac{(W(1-(L_1/L)))^2 x^3}{6EI_1}\right]_0^{L_1}$$

$$= \frac{(W(1-(L_1/L)))^2 L_1^3}{6EI_1}$$

Splitting the beam between $x = L_1$ and L, the moment can be expressed as $R_2(L - x)$:

$$U_2 = \int_{L_1}^{L} \frac{(R_2(L-x))^2}{2EI_2}\,\mathrm{d}x$$

$$= \left[\frac{(WL_1/L)^2(L^2x - Lx^2 + x^3/3)}{2EI_2}\right]_{L_1}^{L}$$

$$= \frac{W^2L_1^2}{2EI_2L^2}\left(L^3 - L^3 + \frac{L^3}{3} - L^2L_1 + LL_1^2 - \frac{L_1^3}{3}\right)$$

$$U = U_1 + U_2 = \frac{(W(1-(L_1/L)))^2 L_1^3}{6EI_1} + \frac{W^2L_1^2}{2EI_2L^2} \times \left(L^3 - L^3 + \frac{L^3}{3} - L^2L_1 + LL_1^2 - \frac{L_1^3}{3}\right)$$

Differentiating the above expression with respect to W gives the deflection at W:

$$\frac{\partial U}{\partial W} = \frac{L_1^3}{6EI_1}\left(2W - \frac{4WL_1}{L} + \frac{2WL_1^2}{L^2}\right) + \frac{2WL_1^2}{2EI_2L^2} \times \left(\frac{L^3}{3} - L^2L_1 + LL_1^2 - \frac{L_1^3}{3}\right) \tag{3.9}$$

$$I_1 = \frac{\pi(d_o^4 - d_i^4)}{64} = \frac{\pi(0.045^4 - 0.02^4)}{64}$$

$$= 1.9343 \times 10^{-7}\,\text{m}^4$$

$$I_2 = \frac{\pi(0.04^4 - 0.02^4)}{64} = 1.1781 \times 10^{-7}\,\text{m}^4$$

Substitution of $W = 125 \times 9.81\,\text{N}$, $L_1 = 0.2\,\text{m}$, $L = 0.5\,\text{m}$, $E = 200 \times 10^9\,\text{N/m}^2$, I_1 and I_2 into equation 3.9 gives a deflection $\partial U/\partial W = 1.05372 \times 10^{-4}\,\text{m}$, and hence a first critical frequency using the approximation of equation 3.1 of

$$\sqrt{\frac{9.81}{1.05372 \times 10^{-4}}} = 305.121\,\text{rad}/s = 2913\,\text{rpm}$$

An alternative to the use of Castigliano's method for determining the deflection of stepped shafts is direct numerical integration. This is outlined by Mischke (1996).

3.5 The ASME design code for transmission shafting

According to Fuchs and Stephens (1980), between 50 and 90 per cent of all mechanical failures are fatigue failures. This challenges the engineer to consider the possibility of fatigue failure at the design stage. Figure 3.21 shows the variation of fatigue strength for a steel with a number of stress cycles. For low-strength steels a 'levelling off' occurs in the graph between 10^6 and 10^7 cycles under non-corrosive conditions, and regardless of the number of stress cycles beyond this the component will not fail due to fatigue. The value of stress corresponding to this levelling off is called the endurance stress or the fatigue limit.

In order to design hollow or solid rotating shafts under combined cyclic bending and steady torsional loading for limited life, the ASME (American Society of Mechanical Engineers) design code for the design of transmission shafting can be used. The ASME procedure ensures that the shaft is properly sized to provide adequate service life, but the designer must ensure that the shaft is stiff enough to limit deflections of power transfer elements such as gears, and to minimise misalignment through seals and bearings. In addition, the shaft's stiffness must be such that it avoids unwanted vibrations through the running range.

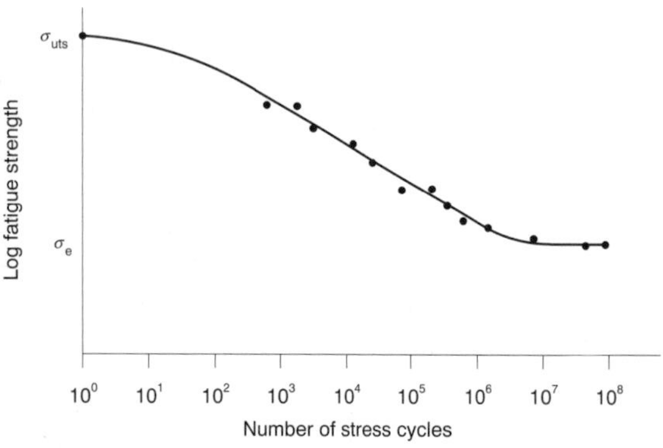

Fig. 3.21 A typical strength cycle diagram for various steels.

The equation for determining the diameter for a solid shaft is

$$d = \left[\frac{32n_s}{\pi}\sqrt{\left(\frac{M}{\sigma_e}\right)^2 + \frac{3}{4}\left(\frac{T}{\sigma_y}\right)^2}\right]^{1/3} \tag{3.10}$$

where

n_s = factor of safety
M = bending moment (N · m)
σ_e = endurance limit of the item (N/m²)
T = torque (N · m)
σ_y = yield strength (N/m²).

There is usually some uncertainty regarding what level a component will actually be loaded to, how strong a material is and how accurate the modelling methods are. Factors of safety are frequently used to account for these uncertainties. The value of a factor of safety is usually based on experience of what has given acceptable performance in the past. The level is also a function of the consequences of component failure and the cost of providing an increased safety factor. As a guide, typical values for the factor of safety based on strength recommended by Vidosek (1957) are:

- 1.25–1.5 for reliable materials under controlled conditions subjected to loads and stresses known with certainty;
- 1.5–2 for well-known materials under reasonably constant environmental conditions subjected to known loads and stresses;
- 2–2.5 for average materials subjected to known loads and stresses;
- 2.5–3 for less well-known materials under average conditions of load, stress and environment;
- 3–4 for untried materials under average conditions of load, stress and environment;
- 3–4 for well-known materials under uncertain conditions of load, stress and environment.

The endurance limit, σ_e, for a mechanical element can be estimated by equation 3.11. Here a series of modifying factors are applied to the endurance limit of a test specimen for various effects such as size, load and temperature:

$$\sigma_e = k_a k_b k_c k_d k_e k_f k_g \sigma'_e \tag{3.11}$$

where

k_a = surface factor
k_b = size factor
k_c = reliability factor
k_d = temperature factor
k_e = duty cycle factor
k_f = fatigue stress concentration factor
k_g = miscellaneous effects factor
σ'_e = endurance limit of test specimen (N/m^2).

If the stress at the location under consideration is greater than σ_e, then the component will fail eventually due to fatigue (i.e. the component has a limited life).

Mischke (1987) has determined the following approximate relationships between the endurance limit of test specimens and the ultimate tensile strength of the material (for steels only):

$$\sigma'_e = 0.504\sigma_{uts} \text{ for } \sigma_{uts} \leq 1400\,\text{MPa} \tag{3.12}$$

$$\sigma'_e = 700\,\text{MPa for } \sigma_{uts} \geq 1400\,\text{MPa} \tag{3.13}$$

The surface finish factor is given by

$$k_a = a\sigma_{uts}^b \tag{3.14}$$

Values for a and b can be found in Table 3.3.

The size factor k_b can be calculated from (Kuguel, 1969):

$$k_b = \left(\frac{d}{7.62}\right)^{-0.1133} \quad (\text{for } d < 50\,\text{mm}) \tag{3.15}$$

$$k_b = 1.85d^{-0.19} \quad (\text{for } d > 50\,\text{mm}) \tag{3.16}$$

The reliability factor k_c is given by Table 3.4.

For temperatures between −57°C and 204°C the temperature factor k_d can be taken as 1. The ASME standard documents values to use outside of this range.

Table 3.3 Surface finish factors

Surface finish	a (MPa)	b
Ground	1.58	−0.085
M/c or cold-drawn	4.51	−0.265
Hot rolled	57.7	−0.718
Forged	272.0	−0.995

Source: Noll and Lipson (1946).

Table 3.4 Reliability factors for use in the ASME transmission shaft equation

Shaft nominal reliability	k_c
0.5	1.0
0.9	0.897
0.99	0.814
0.999	0.753

Source: ANSI/ASME B106.1M-1985.

The duty cycle factor k_e is used to account for cycle loading experienced by the shaft such as stops and starts, transient overloads, shock loading etc., and requires prototype fatigue testing for its quantification. k_e is taken as 1 in the examples presented here.

The fatigue stress concentration factor k_f is used to account for stress concentration regions such as notches, holes, keyways and shoulders. It is given by

$$k_f = \frac{1}{K_f} \tag{3.17}$$

where K_f is the component fatigue stress concentration factor which is given by

$$K_f = 1 + q(k_t - 1) \tag{3.18}$$

where q is the notch sensitivity and k_t is the geometric stress concentration factor.

Values for the notch sensitivity and typical geometric stress concentration factors are given in Figs 3.22–3.25 and Table 3.5.

The miscellaneous factor k_g is used to account for residual stresses, heat treatment, corrosion, surface coatings, vibration, environment and unusual loadings. k_g is taken as 1 here.

Example Using the ASME equation for the design of transmission shafting, determine a sensible minimum nominal diameter for the drive shaft illustrated in Fig. 3.26, consisting of a mid-mounted spur gear and overhung pulley wheel. The shaft is to be manufactured using 817M40 hot-rolled alloy steel, with $\sigma_{uts} = 1000$ MPa, $\sigma_y = 770$ MPa and Brinell hardness approximately 220 BHN (*see* Appendix A for an overview of the British Standard designation of steels and some material properties). The radius of the fillets at the gear and pulley shoulders is 3 mm. The power to be transmitted is 8 kW at 900 rpm. The pitch circle diameter of the 20° pressure angle spur gear

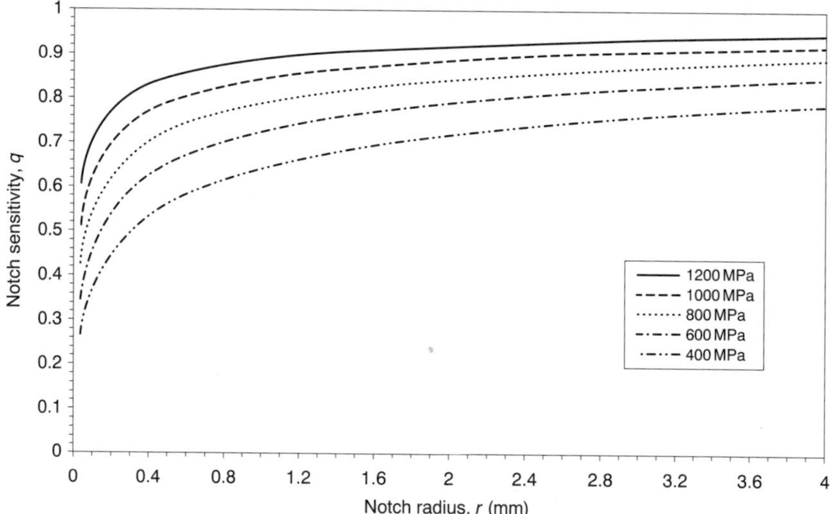

Fig. 3.22 Notch sensitivity index versus notch radius for a range of steels subjected to reversed bending or reversed axial loads. (Data from Kuhn and Hardrath, 1952.)

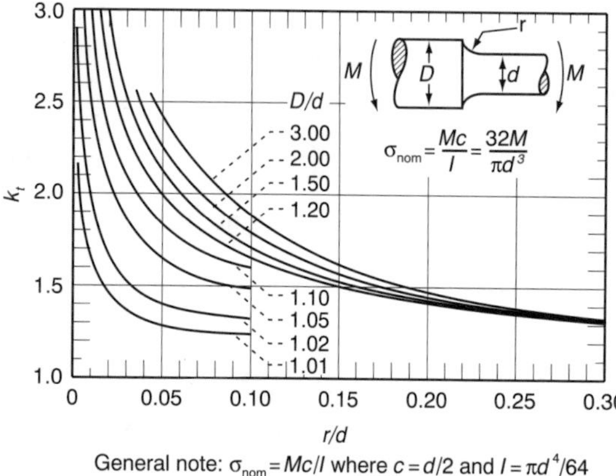

Fig. 3.23 Stress concentration factors for a shaft with a fillet subjected to bending. (Reproduced from Peterson, 1974.)

is 192 mm and the pulley diameter is 250 mm. The masses of the gear and pulley are 8 kg and 10 kg respectively. The ratio of belt tensions should be taken as 2.5.[1] Profiled keys are used to transmit torque through the gear and pulley. A shaft nominal reliability of 90 per cent is desired. Assume the shaft is of constant diameter for the calculation.

Solution In order to use the ASME design equation for transmission shafts, the maximum combination of torque and bending moment must be determined. A sensible approach is to determine the overall bending moment diagram for the shaft, as this information may be of use in other design calculations such as calculating the shaft deflection.

The loading on the shaft can be resolved into both horizontal and vertical planes, and must be considered in determining the resulting bending

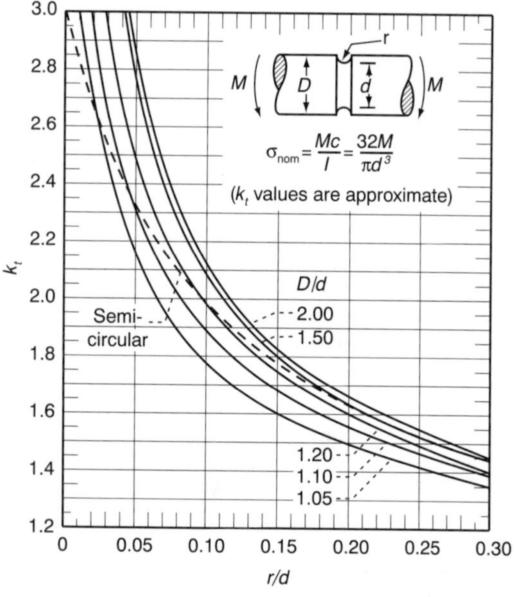

Fig. 3.24 Stress concentration factors for a shaft with a groove subjected to bending. (Reproduced from Peterson, 1974.)

[1] Note that some of the information given in this example concerning pitch circle diameters, pressure angles and ratios of belt tensions has not yet been covered. These terms are dealt with in Chapters 4 and 6 and defined within the solution.

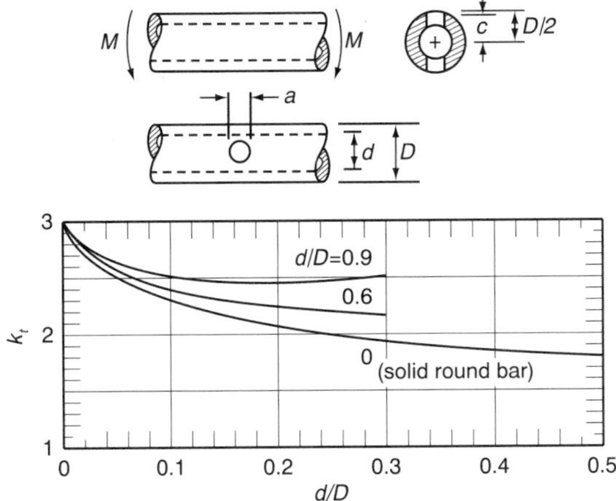

Fig. 3.25 Stress concentration factors for a shaft with a transverse hole subjected to bending. (Reproduced from Peterson, 1974.)

moments on the shaft. Figure 3.27 shows the combined loadings on the shaft from the gear forces, the gear's mass, the belt tensions and the pulley's mass. The mass of the shaft itself has been ignored. The tension on a pulley belt is tighter on the 'pulling' side than on the 'slack' side, and the relationship of these tensions is normally given as a ratio, which in this case is 2.5:1.

The power transmitted through the shaft is 8000 W. The torque is given by

$$\text{Torque} = \frac{\text{Power}}{\omega} = \frac{8000}{(2\pi/60) \times 900} = 84.9\ \text{N} \cdot \text{m}$$

Table 3.5 Fatigue stress concentration factors, k_f, for keyways in steel shafts

Steel	Profiled keyway bending stress	Sled runner keyway bending stress
Annealed, <200 BHN	0.63	0.77
Quenched and drawn, >200 BHN	0.5	0.63

Source: Juvinall (1967).

The ratio of belt tensions is $T_1/T_2 = 2.5$, so $T_1 = 2.5T_2$. The torque on the pulley in terms of the belt tensions is given by

$$\text{Torque} = T_1 r_p - T_2 r_p = 2.5T_2 r_p - T_2 r_p = 1.5T_2 r_p$$

The belt tensions are

$$T_2 = 84.9/(1.5 \times 0.125) = 452.8\ \text{N}$$

$$T_1 = 2.5T_2 = 1131.8\ \text{N}$$

A spur gear experiences both radial and tangential loading, as shown in Fig. 3.24. The tangential load is given by

$$F_t = \frac{\text{Torque}}{r_g} = \frac{84.9}{0.192/2} = 884.4\ \text{N}$$

where r_g is the pitch circle radius. The radial load is given by

$$F_r = F_t \tan\phi = 884.4 \times \tan 20 = 321.9\ \text{N}$$

where ϕ is the pressure angle.

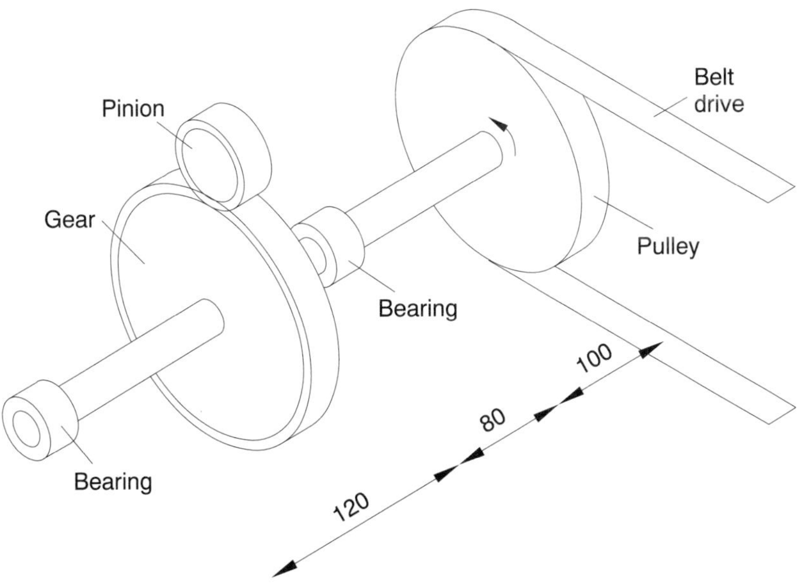

Fig. 3.26 Transmission drive shaft.

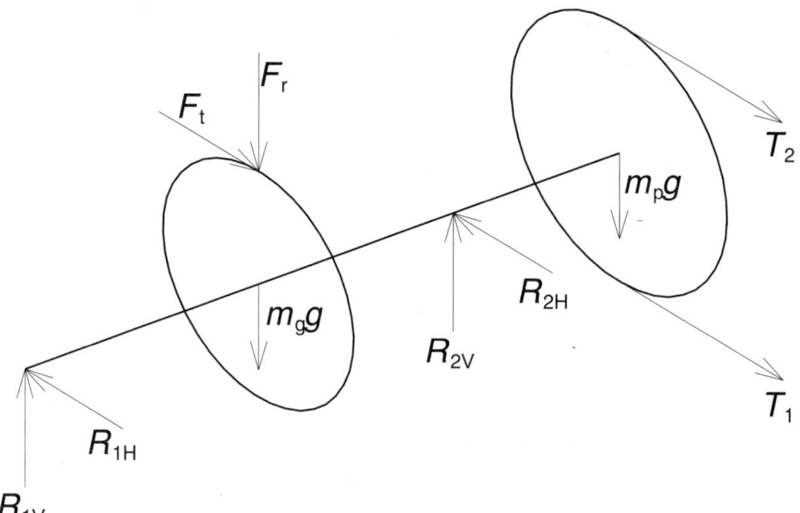

Fig. 3.27 Shaft loading diagram.

Note that when determining the horizontal and vertical bending moment diagrams, both transmitted and gravitational loads should be included.

The vertical loads on the shaft are illustrated in Fig. 3.28.

Moments about A:

$$(F_r + m_g g)L_1 - R_{2V}(L_1 + L_2) + m_p g(L_1 + L_2 + L_3) = 0$$

$$R_{2V} = \frac{(F_r + m_g g)L_1 + m_p g(L_1 + L_2 + L_3)}{L_1 + L_2}$$

$$= \frac{[321.9 + (8 \times 9.81)]0.12 + (10 \times 9.81)0.3}{0.12 + 0.08}$$

$$= 387.4\,\text{N}$$

Resolving vertical forces:

$$R_{1V} + R_{2V} = F_r + m_g g + m_p g$$

Hence

$$R_{1V} = 321.9 + 78.5 + 98.1 - 387.4 = 111.1\,\text{N}$$

The horizontal loads on the shaft are illustrated in Fig. 3.29. The total tension T is

$$T = T_1 + T_2 = 1131.8 + 452.8 = 1585\,\text{N}$$

Moments about A:

$$F_t L_1 - R_{2H}(L_1 + L_2) + T(L_1 + L_2 + L_3) = 0$$

Hence

$$R_{2H} = \frac{F_t L_1 + T(L_1 + L_2 + L_3)}{L_1 + L_2}$$

$$= \frac{(884.4 \times 0.12) + (1585 \times 0.3)}{0.2} = 2908\,\text{N}$$

Resolving horizontal forces:

$$R_{1H} + R_{2H} = F_t + T$$

Hence

$$R_{1H} = 884.4 + 1585 - 2908 = -438.5\,\text{N}$$

The bending moment diagrams can now be determined. You may wish to recall that the bending

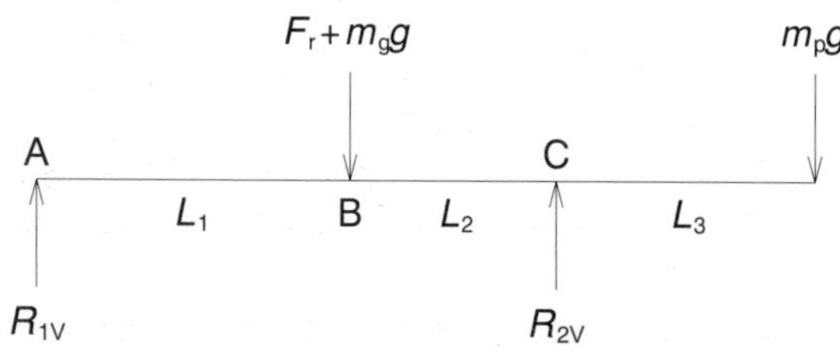

Fig. 3.28 Vertical loading diagram.

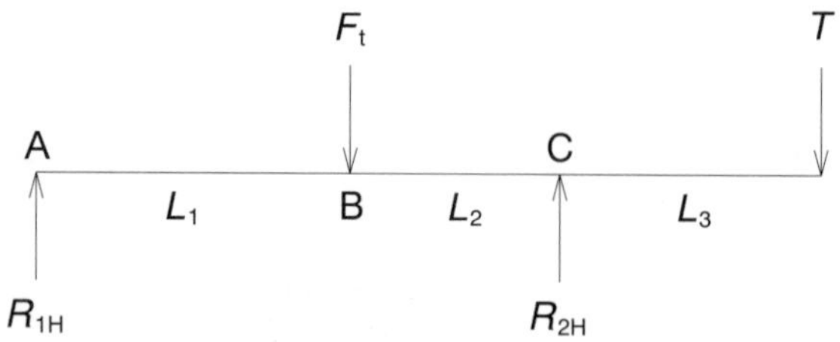

Fig. 3.29 Horizontal loading diagram.

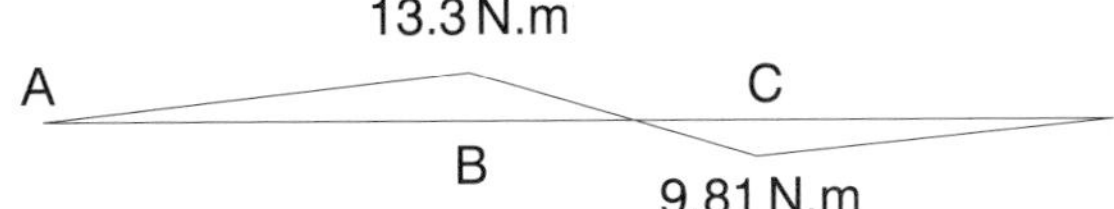

Fig. 3.30 Vertical bending moment diagram.

moment is the algebraic sum of the moments of the external forces to one side of the section about an axis through the section.

The vertical bending moments at B and C (Fig. 3.30) are given by

$$M_{BV} = R_{1V}L_1 = 111.1 \times 0.12 = 13.3\,\text{N}\cdot\text{m}$$

$$M_{CV} = R_{1V}(L_1 + L_2) - [(F_r + m_g g)L_2]$$
$$= 111.1(0.12 + 0.08) - (321.9 + 8 \times 9.81)0.08$$
$$= -9.810\,\text{N}\cdot\text{m}$$

The horizontal bending moments at B and C (Fig. 3.31) are given by

$$M_{BH} = R_{1H}L_1 = -438.5 \times 0.12 = -52.62\,\text{N}\cdot\text{m}$$

$$M_{CH} = R_{1H}(L_1 + L_2) - (F_t L_2)$$
$$= -438.5 \times 0.2 - 884.4 \times 0.08$$
$$= -158.5\,\text{N}\cdot\text{m}$$

The resultant bending moment diagram can be determined by calculating the resultant bending moments at each point:

$$|M_B| = \sqrt{(13.3)^2 + (-52.62)^2} = 54.27\,\text{N}\cdot\text{m}$$

$$|M_C| = \sqrt{(-9.81)^2 + (-158.5)^2} = 158.8\,\text{N}\cdot\text{m}$$

From Fig. 3.32 it can be seen that the maximum bending moment occurs at the right hand bearing.

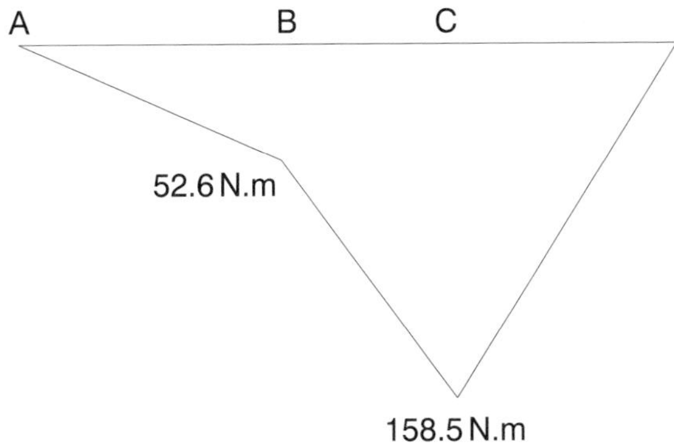

Fig. 3.31 Horizontal bending moment diagram.

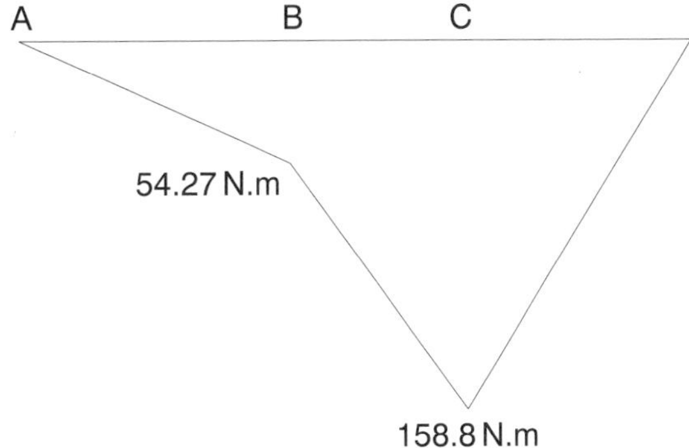

Fig. 3.32 Resultant bending moment diagram.

The value of the resultant bending moment here should be used in the ASME design equation.

The next task is to determine the endurance limit of the shaft and the modifying factors. The shaft material is hot rolled steel and the endurance limit of the test specimen if unknown can be estimated using $\sigma'_e = 0.504\sigma_{uts}$. From materials charts (see Appendix 1), the ultimate tensile strength for 817M40 is 1000 MPa and σ_y is 770 MPa. So

$$\sigma'_e = 0.504 \times 1000 = 504\,\text{MPa}$$

The material is hot rolled, so from equation 3.14 and Table 3.3,

$$k_a = a\sigma_{uts}^b = 57.7(1000)^{-0.718} = 0.405$$

Assuming that the diameter will be about 30 mm, the size factor can be estimated using

$$k_b = (d/7.62)^{-0.1133} = (30/7.62)^{-0.1133} = 0.856$$

If the shaft is significantly different in size to 30 mm, then the calculation should be repeated until convergence between the assumed and the final calculated value for the shaft diameter is achieved.

The desired nominal reliability is 90 per cent, so $k_c = 0.897$.

The operating temperature is not stated, so a value of temperature factor of $k_d = 1$ is assumed. The duty cycle factor is assumed to be $k_e = 1$.

The fillet radius at the shoulders is 3 mm. The ratio of diameters $D/d = (3 + 3 + 30)/30 = 36/30 = 1.2$. $r/d = 3/30 = 0.1$. From Fig. 3.23 the geometric stress concentration factor is $k_t = 1.65$. From Fig. 3.22 the notch sensitivity index for a 1000 MPa strength material with notch radius of 3 mm is $q = 0.9$. Thus

$$K_f = 1 + q(k_t - 1) = 1 + 0.9(1.65 - 1) = 1.59$$

$$k_f = 1/K_f = 0.629$$

The miscellaneous factor is taken as $k_g = 1$.

The endurance limit can now be calculated from $\sigma_e = k_a k_b k_c k_d k_e k_f k_g \sigma'_e$:

$$\sigma_e = 0.405 \times 0.856 \times 0.897 \times 1 \times 1 \times 0.629 \times 1 \times 504 = 98.6\,\text{MPa}$$

As a well-known material has been selected subject to known loads, the factor of safety can be taken as $n_s = 2$. The diameter can now be calculated from the ASME equation:

$$d = \left[\frac{32n_s}{\pi}\sqrt{\left(\frac{M}{\sigma_e}\right)^2 + \frac{3}{4}\left(\frac{T}{\sigma_y}\right)^2}\right]^{1/3}$$

$$= \left[\frac{32 \times 2}{\pi}\sqrt{\left(\frac{158.8}{98.6 \times 10^6}\right)^2 + 0.75\left(\frac{84.9}{770 \times 10^6}\right)^2}\right]^{1/3}$$

$$= 0.032\,\text{m}$$

As this value is close to the assumed value used to evaluate the size and fatigue stress factors, further iteration is not necessary. For manufacturing convenience it may be necessary to modify this diameter to the nearest standard size as used within the company, or taking into account materials readily available from suppliers – in this case 35 mm. Standard sizes can be found in texts such as *Machinery's Handbook* (Schubert, 1982). An advantage of using standard sizes is that standard stock bearings can be selected to fit.

In general, the following principles should be used when designing to avoid fatigue failures.

- Calculations should allow an appropriate safety factor, particularly where stress concentrations occur (e.g. keyways, notches, change of section).
- Provide generous radii at changes of section and introduce stress relief grooves etc.
- Choose materials if possible that have limiting fatigue stresses, e.g. most steels.
- Provide for suitable forms of surface treatment, e.g. shot peening, work hardening, nitriding. Avoid treatments which introduce residual tensile stresses such as electroplating.
- Specify fine surface finishes.
- Avoid corrosive conditions.
- Stress relieving should be used where possible, particularly for welded structures.

Learning objectives checklist

Are you familiar with the general design parameters for shafts? ☐

Can you determine deflections for both constant section and stepped shafts? ☐

Can you determine the first two critical frequencies of a shaft? ☐

Can you use the ASME transmission shaft design equation? ☐

References and sources of information

Books and papers

Beswarick, J. 1994: in Hurst, K. (ed.), *Rotary power transmission design*. McGraw Hill.

Fuchs, H.O. and Stephens, R.I. 1980: *Metal fatigue in engineering*. Wiley.

Hurst, K. (ed.) 1994: *Rotary power transmission design*. McGraw Hill.

Juvinall, R.C. 1967: *Engineering considerations of stress, strain and strength*. McGraw Hill.

Juvinall, R.C. and Marshek, K.M. 1991: *Fundamentals of machine component design*. Wiley.

Kuguel, R. 1969: A relation between theoretical stress concentration factor and fatigue notch factor deduced from the concept of highly stressed volume. *Proceedings of the ASTM* **61**, 732–48.

Kuhn, P. and Hardrath, H.F. 1952: An engineering method for estimating notch size effect on fatigue tests of steel. NACA TN2805.

Mischke, C.R. 1987: Prediction of stochastic endurance strength. *Transactions of the ASME, Journal of Vibration, Acoustic Stress and Reliability in Design*, 113–22.

Mischke, C.R. 1996: Shafts. In Shigley, J.E. and Mischke, C.R. (eds), *Standard handbook of machine design*. McGraw Hill.

Mott, R.L. 1992: *Machine elements in mechanical design*. Merrill.

Noll, C.G. and Lipson, C. 1946: *Allowable working stresses*. Society for Experimental Stress Analysis, Vol. 3.

Peterson, R.E. 1974: *Stress concentration factors*. Wiley.

Reshetov, D.N. 1978: *Machine design*. Mir Publishers.

Schubert, P.B. (ed.) 1982: *Machinery's handbook*. Industrial Press Inc.

Sines, G. and Waisman, J.L. 1959: *Metal fatigue*. McGraw Hill.

Vidosek, J.P. 1957: *Machine design projects*. Ronald Press.

Standards

ANSI/ASME B106.1M-1985. Design of transmission shafting.

BSI 1984: *Manual of British Standards in engineering drawing and design*. BSI Hutchinson.

BS 3550: 1963. Specification for involute splines. BSI.

BS 4235: Part 2: 1977. Specification for metric keys and keyways: Woodruff keys and keyways. BSI.

BS 4235: Part 1: 1992. Specification for metric keys and keyways. Parallel and taper keys. BSI.

Shafts worksheet

1. An air compressor consists of four compressor discs mounted on a steel shaft 5 cm apart, as

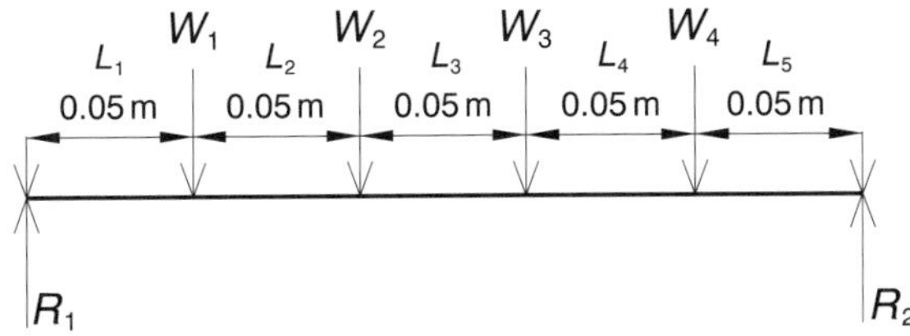

Fig. 3.33 Air compressor shaft.

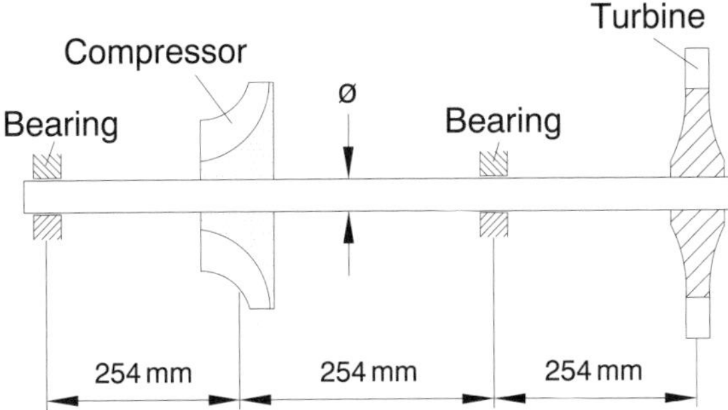

Fig. 3.34 Turbine shaft.

shown in Fig. 3.33. The shaft is simply supported at either end. Each compressor wheel mass is 18 kg. Calculate the deflection at each wheel using Macaulay's method, and calculate, using the Rayleigh–Ritz equation, the first critical frequency of the shaft. The shaft outer diameter is 0.05 m, the inner diameter is 0.03 m.

2. A 22.63 kg compressor impeller wheel is driven by a 13.56 kg turbine, mounted on a common shaft (*see* Fig. 3.34) manufactured from steel, with Young's modulus $E = 207\,\text{GN/m}^2$. The design speed is 10 000 rpm. Determine the shaft diameter so that the first critical speed is 12 000 rpm, giving a safety margin of 2000 rpm. Use Rayleigh's equation and assume rigid bearings and a massless shaft.
3. A research turbocharger shaft is simply supported by bearings, as shown in Fig. 3.35. Assuming there is no deflection at the bearings and that the shaft mass is negligible, calculate the first *two* critical frequencies of the shaft. The compressor mass is 3 kg, the turbine mass 2.5 kg. The steel shaft diameter is 0.02 m. Take Young's modulus for steel as $200\,\text{GN/m}^2$.

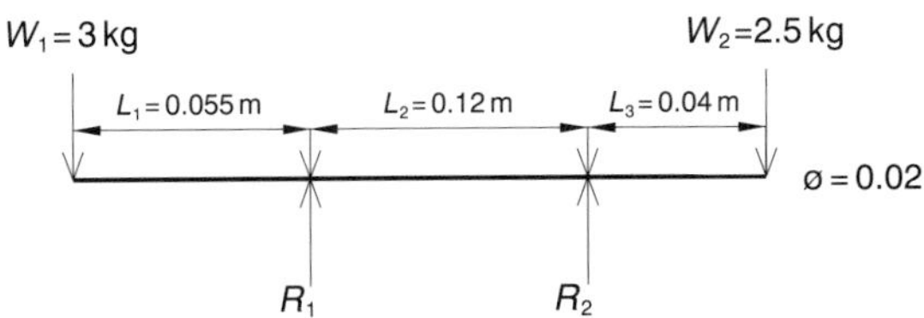

Fig. 3.35 Turbocharger.

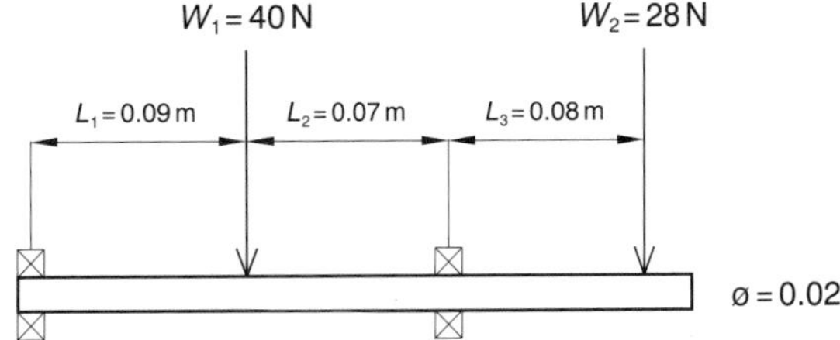

Fig. 3.36 Turbocharger shaft.

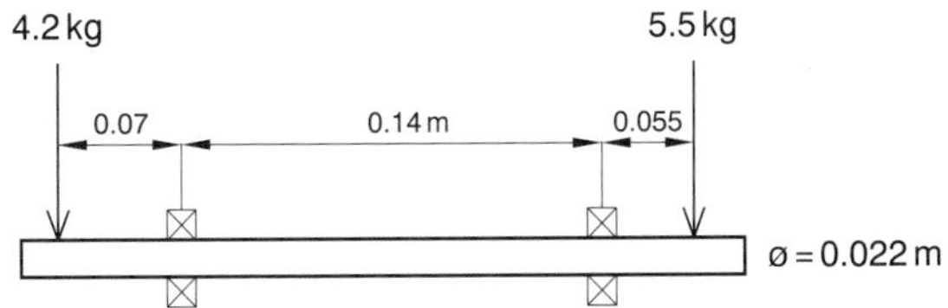

Fig. 3.37 Turbocharger shaft.

4. As part of a preliminary design for a turbocharger determine the first *two* critical frequencies, stating any assumptions made, for the shaft arrangement shown in Fig. 3.36. To assist with your calculation, a_{22} may be taken as 3.2594926×10^{-7} m. Young's modulus should be taken as $200\,\text{GN/m}^2$.
5. As an initial step in the design of an experimental turbocharger, calculate the first *two* critical frequencies, stating any assumptions made, for the steel shaft shown in Fig. 3.37. Young's modulus can be taken as $200\,\text{GN/m}^2$.
6. Determine the first *two* critical speeds for the shaft shown in Fig. 3.38, stating and discussing the practical implications for any assumptions

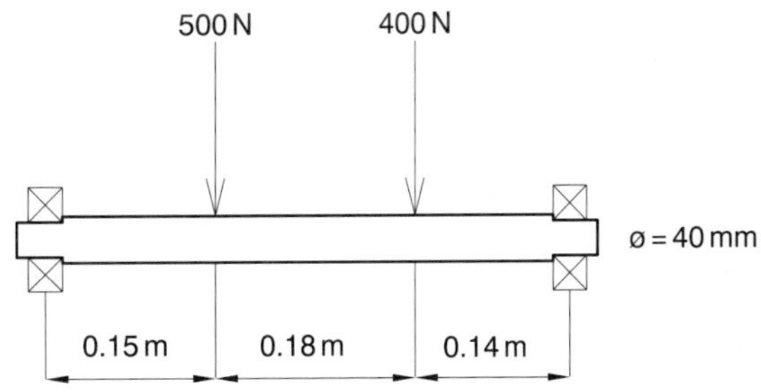

Fig. 3.38 Shaft loading diagram.

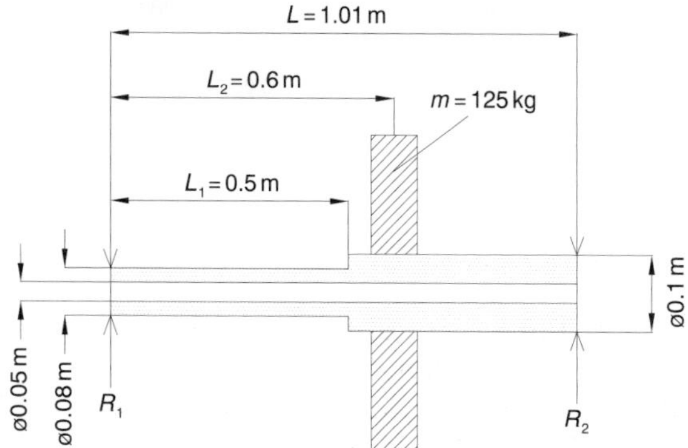

Fig. 3.39 Simple stepped shaft and load.

made. Young's modulus for steel should be taken as 200 GN/m^2.

7. Using Castigliano's theorem for calculating deflections, determine the critical frequency of the steel shaft shown in Fig. 3.39. Neglect the shaft mass.
8. Determine the diameter of the drive shaft for a chain conveyor which has the loading parameters illustrated in Fig. 3.40. A roller chain sprocket of 500 mm pitch diameter, weighing 90 kg, will be mid-mounted between two bearings. A 400 mm, 125 kg roller chain sprocket will be mounted overhung. The drive shaft is to be manufactured using cold-drawn 070M20 steel. Operating temperatures are not expected to exceed 65°C and the operating environment is non-corrosive. The shaft is to be designed for non-limited life of greater than 10^8 cycles, with a 90 per cent survival rate. The shaft will carry a steady driving torque of 1600 N·m and rotate at 36 rpm. A sled runner keyway will be used for the overhung pulley and a profile keyway for the mid-mounted pulley. (Example adapted from ANSI/ASME B106.1M-1985.)

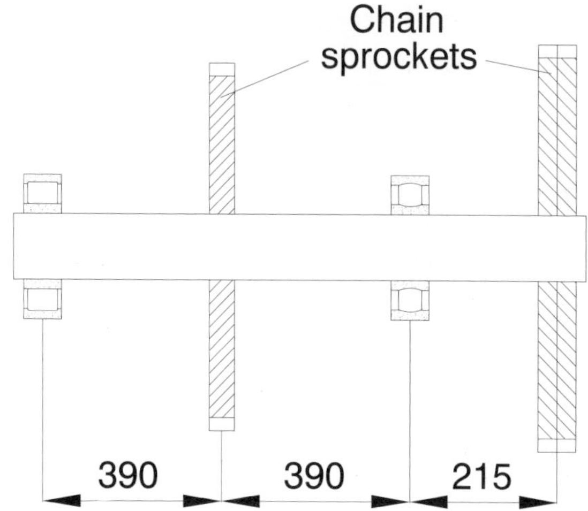

Fig. 3.40 Chain conveyor drive shaft.

9. The geometry and loading for the drive shaft of a snow-track mobile is given in Fig. 3.41. Discuss and outline, with sketches as appropriate, the design decisions that should be taken in scheming out a general arrangement for the shaft. (After Juvinall and Marshek, 1991.)

Answers

Worked solutions to the above problems are given in Appendix B.

1. $y_{w1} = 1.929 \times 10^{-6}$ m, $y_{w2} = 3.100 \times 10^{-6}$ m, $y_{w3} = 3.100 \times 10^{-6}$ m, $y_{w4} = 1.929 \times 10^{-6}$ m, $\omega = 1923.8$ rad/s = 18 371 rpm.
2. $d = 67$ mm.
3. $\omega_1 = 1660$ rad/s, $\omega_2 = 3017$ rad/s.
4. $\omega_1 = 985.3$ rad/s, $\omega_2 = 2949$ rad/s.
5. $\omega_1 = 1153$ rad/s, $\omega_2 = 1703$ rad/s.
6. $\omega_1 = 1521.6$ rad/s, $\omega_2 = 432.8$ rad/s.
7. $\omega_c = 482.3$ rad/s.
8. 74 mm.
9. No unique solution.

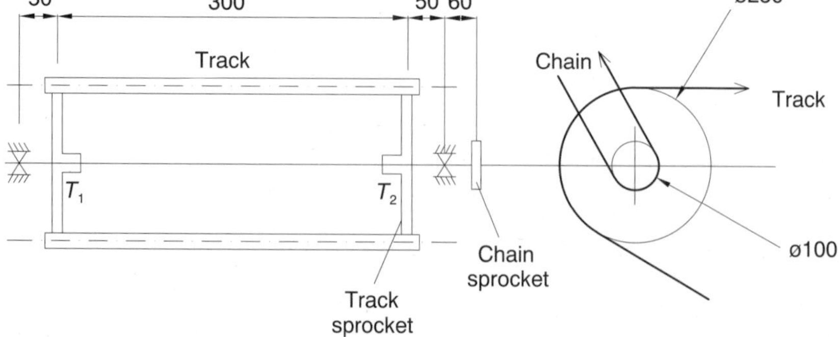

Fig. 3.41 Schematic representation of the geometry and loading for a snow-track mobile drive shaft (dimensions in mm). (Reproduced from Juvinall and Marshek, 1991.)

4 Gears

Gears are toothed cylindrical wheels used for transmitting mechanical power from one rotating shaft to another. Several types of gears are in common use. This chapter introduces the various types of gears and details the design, specification and selection of spur gears in particular.

4.1 Introduction

When transmitting power from a source to the required point of application, a series of devices are available including belts, pulleys, chains, hydraulic and electrical systems, and gears (*see* Table 1.3). Generally, if the distances of power transmission are large, gears are not suitable and chains and belts can be considered (*see* Table 6.8 for a comparison of the merits of belts, chains and gears). However, when a compact, efficient or high-speed drive is required, gear trains offer a competitive and suitable solution. Additional benefits of gear drives include reversibility, configuration at almost any angle between input and output and their suitability to operate in arduous conditions.

Simplistically a speed change can be achieved by running discs (*see* Fig. 4.1) of different diameter together, or cones for turning corners as well. However, the torque capacity of disc or cone drives is limited by the frictional properties of the surfaces. The addition of teeth to the surfaces of the discs or cones makes the drive positive, ensuring synchronisation, and substantially increases the torque capacity.

Primitive gear trains consisted of teeth or pegs located on discs, as illustrated in Fig. 4.2. The disadvantage of these simple teeth is that the velocity ratio is not constant and changes as the teeth go through the meshing cycle, causing noise and vibration problems at elevated speeds. The solution to this problem can be achieved by using a profile on the gear teeth which gives a constant velocity ratio throughout the meshing cycle. Several different geometrical forms can be used, but the full depth involute form is primarily used in current professional engineering practice.

Gears can be divided into several broad classifications.

1. Parallel axis gears:
   (a) spur gears (*see* Fig. 4.3)
   (b) helical gears (*see* Figs 4.4 and 4.5)
   (c) internal gears.
2. Non-parallel, coplanar gears (intersecting axes):
   (a) bevel gears (*see* Fig. 4.6)
   (b) face gears
   (c) conical involute gearing.

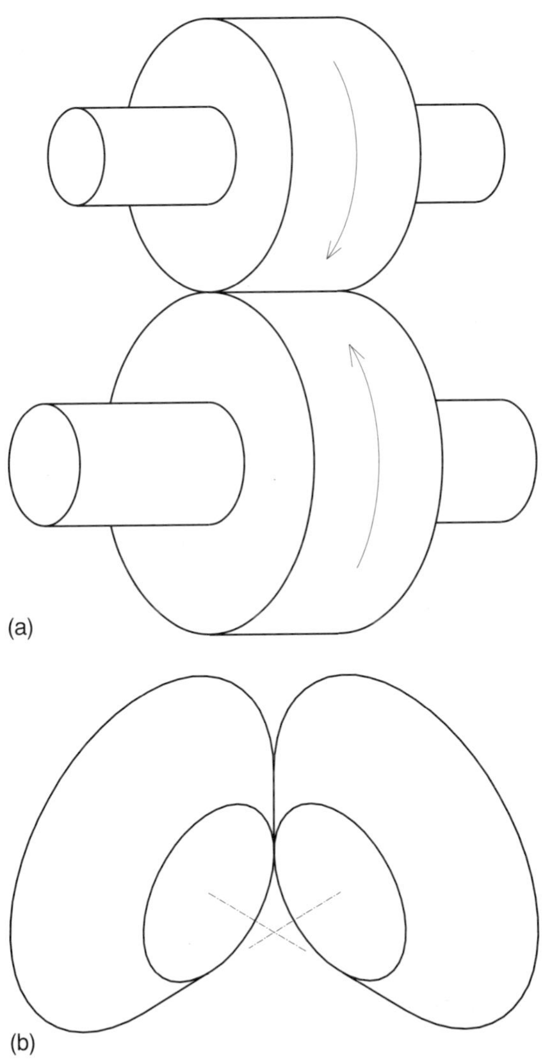

Fig. 4.1 Disc or cone drives.

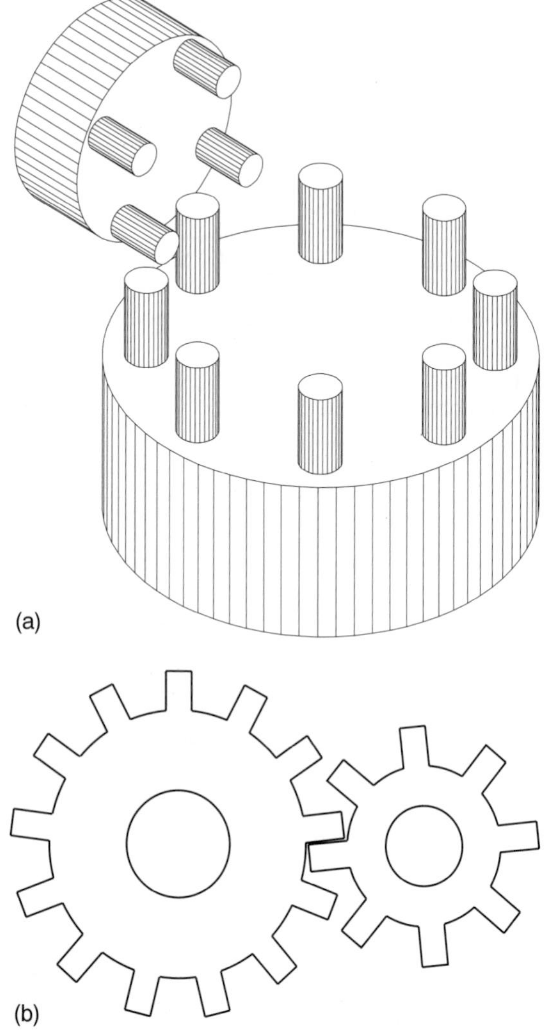

Fig. 4.2 Primitive gears.

3. Non-parallel, non-coplanar gears (non-intersecting axes):
   (a) crossed axis helicals (*see* Fig. 4.7)
   (b) cylindrical worm gearing (*see* Fig. 4.8)
   (c) single enveloping worm gearing
   (d) double enveloping worm gearing
   (e) hypoid gears
   (f) spiroid and helicon gearing
   (g) face gears (off centre).
4. Special gear types:
   (a) square and rectangular gears
   (b) elliptical gears
   (c) scroll gears
   (d) multiple sector gears.

Fig. 4.3 Spur gears. (Photograph courtesy of Hinchliffe Precision Components Ltd.)

Spur gears (Fig. 4.3) are the cheapest of all types for parallel shaft applications. Their straight teeth allow running engagement or disengagement using sliding shaft and clutch mechanisms. Typical applications of spur gears include: manual and automatic motor vehicle gearboxes, machine tool drives, conveyor systems, electric motor gearboxes, timing mechanisms and power tool drives. The majority of power gears are manufactured from hardened and case hardened steel. Other materials

Fig. 4.4 Helical gears. (Photograph courtesy of Hinchliffe Precision Components Ltd.)

Fig. 4.5 Double helical gears.

used include iron, brass and bronze, and polymers such as polyamide (e.g. nylon), polyacetal (e.g. Delrin) and SRBF (e.g. Tufnol). Typical material matches are listed in Table 4.1. Steel gears are often supplied in unhardened form to allow for grub screws, keyways or splines to be added before hardening and finishing.

A helical gear is a cylindrical gear whose tooth traces are helixes, as shown in Fig. 4.4. Common

Fig. 4.6 Bevel gears. (Photograph courtesy of Hinchliffe Precision Components Ltd.)

Fig. 4.7 Crossed axis helical gears. (Photograph courtesy of Hinchliffe Precision Components Ltd.)

helix angles are 15° to 30°. Helical gears are typically used for heavy-duty high-speed (>3500 rpm) power transmission, turbine drives, locomotive gearboxes and machine tool drives. Helical gears are generally

Fig. 4.8 Worm gears. (Photograph courtesy of Hinchliffe Precision Components Ltd.)

Table 4.1 Typical material matches for gears and pinions

Gear	Pinion
Cast iron	Cast iron
Cast iron	Carbon steel
Carbon steel	Alloy steel
Alloy steel	Alloy steel
Alloy steel	Case hardened steel

more expensive than spur gears. Noise levels are lower than for spur gears because teeth in mesh make point contact rather than line contact. The forces arising from meshing of helical gears can be resolved into three component loads: radial, tangential and axial (axial loads are often called thrust loads). Bearings used to support the gear on a shaft must be able to withstand the axial or thrust force component. A neat design solution is the use of double helical gears, as illustrated in Fig. 4.5. These eliminate the need for thrust bearings because the axial force components cancel each other. Whilst spur gears are generally cheaper, a comparable helical gear will be smaller.

Bevel gears have teeth cut on conical blanks (*see* Fig. 4.6) and a gear pair can connect non-parallel intersecting shafts. Bevel gears are used for, e.g. motor transmission differential drives, valve control and mechanical instruments.

A worm gear is a cylindrical helical gear with one or more threads (*see* Fig. 4.8). A worm wheel is a cylindrical gear with flanks cut in such a way as to ensure contact with the flanks of the worm gear. Worm gears are used for steering gear, winch blocks, low-speed gear boxes, rotary tables and remote valve control. Worm gear sets are capable of high-speed reduction and high-load applications where non-parallel, non-interacting shafts are used. The 90° configuration is most common. Frictional heat generation is high in worm gears, so continuous lubrication is required and heat dissipation provision must be made.

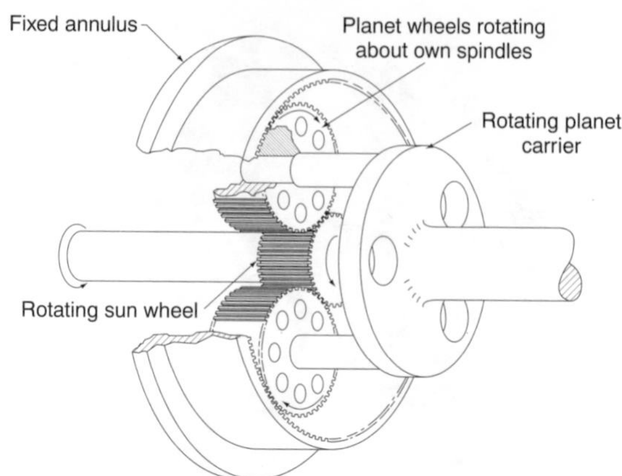

Fig. 4.9 Epicyclic gears. Reproduced from Townsend (1992).

Some gear axes can be allowed to rotate about others. In such cases the gear trains are called planetary or epicyclic (*see* Fig. 4.9). Planetary trains always include a sun gear, a planet carrier or arm, and one or more planet gears.

All gear mechanisms and gear trains demand continuous lubrication which must be pressure fed for high-speed gears in order to counteract centrifugal effects on the oil. Plastic gears, made from, e.g. nylon, can be used in certain applications and have the advantage that there is no need for lubrication, but are only suitable for low-speed applications. Plastic gears can reduce noise levels significantly.

Generally, the pinion of a pair of gears should have the largest number of teeth consistent with adequate resistance to failure by bending stress in the teeth. For a given diameter the larger the number of teeth the finer the pitch, and consequently the weaker they are and the greater the liability to fracture. Table 4.2 lists the range of gear ratios and performance characteristics typically achievable.

When the number of teeth selected for a gear wheel is equal to the product of an integer and the number of teeth in the mating pinion, then

Table 4.2 Useful range of gear ratios

Gear	Ratio range	Pitch line velocity (m/s)	Efficiency (%)
Spur	1:1 to 6:1	25	98–99
Helical	1:1 to 10:1	50	98–99
Double helical	1:1 to 15:1	150	98–99
Bevel	1:1 to 4:1	20	
Worm	5:1 to 75:1	30	20–98
Crossed helical	1:1 to 6:1	30	

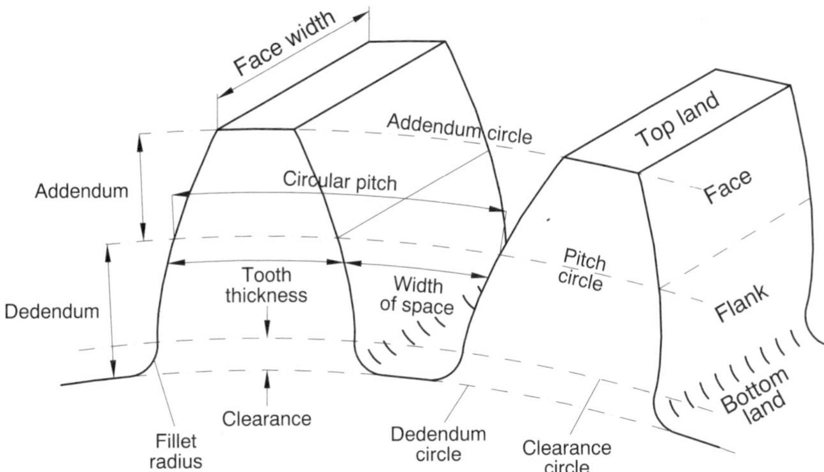

Fig. 4.10 Spur gear schematic showing principal terminology.

the same tooth on the gear wheel will touch the same tooth on the pinion in each revolution of the gear wheel. If there is no common factor between the numbers of teeth on the gear wheel and pinion, then each tooth on the gear wheel will touch each tooth on the pinion in regular succession, with the frequency of contact between a particular pair of teeth being the speed of the pinion divided by the number of teeth on the gear wheel. The avoidance of an integer ratio between the number of teeth on the gear wheel and the pinion can be achieved by the addition of an extra tooth to the gear wheel, provided that there is no operational need for an exact velocity ratio. This extra tooth is called a hunting tooth and can have advantages for wear equalisation (*see* Tuplin (1962) for a fuller discussion).

Various definitions used for describing gear geometry are illustrated in Fig. 4.10 and listed below. For a pair of meshing gears, the smaller gear is called the pinion, the larger the gear wheel.

- **Pitch circle**. This is a theoretical circle on which calculations are based. Its diameter is called the pitch diameter:

$$d = mN \tag{4.1}$$

where d is the pitch diameter (mm), m is the module (mm) and N is the number of teeth. (Care must be taken to distinguish the module from the unit symbol for a metre.)

- **Circular pitch**. This is the distance from a point on one tooth to the corresponding point on the adjacent tooth measured along the pitch circle:

$$p = \pi m = \frac{\pi d}{N} \tag{4.2}$$

where p is the circular pitch (mm), m the module, d the pitch diameter (mm) and N the number of teeth.

- **Module**. This is the ratio of the pitch diameter to the number of teeth:

$$m = \frac{d}{N} \tag{4.3}$$

The unit of the module should be millimetres (mm). Typically the height of a tooth is about 2.25 times the module. Various modules are illustrated in Fig. 4.11.

- **Addendum**, a. This is the radial distance from the pitch circle to the outside of the tooth.
- **Dedendum**, b. This is the radial distance from the pitch circle to the bottom land.

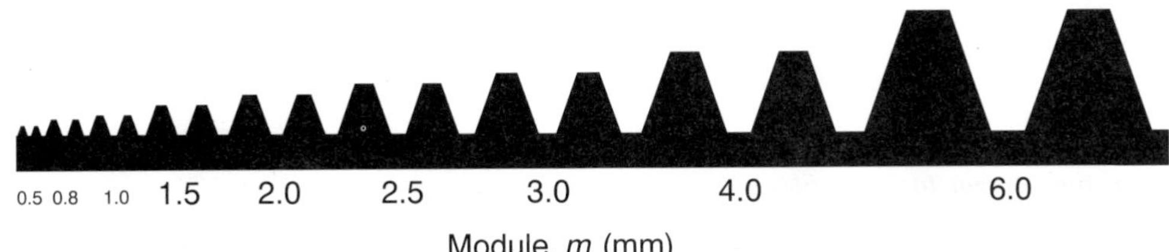

Fig. 4.11 Gear tooth size as a function of the module.

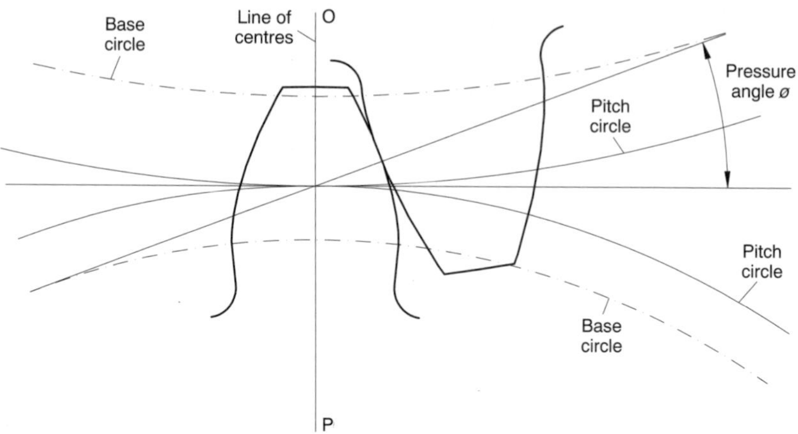

Fig. 4.12 Schematic showing the pressure line and pressure angle.

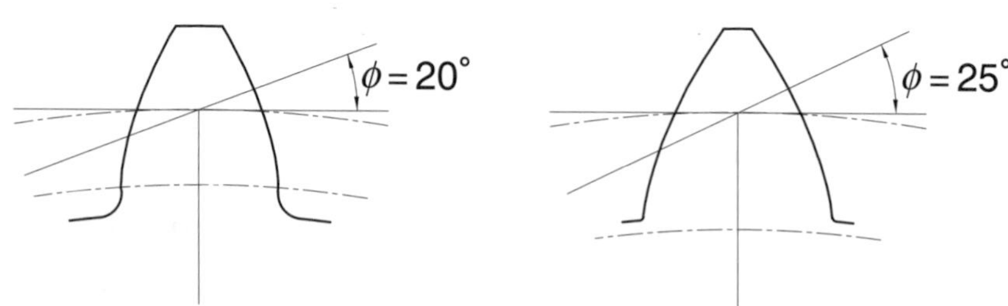

Fig. 4.13 Full depth involute form for varying pressure angles.

- **Backlash**. This is the amount by which the width of a tooth space exceeds the thickness of the engaging tooth measured on the pitch circle.

Prior to use of the metric module, the term diametral pitch was commonly used. The diametral pitch is the ratio of the number of teeth in the gear to the pitch diameter, $P_d = N/d$ (usually in US/English units only, i.e. teeth per inch (tpi)). To convert from pitch diameter P_d (tpi) to the module m (mm) use $m = 25.4/P_d$.

Figure 4.12 shows the line of centres OP connecting the rotation axes of a pair of meshing gears. The angle ϕ is called the pressure angle. The pressure line (also called the generating line or line of action) is defined by the pressure angle. The resultant force vector between a pair of operating meshing gears acts along this line. The actual shape or form of the gear teeth depends on the pressure angle chosen. Figure 4.13 shows two gear teeth of differing pressure angle for a gear with identical number of teeth, pitch and pitch diameter. Current standard pressure angles are 20° and 25°, with the 20° form most widely available.

4.1.1 Construction of gear tooth profiles

The most widely used tooth form for spur gears is the full-depth involute form as illustrated in Fig. 4.14. An involute is one of a class of curves called conjugate curves. When two involute-form gear teeth are in mesh there is a constant velocity ratio between them. From the moment of initial contact to the moment of disengagement, the speed of the gear is in constant proportion to the speed of the pinion. The resulting action of the gears is very smooth. If the velocity ratio was not constant, there would be accelerations and decelerations during the engagement and disengagement, causing vibration noise and potentially damaging torsional oscillations.

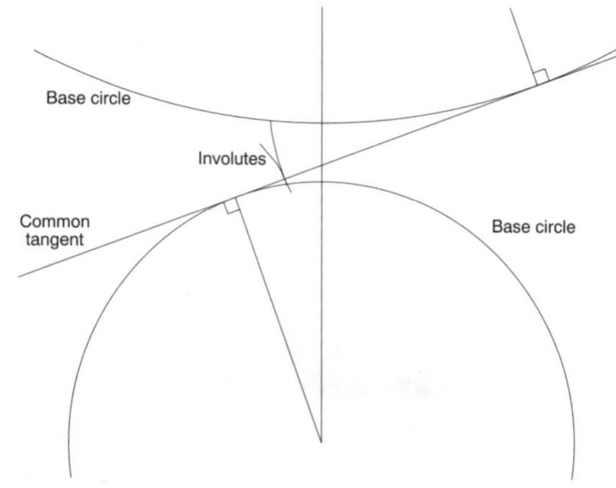

Fig. 4.14 Schematic of the involute form.

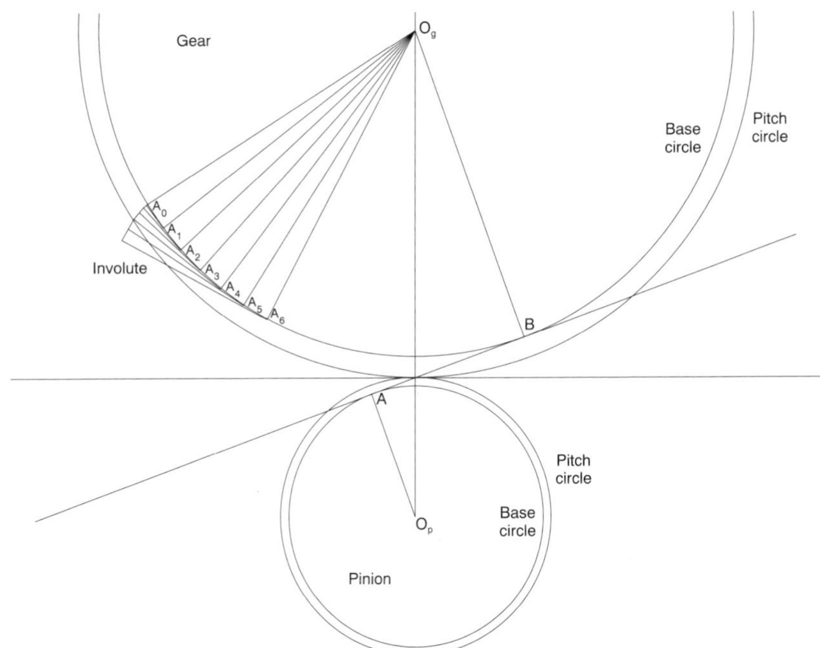

Fig. 4.15 Construction of gear geometry.

The layout and geometry for a pair of meshing spur gears can be determined by the procedure set out below. This procedure assumes access to a CAD draughting package such as Autocad. The procedure can be programmed as a set of procedural 'script' commands to automate the process for any set of initial gear parameters.

1. Calculate the pitch diameter and draw pitch circles tangential to each other (*see* Fig. 4.15):

   $d = mN$

   $d_p = mN_p$

   $d_g = mN_g$

   In the example shown, the module has been selected as $m = 2.5$, the number of teeth in the pinion 20, and in the gear 50. So $d_p = 2.5 \times 20 = 50\,\text{mm}$ and $d_g = 2.5 \times 50 = 125\,\text{mm}$.
2. Draw a line perpendicular to the line of centres through the pitch point (this is the point of tangency of the pitch circles). Draw the pressure line at an angle equal to the pressure angle from the perpendicular. It is called the pressure line because the resultant tooth force is along this line during meshing. Here the pressure angle is 20°.
3. Construct perpendiculars O_pA and O_gB to the pressure line through the centres of each gear. The radial distance of each of these lines are the radii of the base circles of the pinion and gear respectively. Draw the base circles.
4. Draw an involute curve on each base circle. This is illustrated on the gear. First divide the base circle in equal parts A_0, A_1, A_2, A_3, A_4, $A_5, \ldots, A_n$. Construct radial lines O_gA_0, O_gA_1, O_gA_2, $O_gA_3, \ldots, O_gA_n$. Construct perpendiculars to these radial lines. The involute begins at A_0. The second point is obtained by measuring off the distance A_0A_1 on the perpendicular through A_1. The next point is found by measuring off twice the distance A_0A_1 on the perpendicular through A_2 and so on. The curve constructed through these points is the involute for the gear. The involute for the pinion is constructed in the same way on the base circle of the pinion.
5. Calculate the circular pitch:

   $p = \pi m$

   The width of the teeth and the width of the spaces are equal to half the circular pitch. Mark these distances off on the pitch circles (here $p = \pi \times 2.5$).
6. Draw the addendum and dedendum circles for the pinion and gear (*see* Fig. 4.16). Here a tooth system (*see* Section 4.1.3 and Table 4.3) has been selected with

   $a = m$

   $b = 1.25m$

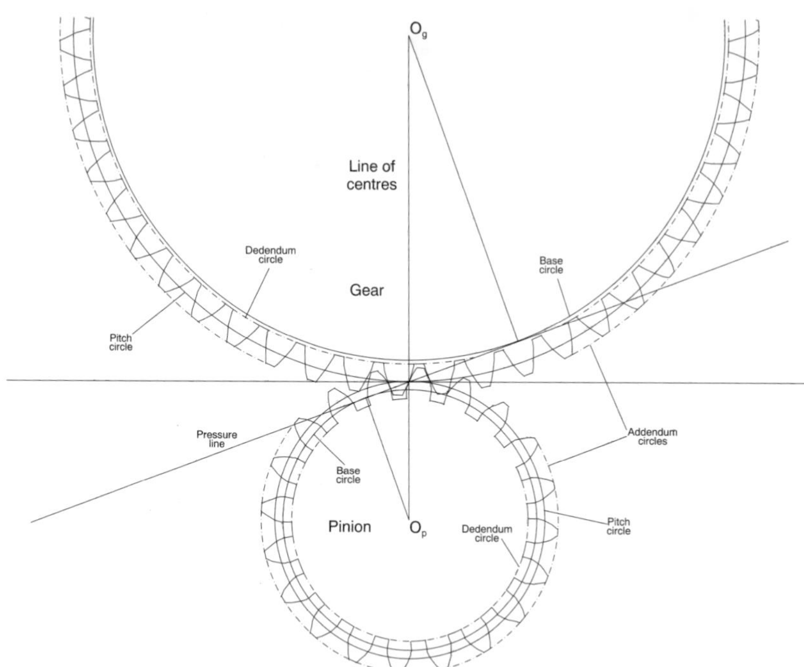

Fig. 4.16 Layout of a pair of meshing spur gears.

7. Mirror the involute profile about a line constructed using a distance half the tooth width along the pitch circle and the gear centre. Using a polar array, generate all of the teeth for the gear. Construct the root fillets as appropriate. Construct the tooth top and bottom lands.

Gears can be purchased as standard items from specialist manufacturers and the exact geometry for the blade teeth is not necessary for engineering drawings and general design purposes (unless you are a designer of gear cutters/forming tools). However, the tooth geometry is necessary for detailed stress/reliability and dynamic analysis.

4.1.2 Gear trains

A gear train is one or more pairs of gears operating together to transmit power. When two gears are in mesh, their pitch circles roll on each other without slippage. The pitch line velocity is given by

$$V = |r_1\omega_1| = |r_2\omega_2| \tag{4.4}$$

where r_1 and r_2 are the pitch radii of gears 1 and 2 and ω_1 and ω_2 are their angular velocities.

The velocity ratio is

$$\left|\frac{\omega_1}{\omega_2}\right| = \frac{r_2}{r_1} \tag{4.5}$$

and can be defined in any of the following ways:

$$\frac{\omega_P}{\omega_G} = \frac{n_P}{n_G} = \frac{N_G}{N_P} = \frac{d_G}{d_P} \tag{4.6}$$

where

ω_P and ω_G are the angular velocities of the pinion and gear respectively (rad/s);
n_P and n_G are the rotational speeds of the pinion and gear respectively (rpm);
N_P and N_G are the number of teeth in the pinion and gear respectively;
d_P and d_G are the pitch diameters of the pinion and gear respectively (mm).

Consider a pinion 1, driving a gear 2. The speed of the driven gear is

$$n_2 = \left|\frac{N_1}{N_2}n_1\right| = \left|\frac{d_1}{d_2}n_1\right| \tag{4.7}$$

where n is rotational speed (rpm), N is the number of teeth and d is the pitch diameter.

Equation 4.7 applies to any gear set (spur, helical, bevel or worm). For spur and parallel helical gears, the convention for direction is positive for anti-clockwise rotation.

Example Consider the gear train shown in Fig. 4.17. Calculate the speed of gear five.

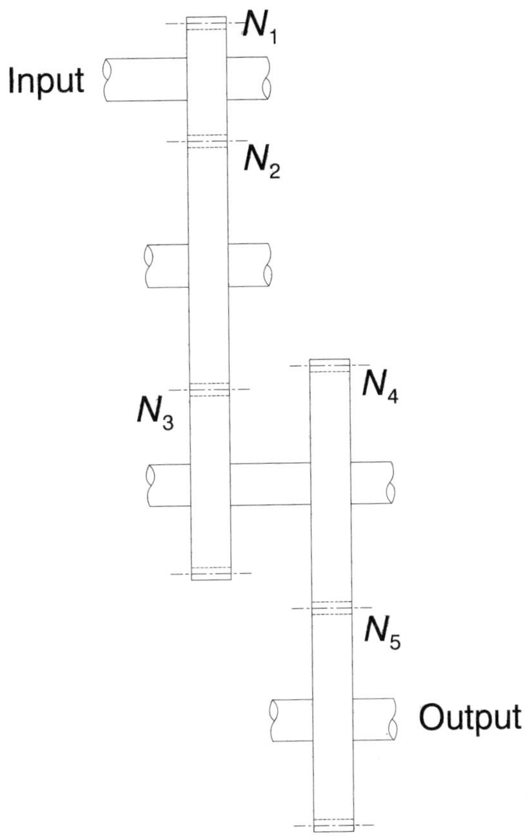

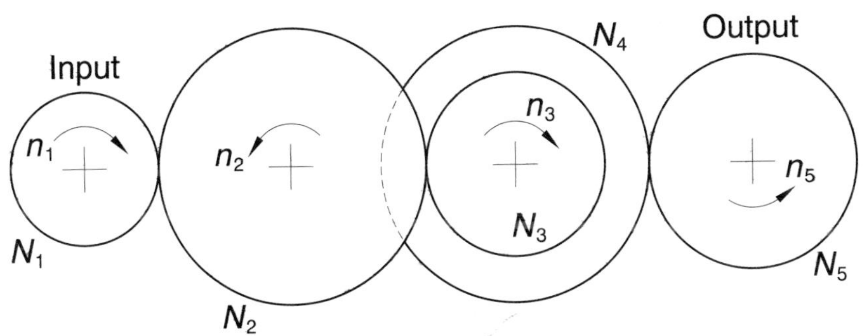

Fig. 4.17 Example gear train.

Solution

$$n_2 = \left| \frac{N_1}{N_2} n_1 \right|$$

$$n_3 = \left| \frac{N_2}{N_3} n_2 \right|$$

$$n_4 = n_3$$

$$n_5 = \left| \frac{N_4}{N_5} n_4 \right|$$

$$n_5 = -\frac{N_4 N_2 N_1}{N_5 N_3 N_2} n_1$$

Example For the double reduction gear train shown in Fig. 4.18, if the input speed is 1750 rpm what is the output speed?

Solution

$$n_2 = \left| \frac{N_1}{N_2} n_1 \right|$$

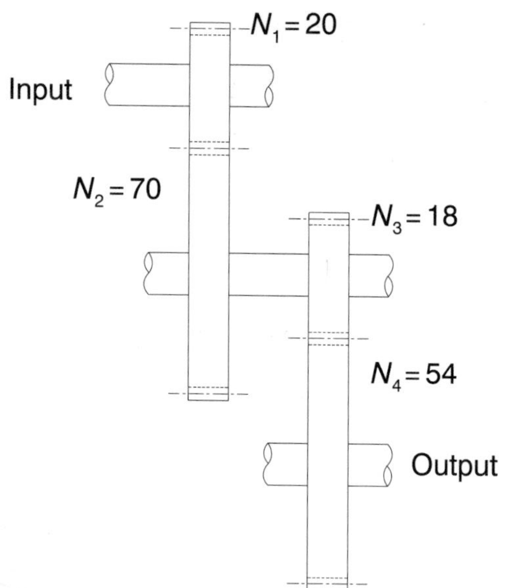

Fig. 4.18 Double reduction gear train.

$$n_3 = n_2$$

$$n_4 = \left| \frac{N_3}{N_4} n_3 \right|$$

$$n_4 = \frac{N_3}{N_4}\frac{N_1}{N_2} n_1 = \left(\frac{18}{54}\right)\left(\frac{20}{70}\right) n_1 = 166.7\,\text{rpm}$$

Example For the double reduction gear train with an idler shown in Fig. 4.19, if the input speed is 2000 rpm what is the output speed?

Solution

$$n_5 = \left| \frac{N_4}{N_5} n_4 \right|$$

$$n_4 = \left| \frac{N_3}{N_4} n_3 \right|$$

$$n_3 = n_2$$

$$n_2 = \left| \frac{N_1}{N_2} n_1 \right|$$

$$n_5 = -\frac{N_4}{N_5}\frac{N_3}{N_4}\frac{N_1}{N_2} n_1$$

$$= -\left(\frac{22}{54}\right)\left(\frac{18}{22}\right)\left(\frac{20}{70}\right) 2000 = -190.4\,\text{rpm}$$

Notice that the presence of the idler gear has caused the gear train output to reverse direction, but has not altered the gear ratio in comparison to the previous example.

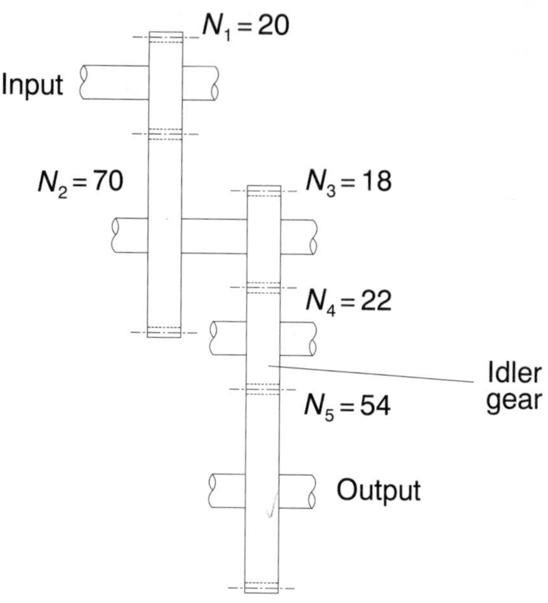

Fig. 4.19 Double reduction gear with idler.

As an illustration of a typical gear train, Fig. 4.20 shows a transmission for small and medium size lorries.

4.1.3 Tooth systems

Tooth systems are standards which define the geometric proportions of gear teeth. Table 4.3 lists the basic tooth dimensions for full depth teeth with pressure angles of 20° and 25°.

Table 4.4 lists preferred values for the module, m, which are used to minimise gear cutting tool requirements, and Table 4.5 lists the preferred standard gear teeth numbers.

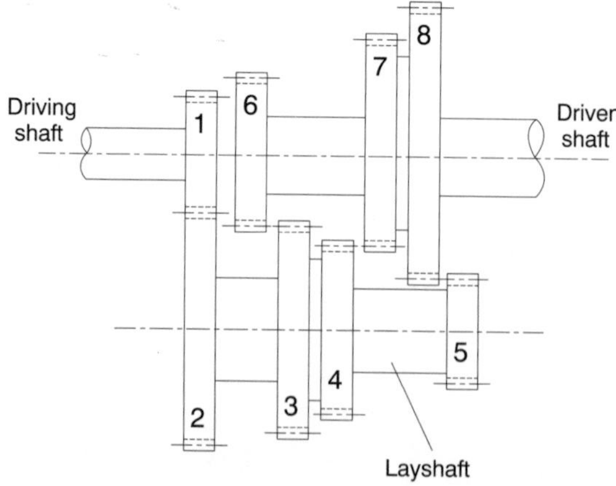

Fig. 4.20 Lorry transmission with four forward speeds.

Table 4.3 Tooth dimension formulas for $\phi = 20°$ and 25°

Dimension	Formula
Addendum	$a = m$
Dedendum	$b = 1.25m$
Working depth	$h_k = 2m$
Whole depth (min.)	$h_t = 2.25m$
Tooth thickness	$t = \pi m/2$
Fillet radius of basic rack	$r_f = 0.3m$
Clearance (min.)	$c = 0.25m$
Clearance for shaved or ground teeth	$c = 0.35m$
Minimum number of pinion teeth, $\phi = 20°$	$N_P = 18$
Minimum number of pinion teeth, $\phi = 25°$	$N_P = 12$
Minimum number of teeth per pair, $\phi = 20°$	$N_P + N_G = 36$
Minimum number of teeth per pair, $\phi = 25°$	$N_P + N_G = 24$
Width of top land (min.)	$t_o = 0.25m$

Note: m = module.

The failure of gears can principally be attributed to tooth breakage, surface failure and scuffing and scoring of the teeth surfaces. Section 4.2 outlines the forces involved on spur gear teeth and Section 4.3 defines methods for estimating gear stresses and the preliminary selection of standard stock gears. In Section 4.4 an overview of gear selection is given.

4.2 Force analysis

Figure 4.21 shows the forces involved for two spur gears in mesh. The force acting at the pressure angle ϕ can be subdivided into two components: the tangential component F_t and the radial component F_r. The radial component serves no useful purpose. The tangential component F_t transmits the load from one gear to the other. If W_t is defined as the transmitted load, $W_t = F_t$. The transmitted load is related to the power transmitted through the gears by the equation

$$W_t = \frac{\text{Power}}{V}$$

where V is the pitch line velocity (m/s). Alternatively it can be defined by

$$W_t = \frac{60 \times 10^3 H}{\pi d n} \tag{4.8}$$

where

W_t is the transmitted load (kN)
H is the power (kW)
d is the gear pitch diameter (mm)
n is the speed (rpm).

4.3 Gear stresses

Gears experience two principle types of stress: bending stress at the root of the teeth due to the transmitted load and contact stresses on the flank of the teeth due to repeated impact, or sustained contact, of one tooth surface against another. A simple method of calculating bending stresses is presented in Section 4.3.1, and this is utilised within a gear selection procedure given in Section 4.3.2. The calculation of contact stresses is defined in Section 4.3.3. A more detailed standard method of determining bending and contact stresses defined by the American Gear Manufacturers Association is presented in Section 4.3.4.

4.3.1 Bending stresses

The calculation of bending stress in gear teeth can be based on the Lewis formula, which in terms of the diametral pitch is given by

$$\sigma = \frac{W_t p}{FY} \tag{4.9}$$

where

W_t = transmitted load (N)
p = diametral pitch (m or mm)
F = face width (m or mm)
Y = Lewis form factor (Table 4.6).

Table 4.4 Preferred values for the module, m

0.5	0.8	1	1.25	1.5	2	2.5	3	4	5	6	8	10	12	16	20	25	32	40	50

Table 4.5 Preferred standard gear teeth numbers

12	13	14	15	16	18	20	22	24	25	28	30
32	34	38	40	45	50	54	60	64	70	72	75
80	84	90	96	100	120	140	150	180	200	220	250

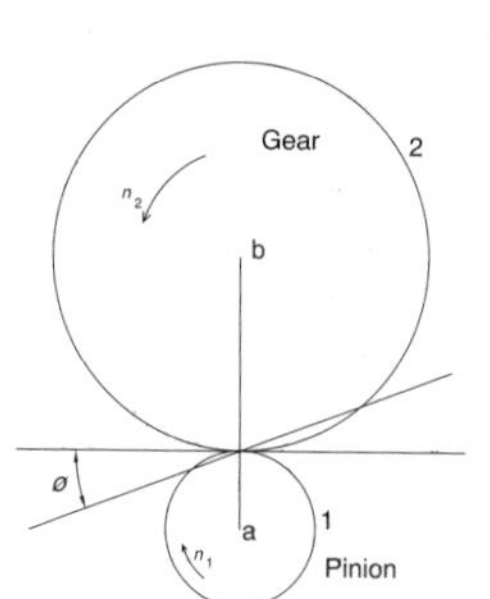

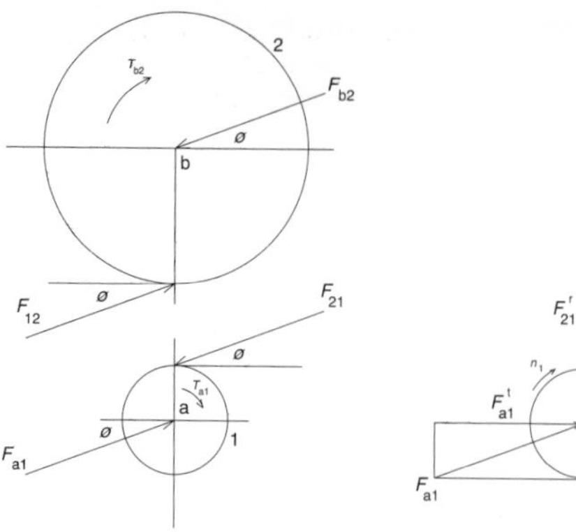

Fig. 4.21 Forces for two spur gears in mesh.

When teeth mesh, the load is delivered to the teeth with some degree of impact. The velocity factor is used to account for this and is given by the Barth equation:

$$K_v = \frac{6.1}{6.1 + V} \tag{4.10}$$

Table 4.6 Values for the Lewis form factor Y defined for two different tooth standards

Number of teeth, N	Y	
	$\phi = 20°$, $a = 0.8m$, $b = m$	$\phi = 20°$, $a = m$, $b = 1.25m$
12	0.33512	0.22960
13	0.34827	0.24317
14	0.35985	0.25530
15	0.37013	0.26622
16	0.37931	0.27610
17	0.38757	0.28508
18	0.39502	0.29327
19	0.40179	0.30078
20	0.40797	0.30769
21	0.41363	0.31406
22	0.41883	0.31997
24	0.42806	0.33056
26	0.43601	0.33979
28	0.44294	0.34790
30	0.44902	0.35510
34	0.45920	0.36731
38	0.46740	0.37727
45	0.47846	0.39093
50	0.48458	0.39860
60	0.49391	0.41047
75	0.50345	0.42283
100	0.51321	0.43574
150	0.52321	0.44930
300	0.53348	0.46364
Rack	0.54406	0.47897

a = addendum, b = dedendum, ϕ = pressure angle, m = module. Source: Mitchener and Mabie (1982).

where V is the pitch line velocity, given by

$$V = \frac{d}{2} \times 10^{-3} n \frac{2\pi}{60} \tag{4.11}$$

where d is in mm and n in rpm. Introducing the velocity factor into the Lewis equation gives

$$\sigma = \frac{W_t}{K_v F m Y} \tag{4.12}$$

Equation 4.12 forms the basis of a simple modern approach to the calculation of bending stresses in gears.

Example A 20° full depth spur pinion is to transmit 1.25 kW at 850 rpm. The pinion has 18 teeth. Determine the Lewis bending stress if the module is 2 and the face width is 25 mm.

Solution Calculating the pinion pitch diameter:

$$d_P = mN_P = 2 \times 18 = 36\,\text{mm}$$

Calculating the pitch line velocity:

$$V = \frac{d_P}{2} \times 10^{-3} \times n \frac{2\pi}{60}$$

$$= \frac{0.036}{2} \times 850 \times 0.1047 = 1.602\,\text{m/s}$$

Calculating the velocity factor:

$$K_v = \frac{6.1}{6.1 + V} = \frac{6.1}{6.1 + 1.602} = 0.7920$$

Calculating the transmitted load:

$$W_t = \frac{60 \times 10^3 H}{\pi d n} = \frac{60 \times 10^3 \times 1.25}{\pi 36 \times 850} = 0.7802\,\text{kN}$$

From Table 4.6, for $N_P = 18$, the Lewis form factor $Y = 0.29327$.

Table 4.7 Permissible bending stresses for various commonly used gear materials

Material	Treatment	σ_{uts} (MPa)	Permissible bending stress σ_p (MPa)
Nylon		65 (20°C)	27
Tufnol		110	31
080M40		540	131
080M40	Induction hardened	540	117
817M40		772	221
817M40	Induction hardened	772	183
045M10		494	117
045M10	Case hardened	494	276
655M13	Case hardened	849	345

The Lewis equation for bending stress gives

$$\sigma = \frac{W_t}{K_v F m Y}$$

$$= \frac{0.7802 \times 10^3}{0.792 \times 0.025 \times 0.002 \times 0.29327}$$

$$= 67.18 \times 10^6 \text{ N/m}^2 = 67.18 \text{ MPa}$$

4.3.2 Simple gear selection procedure

The Lewis formula in the form given by equation 4.12 ($\sigma = W_t/(K_v F m Y)$) can be used in a provisional spur gear selection procedure for a given transmission power, input and output speeds. The procedure is outlined below.

1. Select the number of teeth for the pinion and the gear to give the required gear ratio (observe the guidelines presented in Table 4.2 for maximum gear ratios). Note that the minimum number of teeth permissible when using a pressure angle of 20° is 18 (Table 4.3). Use either the standard teeth numbers as listed in Table 4.5 or as listed in a stock gear catalogue.
2. Select a material. This will be limited to those listed in the stock gear catalogues.
3. Select a module, m, from Table 4.4, or as listed in a stock gear catalogue (*see* Tables 4.8–4.11 which give examples of a selection of stock gears available).
4. Calculate the pitch diameter, $d = mN$.
5. Calculate the pitch line velocity, $V = (d/2) \times n \times (2\pi/60)$. Ensure this does not exceed the guidelines given in Table 4.2.
6. Calculate the dynamic factor, $K_v = 6/(6 + V)$.
7. Calculate the transmitted load, $W_t = \text{Power}/V$.
8. Calculate an acceptable face width using the Lewis formula, $F = W_t/(K_v m Y \sigma_p)$. The Lewis form factor can be obtained from Table 4.6.

The permissible bending stress, σ_p, can be taken as σ_{uts}/factor of safety, where the factor of safety is set by experience but may range from 2 to 5. Alternatively use values of σ_p as listed for the appropriate material in a stock gear catalogue (Table 4.7 lists values for a few gear materials).

The design procedure consists of proposing teeth numbers for the gear and pinion, selecting a suitable material, selecting a module and calculating the various parameters as listed, resulting in a value for the face width. If the face width is greater than those available in the stock gear catalogue or if the pitch line velocity is too high, repeat the process for a different module. If this does not provide a sensible solution try a different material etc. The process can be optimised, taking cost and other performance criteria into account if necessary, and can easily be programmed.

Example A gearbox is required to transmit 18 kW from a shaft rotating at 2650 rpm. The desired output speed is approximately 12 000 rpm. For space limitation and standardisation reasons a double step-up gearbox is requested with equal ratios. Using the limited selection of gears presented in Tables 4.8–4.11 select suitable gears for the gear wheels and pinions.

Solution

Overall ratio = 12 000/2650 = 4.528

First stage ratio = $\sqrt{4.528} = 2.128$

This could be achieved using a gear with 38 teeth and pinion with 18 teeth (ratio = 38/18 = 2.11).

The gear materials listed in Tables 4.8–4.11 are 817M40 and 655M13 steels (*see* Appendix A for an introduction to the British Standard designation system for steels). From Table 4.7 the 655M13 is

Table 4.8 Spur gears: 1.0 module, heavy duty steel 817M40, 655M13, face width 15 mm

Part number	Teeth	PCD (mm)	O/D (mm)	Boss dia. (mm)	Bore dia. (mm)
SG1-9	9	10.00	12.00	12	6
SG1-10	10	11.00	13.00	13	6
SG1-11	11	12.00	14.00	14	6
SG1-12	12	12.00	14.00	14	6
SG1-13	13	13.00	15.00	15	6
SG1-14	14	14.00	16.00	16	6
SG1-15	15	15.00	17.00	17	6
SG1-16	16	16.00	18.00	18	8
SG1-17	17	17.00	19.00	18	8
SG1-18	18	18.00	20.00	18	8
SG1-19	19	19.00	21.00	18	8
SG1-20	20	20.00	22.00	20	8
SG1-21	21	21.00	23.00	20	8
SG1-22	22	22.00	24.00	20	8
SG1-23	23	23.00	25.00	20	8
SG1-24	24	24.00	26.00	20	8
SG1-25	25	25.00	27.00	20	8
SG1-26	26	26.00	28.00	25	8
SG1-27	27	27.00	29.00	25	8
SG1-28	28	28.00	30.00	25	8
SG1-29	29	29.00	31.00	25	8
SG1-30	30	30.00	32.00	25	8
SG1-31	31	31.00	33.00	25	8
SG1-32	32	32.00	34.00	25	8
SG1-33	33	33.00	35.00	25	8
SG1-34	34	34.00	36.00	25	8
SG1-35	35	35.00	37.00	25	8
SG1-36	36	36.00	38.00	30	8
SG1-37	37	37.00	39.00	30	8
SG1-38	38	38.00	40.00	30	8
SG1-39	39	39.00	41.00	30	8
SG1-40	40	40.00	42.00	30	8
SG1-41	41	41.00	43.00	30	8
SG1-42	42	42.00	44.00	30	8
SG1-43	43	43.00	45.00	30	8
SG1-44	44	44.00	46.00	30	8
SG1-45	45	45.00	47.00	30	8
SG1-46	46	46.00	48.00	30	8
SG1-47	47	47.00	49.00	30	8
SG1-48	48	48.00	50.00	30	8
SG1-49	49	49.00	51.00	30	8
SG1-50	50	50.00	52.00	35	10
SG1-51	51	51.00	53.00	35	10
SG1-52	52	52.00	54.00	35	10
SG1-53	53	53.00	55.00	35	10
SG1-54	54	54.00	56.00	35	10
SG1-55	55	55.00	57.00	35	10
SG1-56	56	56.00	58.00	35	10
SG1-57	57	57.00	59.00	35	10
SG1-58	58	58.00	60.00	35	10
SG1-59	59	59.00	61.00	35	10
SG1-60	60	60.00	62.00	35	10
SG1-61	61	61.00	63.00	35	10
SG1-62	62	62.00	64.00	35	10
SG1-63	63	63.00	65.00	35	10

Table 4.8 Continued

Part number	Teeth	PCD (mm)	O/D (mm)	Boss dia. (mm)	Bore dia. (mm)
SG1-64	64	64.00	66.00	35	10
SG1-65	65	65.00	67.00	35	10
SG1-66	66	66.00	68.00	35	10
SG1-68	68	68.00	70.00	35	10
SG1-70	70	70.00	72.00	35	10
SG1-72	72	72.00	74.00	35	10
SG1-74	74	74.00	76.00	45	10
SG1-76	76	76.00	78.00	45	10
SG1-78	78	78.00	80.00	45	10
SG1-80	80	80.00	82.00	45	10
SG1-84	84	84.00	86.00	45	10
SG1-88	88	88.00	90.00	45	10
SG1-90	90	90.00	92.00	45	10
SG1-96	96	96.00	98.00	45	10
SG1-100	100	100.00	102.00	45	10
SG1-112	112	112.00	114.00	45	10
SG1-120	120	120.00	122.00	45	10
SG1-130	130	130.00	132.00	45	10
SG1-150	150	150.00	152.00	45	10

Data adapted from HPC Gears Ltd. PCD = pitch circle diameter; O/D = outer diameter.

the stronger steel and this is selected for this example prior to a more detailed consideration. For 655M13 case hardened steel gears, the permissible stress $\sigma_p = 345$ MPa.

Calculations for gear 1: $Y_{38} = 0.37727$, $n =$ 2650 rpm.

m	1.5	2.0
d (mm)	57	76
V (m/s)	7.9	10.5
W_t (N)	2276	1707
K_v	0.4357	0.3675
F	27	18

$m = 1.5$ gives face width greater than catalogue value of 20, so try $m = 2$.

Calculations for pinion 1: $Y_{18} = 0.29327$, $n =$ 5594 rpm.

m	1.5	2.0
d (mm)	27	36
V (m/s)	7.9	10.5
W_t (N)	2276	1707
K_v	0.4357	0.3676
F (mm)	34.4	23

$m = 1.5$ gives face width greater than catalogue value of 20, so try $m = 2$.

Calculations for gear 2: $Y_{38} = 0.37727$, $n =$ 5594 rpm.

m	2.0
d (mm)	76
V (m/s)	22.26
W_t (N)	808.6
K_v	0.215
F (mm)	14.4

$m = 2$ gives value for face width lower than catalogue specification, so the design is OK.

Calculations for pinion 2: $Y_{18} = 0.29327$, $n =$ 11 810 rpm.

m	2.0
d (mm)	36
V (m/s)	22.26
W_t (N)	808.6
K_v	0.215
F (mm)	18.6

$m = 2$ gives value for face width lower than catalogue specification, so the design is OK.

4.3.3 Wear failure

As well as considering failure due to bending stresses in gears, the failure due to wear on the surface of gear teeth should be considered. Possible surface failures are pitting, which is a surface fatigue failure due to many repetitions of high contact stresses, scoring due to failure of lubrication and abrasion due to the presence of foreign particles.

The surface compressive, Hertzian or contact stress for a gear is defined by

$$\sigma_c = -C_p \left[\frac{W_t}{C_v F \cos\phi} \left(\frac{1}{r_1} + \frac{1}{r_2} \right) \right]^{0.5} \qquad (4.13)$$

Table 4.9 Spur gears: 1.5 module, heavy duty steel 817M40, 655M13, face width 20 mm

Part number	Teeth	PCD (mm)	O/D (mm)	Boss dia. (mm)	Bore dia. (mm)
SG1.5-9	9	15.00	18.00	18	8
SG1.5-10	10	16.50	19.50	19.5	8
SG1.5-11	11	18.00	21.00	21	8
SG1.5-12	12	18.00	21.00	21	8
SG1.5-13	13	19.50	22.50	20	8
SG1.5-14	14	21.00	24.00	20	8
SG1.5-15	15	22.50	25.50	20	8
SG1.5-16	16	24.00	27.00	25	10
SG1.5-17	17	25.50	28.50	25	10
SG1.5-18	18	27.00	30.00	25	10
SG1.5-19	19	28.50	31.50	25	10
SG1.5-20	20	30.00	33.00	25	10
SG1.5-21	21	31.50	34.50	25	10
SG1.5-22	22	33.00	36.00	30	10
SG1.5-23	23	34.50	37.50	30	10
SG1.5-24	24	36.00	39.00	30	10
SG1.5-25	25	37.50	40.50	30	10
SG1.5-26	26	39.00	42.00	30	10
SG1.5-27	27	40.50	43.50	30	10
SG1.5-28	28	42.00	45.00	30	10
SG1.5-29	29	43.50	46.50	30	10
SG1.5-30	30	45.00	48.00	30	10
SG1.5-31	31	46.50	49.50	30	10
SG1.5-32	32	48.00	51.00	30	10
SG1.5-33	33	49.50	52.50	30	10
SG1.5-34	34	51.00	54.00	30	10
SG1.5-35	35	52.50	55.50	50	15
SG1.5-36	36	54.00	57.00	50	15
SG1.5-37	37	55.50	58.50	50	15
SG1.5-38	38	57.00	60.00	50	15
SG1.5-39	39	58.50	61.50	50	15
SG1.5-40	40	60.00	63.00	50	15
SG1.5-41	41	61.50	64.50	50	15
SG1.5-42	42	63.00	66.00	50	15
SG1.5-43	43	64.50	67.50	50	15
SG1.5-44	44	66.00	69.00	50	15
SG1.5-45	45	67.50	70.50	50	15
SG1.5-46	46	69.00	72.00	50	15
SG1.5-47	47	70.50	73.50	50	15
SG1.5-48	48	72.00	75.00	50	15
SG1.5-49	49	73.50	76.50	50	15
SG1.5-50	50	75.00	78.00	50	15
SG1.5-51	51	76.50	79.50	50	15
SG1.5-52	52	78.00	81.00	50	15
SG1.5-53	53	79.50	82.50	60	15
SG1.5-54	54	81.00	84.00	60	15
SG1.5-55	55	82.50	85.50	60	15
SG1.5-56	56	84.00	87.00	60	15
SG1.5-57	57	85.50	88.50	60	15
SG1.5-58	58	87.00	90.00	60	15
SG1.5-59	59	88.50	91.50	60	15
SG1.5-60	60	90.00	93.00	60	15
SG1.5-62	62	93.00	96.00	60	15
SG1.5-64	64	96.00	99.00	60	15
SG1.5-65	65	97.50	100.50	60	15

Table 4.9 Continued

Part number	Teeth	PCD (mm)	O/D (mm)	Boss dia. (mm)	Bore dia. (mm)
SG1.5-66	66	99.00	102.00	60	15
SG1.5-68	68	102.00	105.00	60	15
SG1.5-70	70	105.00	108.00	60	15
SG1.5-71	71	106.50	109.50	60	15
SG1.5-72	72	108.00	111.00	60	15
SG1.5-73	73	109.50	112.50	60	15
SG1.5-74	74	111.00	114.00	60	15
SG1.5-75	75	112.50	115.50	60	15
SG1.5-76	76	114.00	117.00	60	15
SG1.5-78	78	117.00	120.00	75	15
SG1.5-80	80	120.00	123.00	75	15
SG1.5-86	86	129.00	132.00	75	15
SG1.5-90	90	135.00	138.00	75	15
SG1.5-96	96	144.00	147.00	75	15
SG1.5-98	98	147.00	150.00	75	15
SG1.5-100	100	150.00	153.00	75	15
SG1.5-105	105	157.50	160.50	75	15
SG1.5-110	110	165.00	168.00	75	15
SG1.5-115	115	175.50	178.50	75	15
SG1.5-120	120	180.00	183.00	75	15

Data adapted from HPC Gears Ltd. PCD = pitch circle diameter; O/D = outer diameter.

where

$$r_1 = \frac{d_P \sin\phi}{2} \qquad r_2 = \frac{d_G \sin\phi}{2} \tag{4.14}$$

and

C_v is the velocity factor = K_v (*see* equation 4.10)
C_p is the elastic coefficient (*see* equation 4.22 and Table 4.12).

If the units of C_P are in $\sqrt{\text{MPa}}$, W_t in newtons, r_1, r_2 in metres, equation 4.13 gives σ_c in kPa.

Example A speed reducer has a 22-tooth spur pinion made of steel, driving a 60-tooth gear made of cast iron. The transmitted power is 10 kW. The pinion speed is 1200 rpm, module 4 and face width 50 mm. Determine the contact stress.

Solution For $N_P = 22$, $N_G = 60$, $n = 1200$ rpm, $m = 4$, $F = 50$ mm, $H = 10$ kW:

$$d_P = mN_P = 4 \times 22 = 88\text{ mm}$$

$$d_G = 4 \times 60 = 240\text{ mm}$$

$$V = \frac{0.088}{2} \times 1200 \times \frac{2\pi}{60} = 5.529\text{ m/s}$$

$$C_v = \frac{6.1}{6.1 + 5.529} = 0.5245$$

$$W_t = \frac{60 \times 10^3 H}{\pi d n} = 1.809\text{ kN}$$

From Table 4.12, for a steel pinion and cast iron gear:

$$C_P = 174\sqrt{\text{MPa}}$$

From equation 4.14:

$$r_1 = \frac{88 \times 10^{-3} \sin 20}{2} = 0.01504$$

$$r_2 = \frac{240 \times 10^{-3} \sin 20}{2} = 0.04104$$

From equation 4.13:

$$\sigma_c = -174\left[\frac{1.809 \times 10^3}{0.5245 \times 0.05 \cos 20}(90.82)\right]^{0.5}$$

$$= -174 \times 2582\text{ kPa}$$

$$= -449\,200\text{ kPa} \approx -449\text{ MPa}$$

4.3.4 AGMA equations for bending and contact stress

The calculation of bending and contact stresses in spur and helical can be determined using standardised methods presented by the British Standards Institution (BSI), the International Organisation for Standardisation (ISO), the Deutsches Institut

Table 4.10 Spur gears: 2.0 module, heavy duty steel 817M40, 655M13, face width 25 mm

Part number	Teeth	PCD (mm)	O/D (mm)	Boss dia. (mm)	Bore dia. (mm)
SG2-9	9	20.00	24.00	24	12
SG2-10	10	22.00	26.00	26	12
SG2-11	11	24.00	28.00	28	12
SG2-12	12	24.00	28.00	28	12
SG2-13	13	26.00	30.00	30	12
SG2-14	14	28.00	32.00	30	12
SG2-15	15	30.00	34.00	30	12
SG2-16	16	32.00	36.00	30	12
SG2-17	17	34.00	38.00	35	15
SG2-18	18	36.00	40.00	35	15
SG2-19	19	38.00	42.00	35	15
SG2-20	20	40.00	44.00	35	15
SG2-21	21	42.00	46.00	35	15
SG2-22	22	44.00	48.00	35	15
SG2-23	23	46.00	50.00	35	15
SG2-24	24	48.00	52.00	35	15
SG2-25	25	50.00	54.00	35	15
SG2-26	26	52.00	56.00	35	15
SG2-27	27	54.00	58.00	50	20
SG2-28	28	56.00	60.00	50	20
SG2-29	29	58.00	62.00	50	20
SG2-30	30	60.00	64.00	50	20
SG2-31	31	62.00	66.00	50	20
SG2-32	32	64.00	68.00	50	20
SG2-33	33	66.00	70.00	50	20
SG2-34	34	68.00	72.00	50	20
SG2-35	35	70.00	74.00	50	20
SG2-36	36	72.00	76.00	50	20
SG2-37	37	74.00	78.00	50	20
SG2-38	38	76.00	80.00	50	20
SG2-39	39	78.00	82.00	50	20
SG2-40	40	80.00	84.00	50	20
SG2-41	41	82.00	86.00	50	20
SG2-42	42	84.00	88.00	50	20
SG2-43	43	86.00	90.00	60	20
SG2-44	44	88.00	92.00	60	20
SG2-45	45	90.00	94.00	60	20
SG2-46	46	92.00	96.00	60	20
SG2-47	47	94.00	98.00	60	20
SG2-48	48	96.00	100.00	60	20
SG2-49	49	98.00	102.00	60	20
SG2-50	50	100.00	104.00	60	20
SG2-51	51	102.00	106.00	60	20
SG2-52	52	104.00	108.00	60	20
SG2-53	53	106.00	110.00	60	20
SG2-54	54	108.00	112.00	60	20
SG2-55	55	110.00	114.00	60	20
SG2-56	56	112.00	116.00	60	20
SG2-57	57	114.00	118.00	60	20
SG2-58	58	116.00	120.00	60	20
SG2-59	59	118.00	122.00	60	20
SG2-60	60	120.00	124.00	60	20
SG2-62	62	124.00	128.00	75	20
SG2-64	64	128.00	132.00	75	20
SG2-65	65	130.00	134.00	75	20

Table 4.10 Continued

Part number	Teeth	PCD (mm)	O/D (mm)	Boss dia. (mm)	Bore dia. (mm)
SG2-66	66	132.00	136.00	75	20
SG2-68	68	136.00	140.00	75	20
SG2-70	70	140.00	144.00	75	20
SG2-71	71	142.00	146.00	75	20
SG2-72	72	144.00	148.00	75	20
SG2-73	73	146.00	150.00	75	20
SG2-74	74	148.00	152.00	75	20
SG2-75	75	150.00	154.00	75	20
SG2-76	76	152.00	156.00	75	20
SG2-78	78	156.00	160.00	100	20
SG2-80	80	160.00	164.00	100	20
SG2-86	86	172.00	176.00	100	20
SG2-90	90	180.00	184.00	100	20
SG2-96	96	192.00	196.00	100	20
SG2-98	98	196.00	200.00	100	20
SG2-100	100	200.00	204.00	100	20
SG2-105	105	210.00	214.00	100	20
SG2-110	110	220.00	224.00	100	20
SG2-115	115	230.00	234.00	100	20
SG2-120	120	240.00	244.00	100	20

Data adapted from HPC Gears Ltd. PCD = pitch circle diameter; O/D = outer diameter.

für Normung (DIN), and the American Gear Manufacturers Association (AGMA). In the opinion of the author, the AGMA standards are currently the most approachable and are utilised here. The procedures make use of a series of geometry and design factors which can be determined from design charts and tables.

The AGMA formula for bending stress is

$$\sigma = \frac{W_t K_a}{K_v} \frac{1}{Fm} \frac{K_s K_m}{J} \qquad (4.15)$$

where

σ is the bending stress
W_t is the transmitted tangential (*see* equation 4.8)
K_a is an application factor (usually taken as $K_a = 1$)
K_v is a dynamic factor (*see* equation 4.23)
m is the module
F is the face width
K_s is a size factor (usually taken as $K_s = 1$)
K_m is a load distribution factor (*see* Table 4.13)
J is a geometry factor (*see* Table 4.14).

The AGMA equation for pitting resistance is

$$\sigma_c = C_p \left(\frac{W_t C_a}{C_v} \frac{C_s}{Fd} \frac{C_m C_f}{I} \right)^{0.5} \qquad (4.16)$$

where

σ_c is the absolute value of contact stress
C_p is the elastic coefficient (*see* equation 4.22 or Table 4.12)
C_a is an application factor (usually taken as $C_a = 1$)
C_v is a dynamic factor (*see* equation 4.23)
C_s is a size factor (usually taken as $C_s = 1$)
d is the pitch diameter of the pinion
C_m is a load distribution factor (*see* Table 4.13)
C_f is a surface condition factor (usually taken as $C_f = 1$)
I is a geometry factor (*see* equation 4.19).

For equation 4.16, if W_t is in newtons, F and d in metres and C_P in $\sqrt{\text{MPa}}$, then the units of σ_c are in kPa.

Based on safe working practices, the AGMA has defined allowable stress equations for gears for the allowable bending stress and for the allowable contact stress.

The AGMA equation for determining a safe value for the allowable bending stress is

$$\sigma_{all} = \frac{S_t K_L}{K_T K_R} \qquad (4.17)$$

where

S_t is the AGMA bending strength (*see* Table 4.16)

Table 4.11 Spur gears: 3.0 module, heavy duty steel 817M40, 655M13, face width 35 mm

Part number	Teeth	PCD (mm)	O/D (mm)	Boss dia. (mm)	Bore dia. (mm)
SG3-9	9	30.00	36.00	36	15
SG3-10	10	33.00	39.00	39	15
SG3-11	11	36.00	42.00	42	15
SG3-12	12	36.00	42.00	42	20
SG3-13	13	39.00	45.00	45	20
SG3-14	14	42.00	48.00	45	20
SG3-15	15	45.00	51.00	45	20
SG3-16	16	48.00	54.00	45	20
SG3-17	17	51.00	57.00	45	20
SG3-18	18	54.00	60.00	45	20
SG3-19	19	57.00	63.00	45	20
SG3-20	20	60.00	66.00	60	25
SG3-21	21	63.00	69.00	60	25
SG3-22	22	66.00	72.00	60	25
SG3-23	23	69.00	75.00	60	25
SG3-24	24	72.00	78.00	60	25
SG3-25	25	75.00	81.00	60	25
SG3-26	26	78.00	84.00	60	25
SG3-27	27	81.00	87.00	60	25
SG3-28	28	84.00	90.00	60	25
SG3-29	29	87.00	93.00	60	25
SG3-30	30	90.00	96.00	60	25
SG3-31	31	93.00	99.00	60	25
SG3-32	32	96.00	102.00	60	25
SG3-33	33	99.00	105.00	60	25
SG3-34	34	102.00	108.00	75	25
SG3-35	35	105.00	111.00	75	25
SG3-36	36	108.00	114.00	75	25
SG3-37	37	111.00	117.00	75	25
SG3-38	38	114.00	120.00	75	25
SG3-39	39	117.00	123.00	75	25
SG3-40	40	120.00	126.00	75	25
SG3-41	41	123.00	129.00	75	25
SG3-42	42	126.00	132.00	75	25
SG3-43	43	129.00	135.00	75	25
SG3-44	44	132.00	138.00	100	25
SG3-45	45	135.00	141.00	100	25
SG3-46	46	138.00	144.00	100	25
SG3-47	47	141.00	147.00	100	25
SG3-48	48	144.00	150.00	100	25
SG3-49	49	147.00	153.00	100	25
SG3-50	50	150.00	156.00	100	25
SG3-51	51	153.00	159.00	100	25
SG3-52	52	156.00	162.00	100	25
SG3-53	53	159.00	165.00	100	25
SG3-54	54	162.00	168.00	100	25
SG3-55	55	165.00	171.00	100	25
SG3-56	56	168.00	174.00	100	25
SG3-57	57	171.00	177.00	127	25
SG3-58	58	174.00	180.00	127	25
SG3-59	59	177.00	183.00	127	25
SG3-60	60	180.00	186.00	127	25
SG3-62	62	186.00	192.00	127	25
SG3-63	63	189.00	195.00	127	25
SG3-64	64	192.00	198.00	127	25

Table 4.11 Continued

Part number	Teeth	PCD (mm)	O/D (mm)	Boss dia. (mm)	Bore dia. (mm)
SG3-65	65	195.00	201.00	127	25
SG3-68	68	204.00	210.00	127	25
SG3-70	70	210.00	216.00	150	30
SG3-72	72	216.00	222.00	150	30
SG3-75	75	225.00	231.00	150	30
SG3-76	76	228.00	234.00	150	30
SG3-78	78	234.00	240.00	150	30
SG3-80	80	240.00	246.00	150	30
SG3-82	82	246.00	252.00	150	30
SG3-84	84	252.00	258.00	150	30
SG3-86	86	258.00	264.00	150	30
SG3-90	90	270.00	276.00	150	30
SG3-92	92	276.00	282.00	150	30
SG3-94	94	282.00	288.00	150	30
SG3-95	95	285.00	291.00	150	30
SG3-96	96	288.00	294.00	150	30

Data adapted from HPC Gears Ltd. PCD = pitch circle diameter; O/D = outer diameter.

Table 4.12 Values of the elastic coefficient C_p ($\sqrt{\text{MPa}}$)

Pinion	E (GPa)	Gear: Steel	Malleable iron	Nodular iron	Cast iron	Aluminium bronze	Tin bronze
Steel	200	191	181	179	174	162	158
Malleable iron	170	181	174	172	168	158	154
Nodular iron	170	179	172	170	166	156	152
Cast iron	150	174	168	166	163	154	149
Aluminium bronze	120	162	158	156	154	145	141
Tin bronze	110	158	154	152	149	141	137

Table 4.13 Load distribution factors, C_m, K_m for spur gears

Condition of support	Face width F (mm): ≤ 50	≤ 150	≤ 225	≥ 400
Low bearing clearances; minimum deflections; precision gears; accurate mounting	1.3	1.4	1.5	1.8
Less accurate gears; contact across full face; less rigid mounting	1.6	1.7	1.8	2.0
Accuracy and mounting with less than full face contact			≥ 2	

After AGMA 225.01.

Table 4.14 Values for the AGMA geometry factor J for teeth with $\phi = 20°$, $a = m$, $b = 1.25m$, $r_f = 0.3m$

	Number of teeth in mating gear							
Number of teeth	1	17	25	35	50	85	300	1000
18	0.24486	0.32404	0.33214	0.33840	0.34404	0.35050	0.35594	0.36112
19	0.24794	0.33029	0.33878	0.34537	0.35134	0.35822	0.36405	0.36963
20	0.25072	0.33600	0.34485	0.35176	0.35804	0.36532	0.37151	0.37749
21	0.25323	0.34124	0.35044	0.35764	0.36422	0.37186	0.37841	0.38475
22	0.25552	0.34607	0.35559	0.36306	0.36992	0.37792	0.38479	0.39148
24	0.25951	0.35468	0.36477	0.37275	0.38012	0.38877	0.39626	0.40360
26	0.26289	0.36211	0.37272	0.38115	0.38897	0.39821	0.40625	0.41418
28	0.26580	0.36860	0.37967	0.38851	0.39673	0.40650	0.41504	0.42351
30	0.26831	0.37432	0.38580	0.39500	0.40359	0.41383	0.42283	0.43179
34	0.27247	0.38394	0.39611	0.40594	0.41517	0.42624	0.43604	0.44586
38	0.27575	0.39170	0.40446	0.41480	0.42456	0.43633	0.44680	0.45735
45	0.28013	0.40223	0.41579	0.42685	0.43735	0.45010	0.46152	0.47310
50	0.28252	0.40808	0.42208	0.43555	0.44448	0.45778	0.46975	0.48193
60	0.28613	0.41702	0.43173	0.44383	0.45542	0.46960	0.48243	0.49557
75	0.28979	0.42620	0.44163	0.45440	0.46668	0.48179	0.49554	0.50970
100	0.29353	0.43561	0.45180	0.46527	0.47827	0.49437	0.50909	0.52435
150	0.29738	0.44530	0.46226	0.47645	0.49023	0.50736	0.52312	0.53954
300	0.30141	0.45526	0.47304	0.48798	0.50256	0.52078	0.53765	0.55533
Rack	0.30571	0.46554	0.48415	0.49988	0.51529	0.53467	0.55272	0.57173

Source: Mitchener and Mabie (1982).

K_L is a life factor (*see* Fig. 4.22)
K_T is a temperature factor (usually taken as $K_T = 1$)
K_R is a reliability factor (*see* Table 4.15).

The AGMA equation for determining a safe value for the allowable contact stress is

$$\sigma_{c,all} = \frac{S_c C_L C_H}{C_T C_R} \qquad (4.18)$$

where

S_c is the AGMA surface fatigue strength (*see* Table 4.17)
C_L is a life factor (*see* Fig. 4.23)
C_H is a hardness ratio factor (*see* equation 4.26 (for gear only))
C_T is the temperature factor (usually taken as $C_T = 1$)
C_R is a reliability factor (*see* Table 4.15).

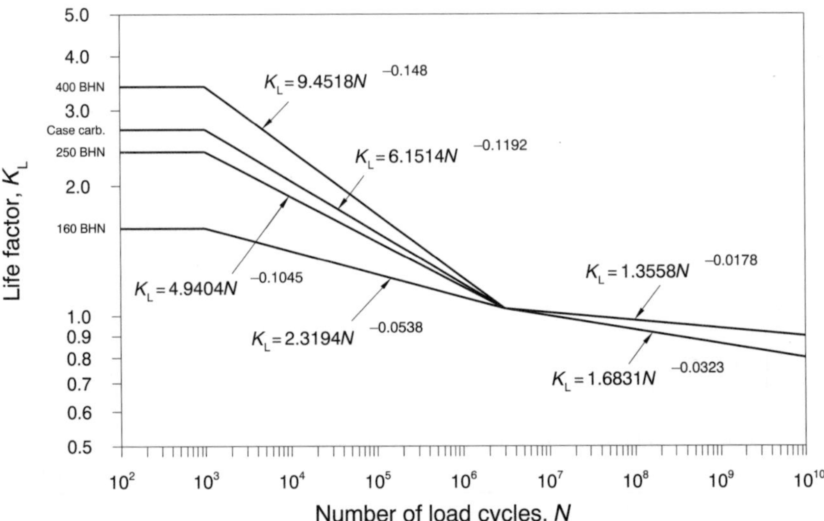

Fig. 4.22 Pitting resistance life factor K_L. (Reproduced from ANSI/AGMA 2001-B88.)

Table 4.15 AGMA reliability factors C_R, K_R

Reliability	C_R, K_R
0.90	0.85
0.99	1.00
0.999	1.25
0.9999	1.50

After ANSI/AGMA 2001-B88.

The bending strength geometry factor, J, for spur gears can be determined from Table 4.14.

For spur and helical gears the surface strength geometry factor, I, can be determined from

$$I = \frac{\cos\phi_t \sin\phi_t}{2m_N}\frac{m_G}{m_G+1} \quad \text{(external gears)} \tag{4.19}$$

$$I = \frac{\cos\phi_t \sin\phi_t}{2m_N}\frac{m_G}{m_G-1} \quad \text{(internal gears)} \tag{4.20}$$

where $m_N = 1$ for spur gears. (ϕ_t is the transverse pressure angle used for helical gears. Replace ϕ_t by ϕ for spur gears.) For helical gears, $m_N = p_N/0.95Z$, where p_N is the base pitch and Z is the length of the line of action in the transverse plane.

The speed ratio m_G is

$$m_G = \frac{N_G}{N_P} \tag{4.21}$$

The elastic coefficient C_p can be determined from Table 4.12, or calculated using

$$C_p = \left[\frac{1}{\pi\left(\frac{1-\nu_P^2}{E_P}+\frac{1-\nu_G^2}{E_G}\right)}\right]^{0.5} \tag{4.22}$$

where

ν_P is Poisson's ratio for the pinion
ν_G is Poisson's ratio for the gear
E_P is Young's modulus for the pinion
E_G is Young's modulus for the gear.

The dynamic factor $C_v = K_v$ can be determined by

$$C_v = K_v = \left(\frac{A}{A+\sqrt{200V}}\right)^B \tag{4.23}$$

where

$$B = \frac{(12-Q_v)^{2/3}}{4} \tag{4.24}$$

and

$$A = 50 + 56[1 - B] \tag{4.25}$$

The dynamic factor is used to account for the effect of tooth spacing and profile errors, the magnitude of the pitch line velocity, the inertia and stiffness of the rotating components and the transmitted load per unit face width. The AGMA (and also BSI and ISO) has defined a set of quality control numbers, Q_v to quantify these parameters. Classes of $3 \le Q_v \le 7$ include most commercial quality gears; classes of $8 \le Q_v \le 12$ are of precision quality. Generally, the higher the peripheral speed and the smaller the specific tooth load, the more accurate the gear manufacture must be and a gear with a higher quality control number should be used.

The application factors C_a and K_a are used at the discretion of the engineer to account for high in-service loads.

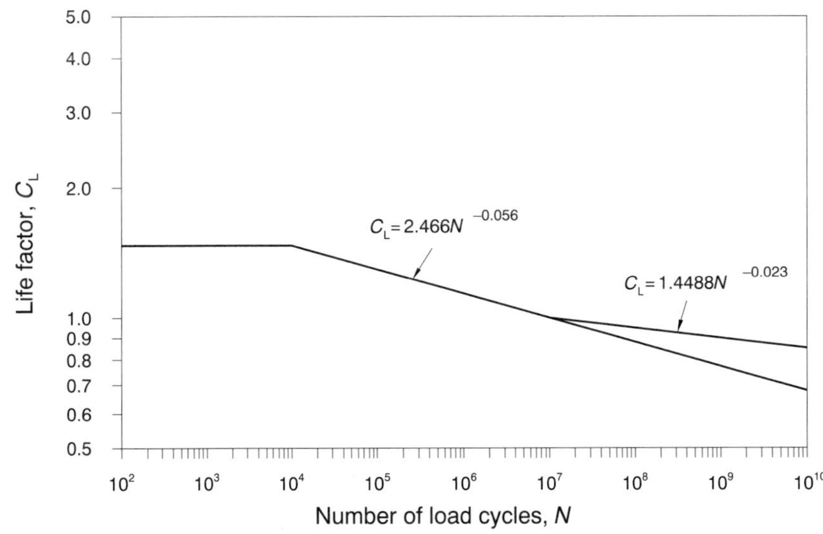

Fig. 4.23 Bending strength life factor C_L. (Reproduced from ANSI/AGMA 2001-B88.)

Table 4.16 Selected AGMA bending strengths

Material		Heat treatment	Minimum hardness at surface (BHN)	S_t (MPa)
Steel				
AGMA class	A1	Through hardened and tempered	≤180	170–230
	A2		240	210–280
	A3		300	250–325
	A4		360	280–360
	A5		400	290–390
BS 080M40		Induction hardened	155	205–215
BS 817M40		Induction hardened	220	260–270
BS 045M10		Case hardened	145	195–205
BS 655M13		Case hardened	245	280–290
Cast iron				
AGMA class	20	As cast	–	35
	30	As cast	175	69
	40	As cast	200	90

After ANSI/AGMA 2001-B88.

The surface condition factor C_f is usually taken as $C_f = 1$. Higher values can be used if surface defects are present.

The size factors C_s and K_s are usually taken as $C_s = 1$ and $K_s = 1$. Higher values can be used if there is any non-uniformity in the material properties.

The load distribution factor (C_m, K_m) is used to account for misalignment of rotational axes, deviations in lead, elastic deflections of shafts, bearing and housing caused by loads. Values of C_m and K_m can be determined from Table 4.13.

The hardness ratio factor is used only for the gear. Its purpose is to adjust the surface strengths to account for different hardness of the gear and pinion. The hardness ratio factor for a gear is given by

$$C_H = 1 + A(m_G - 1) \qquad (4.26)$$

where

$$A = 8.98 \times 10^{-3}\left(\frac{H_{BP}}{H_{BG}}\right) - 8.29 \times 10^{-3} \qquad (4.27)$$

where H_{BP} and H_{BG} are the Brinell hardness values of the pinion and gear respectively.

The life factors K_L and C_L are used to modify the AGMA strengths for lives other than 10^7 cycles. K_L can be determined from Fig. 4.22 and C_L from Fig. 4.23.

The reliability factors C_R and K_R can be determined from Table 4.15. For values other than those listed use logarithmic interpolation.

Values for the bending strength and surface fatigue strength for various materials are listed in Tables 4.16 and 4.17. Alternatively, estimates for S_t and S_c can be obtained from

$$S_t \approx 20.55 + 1.176H_B - 9.584 \times 10^{-4}H_B^2 \qquad (4.28)$$

$$S_c \approx 182.7 + 2.382H_B \qquad (4.29)$$

where

S_t is the bending strength (MPa)
S_c is the contact strength (MPa)
H_B is the Brinell hardness.

Example A gear drive consists of a 20° steel spur pinion with 18 teeth driving a 48-tooth cast iron gear. The pinion speed is 350 rpm, face width 50 mm and module 4. The gears are to be made to number 7 AGMA quality standards and are to be accurately and rigidly mounted, and 3 kW is to be transmitted. Calculate the AGMA contact stress.

Solution For $\phi = 20°$, $N_P = 18$, $N_G = 48$, $n = 350$ rpm, $F = 50$ mm, $m = 4$, $Q_v = 7$, $H = 3$ kW:

$$d_P = mN_P = 4 \times 18 = 72\text{ mm}$$

We need C_p, C_a, C_v, C_s, C_m, C_f, I.

Table 4.17 Selected AGMA surface fatigue strengths

Material		Heat treatment	Minimum hardness at surface (BHN)	S_c (MPa)
Steel				
AGMA class	A1	Through hardened	≤180	590–660
	A2		240	720–790
	A3		300	830–930
	A4		360	1000–1100
	A5		400	1100–1200
BS 080M40		Induction hardened	155	530–575
BS 817M40		Induction hardened	220	675–740
BS 045M10		Case hardened	145	505–550
BS 655M13		Case hardened	245	730–800
Cast iron				
AGMA class	20	As cast	–	340–410
	30	As cast	175	450–520
	40	As cast	200	520–590

After ANSI/AGMA 2001-B88.

From Table 4.12 for a steel pinion and a cast iron gear, $C_p = 174\sqrt{\text{MPa}}$.

Default conditions:

$$C_a = C_s = C_f = 1$$

From Table 4.13 for accurate mountings, with $F = 50\,\text{mm}$, $C_m = 1.3$:

$$C_v = \left(\frac{A}{A + \sqrt{200V}}\right)^B$$

$$B = \frac{(12-7)^{0.6667}}{4} = 0.731$$

$$A = 65.06$$

$$V = \frac{0.072}{2} \times 350 \times \frac{2\pi}{60} = 1.319\,\text{m/s}$$

$$C_v = 0.8497$$

$$W_t = \frac{60 \times 10^3 H}{\pi d n} = 2.274\,\text{kN}$$

$$m_G = \frac{N_G}{N_P} = 2.667$$

$$I = \frac{\cos 20 \sin 20}{2} \times \frac{2.667}{2.667 + 1} = 0.1169$$

$$\sigma_c = C_p\left(\frac{W_t C_a}{C_v}\frac{C_s}{Fd}\frac{C_m C_f}{I}\right)^{0.5}$$

$$= 174\left(\frac{2274 \times 1}{0.8497} \times \frac{1}{0.05 \times 0.072} \times \frac{1.3 \times 1}{0.1169}\right)^{0.5}$$

$$= 174 \times 2876\,\text{kPa} \approx 500\,\text{MPa}$$

Factors of safety can be defined as ratios of a known safe working stress in a component to the expected stress experienced in the component. A factor of safety normally gives an indication of whether the component can withstand the occasional unexpected high load and have a long safe working life.

The factor of safety for surface fatigue failure is defined by

$$n_c = \frac{\sigma_{c,\text{all}}}{\sigma_c} \tag{4.30}$$

This is the ratio of the AGMA allowable contact stress (*see* equation 4.18) to the AGMA contact stress calculated for the gear by equation 4.16. n_c is sometimes called the factor of safety based on contact stress.

The factor of safety based on bending strength is defined by

$$n_s = \frac{\sigma_{\text{all}}}{\sigma} \tag{4.31}$$

where σ_{all} is the allowable AGMA bending stress (given by equation 4.17), and σ is the AGMA bending stress (given by equation 4.15).

As a guideline, factors of safety between 1 and 1.5 are suitable when the loading conditions are well known and understood. For general machine design, factors of safety between 1.5 and 2 should be used. The loading within gearboxes are rarely known exactly and a factor of safety of 2 should be specified.

Example A 22-tooth, 20° pressure angle spur pinion rotates at 450 rpm and transmits 1 kW to a 32-tooth gear. The module is 2.5, face width 34 mm, AGMA quality standard 5 with average mounting conditions. The pinion and the gear material is number 30 cast iron. Find the factor of safety for the gear set based on the contact stress if the life is to be 1×10^5 cycles corresponding to a 90 per cent reliability.

Solution For $\phi = 20°$, $N_P = 22$, $N_G = 32$, $n = 450$ rpm, $F = 34$ mm, $m = 2.5$, $Q_v = 5$, $H = 1$ kW:

$$d_P = mN_P = 2.5 \times 22 = 55\,\text{mm}$$

We need C_p, C_a, C_v, C_s, C_m, C_f, I.

From Table 4.12 for a cast iron pinion and a cast iron gear, $C_p = 163\sqrt{\text{MPa}}$.

Default conditions:

$$C_a = C_s = C_f = 1$$

From Table 4.13 for average mountings, with $F = 34$ mm, $C_m = 1.6$:

$$C_v = \left(\frac{A}{A + \sqrt{200V}}\right)^B$$

$$B = \frac{(12 - 5)^{0.66667}}{4} = 0.9148$$

$$A = 54.77$$

$$V = \frac{0.055}{2} \times 450 \times \frac{2\pi}{60} = 1.296\,\text{m/s}$$

$$C_v = 0.7900$$

$$W_t = \frac{60 \times 10^3 H}{\pi dn} = 0.7717\,\text{kN}$$

$$m_G = \frac{N_G}{N_P} = 1.455$$

$$I = \frac{\cos 20 \sin 20}{2} \times \frac{1.455}{1.455 + 1} = 0.09523$$

$$\sigma_c = C_p\left(\frac{W_t C_a}{C_v}\frac{C_s}{Fd}\frac{C_m C_f}{I}\right)^{0.5}$$

$$= 163\left(\frac{771.7 \times 1}{0.7900} \times \frac{1}{0.034 \times 0.055} \times \frac{1.6 \times 1}{0.09523}\right)^{0.5}$$

$$= 482\,891.3\,\text{kPa} \approx 483\,\text{MPa}$$

From Table 4.17, based on the pinion or the gear (same material), $S_c = 450$ MPa:

$$C_L = 2.466N^{-0.056} = 2.466(10^5)^{-0.056} = 1.294$$

$$C_H = 1 + 0.00069(1.455 - 1) = 1.0003$$

$C_T = 1$. From Table 4.15, $C_R = 0.85$:

$$\sigma_{c,\text{all}} = \frac{S_c C_L C_H}{C_T C_R} = 685.3\,\text{MPa}$$

So the factor of safety based on the ratio of the allowable AGMA contact stress to the AGMA contact stress for the pinion is

$$n_c = \frac{\sigma_{c,\text{all}}}{\sigma_c} = \frac{685.3}{483} = 1.4$$

4.4 Gear selection procedure

Figure 4.24 illustrates a general selection procedure for gears. Normally the required rotational speeds for the gear and pinion are known from the design requirements, as is the power the drive must transmit. A designer has to decide on the type of gears to be used, their arrangements on the shafts, materials, heat treatments and gear geometry (number of teeth, module, tooth form, face width, quality). There is no one solution to a gear design requirement. Several 'good' or optimum designs are possible. The designer's judgement and creativity are important aspects. Whilst Fig. 4.24 shows the general selection procedure, the steps given below outline a general calculation procedure.

Specify the initial geometry. Figure 4.11 can be used as a visual guide to the size of gear teeth:

$$d_P = mN_P \quad d_G = mN_G$$

Calculate the pitch line speed:

$$V = \frac{\pi dn}{60}$$

Calculate the transmitted load:

$$W_t = \frac{\text{Power}}{V}$$

Determine the velocity factor:

$$K_v = \left(\frac{A}{A + \sqrt{200V}}\right)^B$$

Determine the factors K_a, K_s, K_m, C_p, C_f, I, J, K_L, K_T, K_R, C_H.

Specify the face width or calculate from

$$F = \frac{W_t}{K_v m Y \sigma}$$

As a general rule, spur gears should be designed with a face width between three and five times the circular pitch.

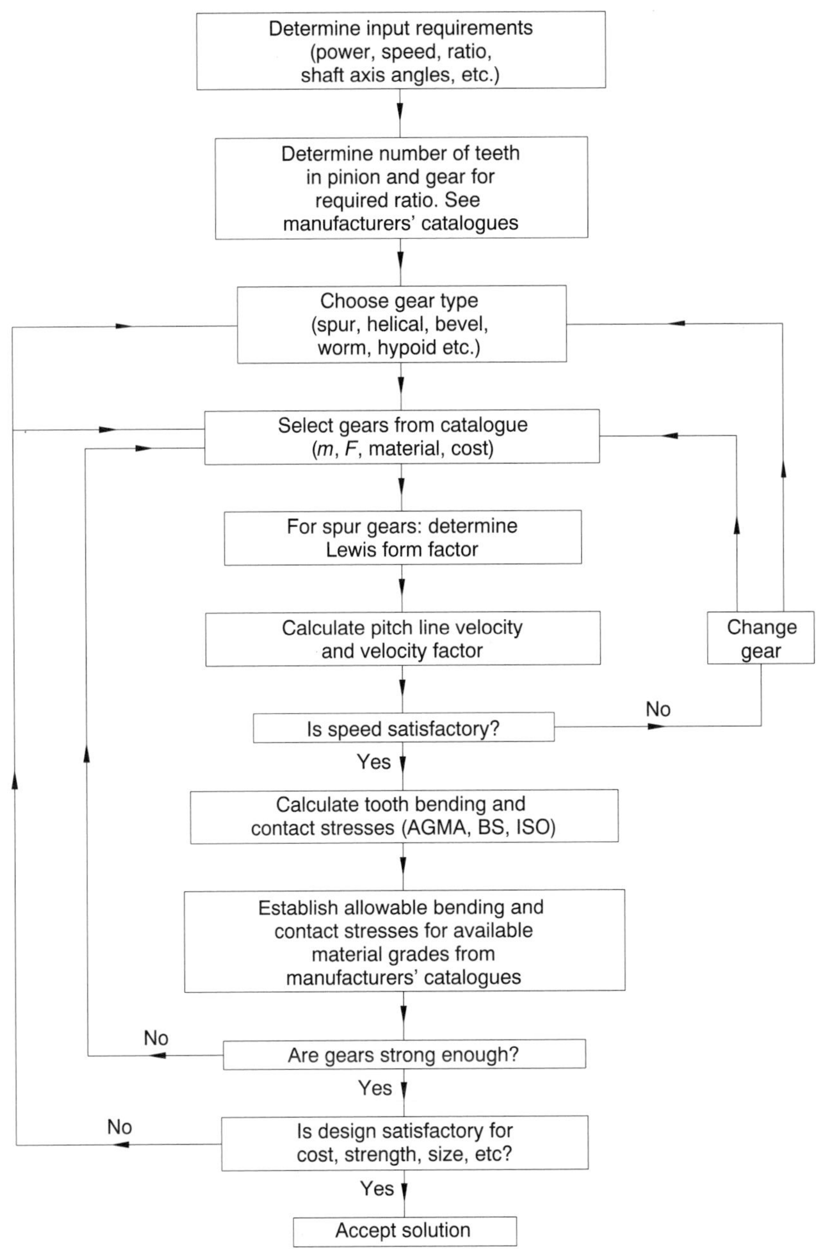

Fig. 4.24 General selection procedure for gears.

Calculate the AGMA bending stress (equation 4.15):

$$\sigma = \frac{W_t K_a}{K_v} \frac{1}{Fm} \frac{K_s K_m}{J}$$

Calculate the AGMA contact stress (equation 4.16):

$$\sigma_c = C_p \left(\frac{W_t C_a}{C_v} \frac{C_s}{Fd} \frac{C_m C_f}{I} \right)^{0.5}$$

Calculate the allowable bending stress (equation 4.17):

$$\sigma_{all} = \frac{S_t K_L}{K_T K_R}$$

Calculate the allowable contact stress (equation 4.18):

$$\sigma_{c,all} = \frac{S_c C_L C_H}{C_T C_R}$$

If values are unsatisfactory, either change initial geometry or change materials.

This procedure can be reordered, say, if a particular gear material is to be used and stress is to be limited to a certain value, or if a particular factor of safety is required etc.

Learning objectives checklist

- Are you familiar with gear nomenclature? ☐
- Can you construct an involute profile? ☐
- Can you select a suitable gear type for different applications? ☐
- Can you calculate contact stresses, bending stresses and factors of safety for a given set of gear parameters? ☐
- Are you familiar with general gear design and selection procedures? ☐

References and sources of information

Books and papers

David Brown Special Products Ltd 1995: *David Brown basic gear book*.

Engineering and Sciences Data Unit 1968: ESDU 68040. Design of parallel axis straight spur and helical gears – choice of materials and preliminary estimate of major dimensions.

Engineering and Sciences Data Unit 1981: ESDU 77002. Design of parallel axis straight spur and helical gears: geometric design.

Engineering and Sciences Data Unit 1988: ESDU 88033. The design of spur and helical involute gears. A procedure compatible with BS436: Part 3: 1986 – Method for calculation of contact and root bending stress limitations for metallic involute gears.

Hofmann, D.A., Kohler, H.K. and Munro, R.G. 1991: *Gear technology teaching pack*. British Gear Association.

HPC Gears Ltd. *Gears catalogue*.

Mitchener, R.G. and Mabie, H.H. 1982: The determination of the Lewis form factor and the AGMA geometry factor J for external spur gear teeth. *ASME Journal of Mechanical Design* **104**, 148–58.

Townsend, D.P. 1992: *Dudley's gear handbook*. McGraw Hill.

Tuplin, W.A. 1962: *Gear design*. The Machinery Publishing Co.

Standards

AGMA 225.01 (1967). Strength of spur, helical, herringbone and bevel gear teeth. AGMA.

AGMA 908-B89 (1989). Geometry factors for determining the pitting resistance and bending strength of spur, helical and herringbone gear teeth. AGMA.

ANSI/AGMA 2001-B88 (1988). Fundamental rating factors and calculation methods for involute spur and helical gear teeth. AGMA.

BS 436: Part 1: 1967. Specification for spur and helical gears. Part 1: Basic rack form, pitches and accuracy (diametral pitch series). BSI.

BS 436: Part 2: 1970. Specification for spur and helical gears. Part 2: Basic rack form, modules and accuracy (1 to 50 metric module). BSI.

BS 2519: Part 1: 1976. Glossary for gears. Geometrical definitions. BSI.

BS 2519: Part 2: 1976. Glossary for gears. Notation. BSI.

BS 436: Part 3: 1986. Spur and helical gears. Part 3. Method for calculation of contact and root bending stress limitations for metallic involute gears. BSI.

BS 6168: 1987. Specification for non-metallic spur gears. BSI.

BS ISO 6336-1: 1996. Calculation of load capacity of spur and helical gears. Basic principles, introduction and general influence factors. BSI.

BS ISO 6336-2: 1996. Calculation of load capacity of spur and helical gears. Calculation of surface durability (pitting). BSI.

BS ISO 6336-3: 1996. Calculation of load capacity of spur and helical gears. Calculation of tooth bending strength. BSI.

BS ISO 6336-5: 1996. Calculation of load capacity of spur and helical gears. Strength and quality of materials. BSI.

Web sites

http://www.powertransmission.com/index.htm

http://www.reliance.co.uk

http://www.the-net-effect.com/hpc/

Gears worksheet

1. For the gear train illustrated in Fig. 4.25, determine the output speed and direction of rotation if the input shaft rotates at 1490 rpm clockwise. Gears A to D have a module of 1.5 and gears E to H a module of 2.
2. A spur pinion has a module of 2.5 and 18 teeth cut on the 20° full depth system and is to transmit 4 kW power at 1100 rpm. Using the Lewis formula, determine the resulting bending stress if the face width is 30 mm.
3. A 20° full depth spur pinion is to transmit 1.75 kW at 1200 rpm. If the pinion has 18 teeth with a module of 2, determine a suitable value for the face width, based on the Lewis formula, if the bending stress should not exceed 75 MPa.
4. Based on 20° full depth teeth, a spur pinion is to be designed to transmit 2.5 kW at 1000 rpm. The pinion has 19 teeth. Determine suitable values for the face width and module, based on the Lewis formula, if the bending stress is not to exceed 75 MPa.

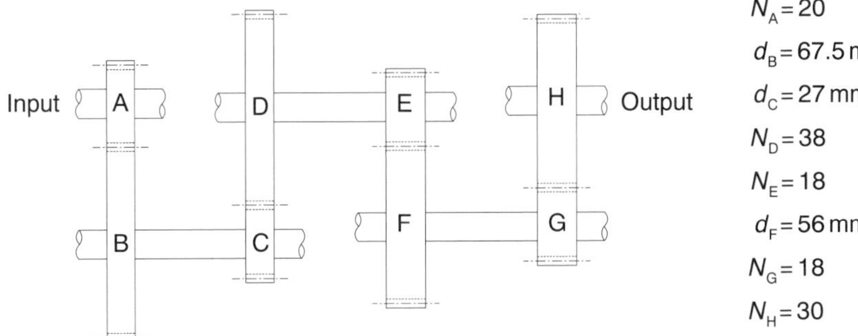

Fig. 4.25 Gear train.

5. Using the Lewis formula, estimate the power rating of a 20° full depth spur pinion having a module of 6, 21 teeth and face width 50 mm, if the maximum bending stress is 117 MPa. The design speed is 850 rpm.
6. A gearbox is required to transmit 7 kW from a shaft rotating at 2000 rpm. The desired output speed is approximately 350 rpm. Select suitable gears.
7. A gearbox is required to transmit 12 kW from a shaft rotating at 2970 rpm. The desired output speed is approximately 13 500 rpm. Select and specify appropriate gears for the gearbox.
8. A gearbox is required to transmit 1 kW from a shaft rotating at 2650 rpm. The desired output speed is approximately 200 rpm. Select and specify appropriate gears for the gearbox.
9. A 20-tooth, 20° pressure angle, module 4 cast iron spur pinion is used to drive a 32-tooth cast iron gear. Using equation 4.13, determine the contact stress if 10.5 kW is transmitted. The pinion speed is 950 rpm and the face width is 50 mm.
10. A 20° gear drive consists of a steel spur pinion with 20 teeth, driving a 36-tooth cast iron gear, transmitting 0.15 kW. The pinion speed is 100 rpm, face width 20 mm and module 1.5. The gears are manufactured with an AGMA quality value of 6, and are installed with less than perfect mounting rigidity but contact across the whole face. Calculate the AGMA contact stress.
11. An 18-tooth, 20° pressure angle spur pinion rotates at 1800 rpm and transmits 3 kW to a 52-tooth gear. The module is 2.5, face width 30 mm, AGMA quality standard 6 and the mounting is of exceptional rigidity. The pinion material is steel, Brinell hardness 240. The gear material is number 30 cast iron. Find the factor of safety based on the surface fatigue strength if the life is to be more than 1×10^6 cycles corresponding to a 90 per cent reliability.
12. A 20° speed reducer with a gear ratio of 4.42 has 19 teeth on the steel pinion. The pinion speed is 750 rpm and transmits 3.75 kW to the gear. The face width is 40 mm, the module is 3. The gear is cast iron. Calculate the AGMA bending stress in the pinion and the gear based on: $Q_v = 6$, $K_m = 1.3$, $K_a = 1$, $K_s = 1$.
13. A double step-up gearbox is required to transmit 30 kW from an input shaft rotating at 2900 rpm. A proposed design for the first gear set consists of 655M13 case hardened 20° pressure angle gears of AGMA quality standard 8. The gear and pinion have 38 and 20 teeth respectively, a module of 2.5 and face width of 30 mm. The gear set is to be accurately mounted. Determine the AGMA bending stress and the factor of safety for the proposed gear set if the life is to be more than 1×10^8 cycles with a reliability of 99.9 per cent. Comment on the suitability of the proposed design.
14. A 19-tooth steel (Brinell hardness 300) 20° spur pinion transmits 14.5 kW to a 77-tooth gear. The gear material is AGMA number 30 cast iron. The pinion speed is 375 rpm, the face width is 75 mm and the module is 6 mm. Determine the factor of safety for the drive based on bending for a life of 10^8 cycles and 99% reliability if $Q_v = 6$, $K_m = 1.6$, $K_a = 1$, $K_s = 1$.

Answers

Worked solutions to the above problems are given in Appendix B.

1. 121 rpm.
2. 100 MPa.
3. 25 mm.
4. No unique solution.

5. 32.2 kW.
6. No unique solution.
7. No unique solution.
8. No unique solution.
9. 541 MPa.
10. 896.4 MPa.
11. 1.239.
12. 65 MPa.
13. 129.8 MPa, 1.658.
14. 0.93.

5 Seals

Seals are devices used to prevent or limit leakage of fluids or particulates. The aims of this chapter are to introduce the variety of seal configurations, give guidelines for the selection of seals and introduce calculation methods for the quantification of some seal leakage rates.

5.1 Introduction

The purpose of a seal is to prevent or limit flow between components. Seals are an important aspect of machine design where pressurised fluids must be contained within an area of a machine such as a hydraulic cylinder, contaminants excluded or lubricants contained within a region. Seals fall into two general categories:

1. static seals, where sealing takes place between two surfaces which do not move relative to each other;
2. dynamic seals, where sealing takes place between two surfaces which move relative to each other by, for example, rotary or reciprocatory motion.

Some of the considerations in selecting the type of seal include:

1. the nature of the fluid to be contained or excluded;
2. pressure levels either side of the seal;
3. the nature of any relative motion between the seal and mating components;
4. the level of sealing required;
5. operating temperatures;
6. life expectancy, serviceability;
7. total cost – component and lifetime.

The variety of seals and sealing systems is extensive. Figure 5.1 shows a general classification diagram for seals detailing the principal types. Several of these will be introduced in Sections 5.2 and 5.3. It should be noted, however, that once a general sealing requirement and possible solution has been identified, the best source of specific information is usually the seal manufacturers, or for sealing bearings, the bearing manufacturers. Figure 5.2 provides a general guide to the selection of seal type.

5.2 Static seals

Static seals are designed to provide a complete physical barrier to leakage flow. To achieve this the seal material must be resilient enough to flow into and fill any irregularities in the surfaces being sealed and at the same time remain rigid

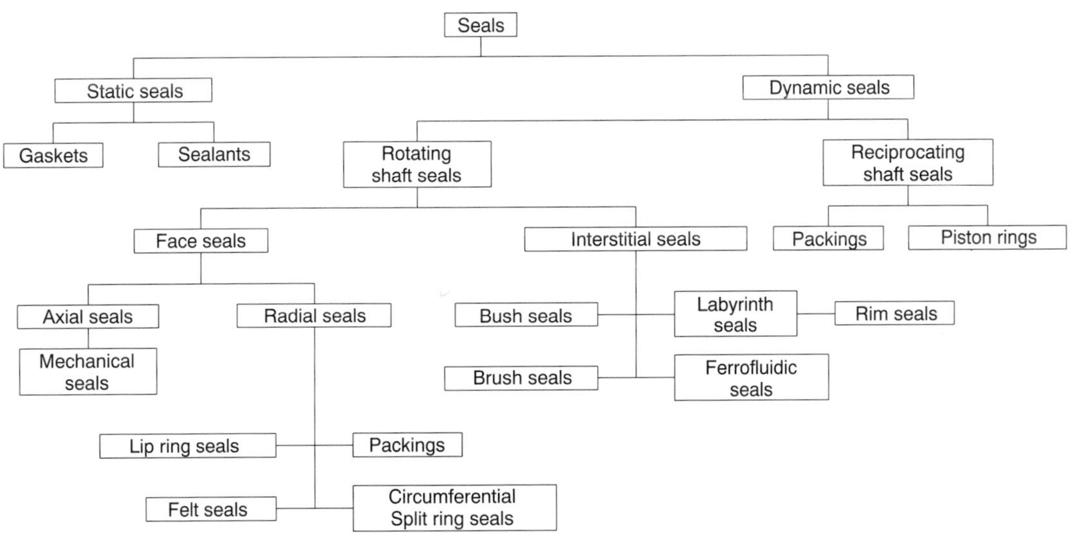

Fig. 5.1 General seal classification chart. (Reproduced with alterations from Buchter, 1979.)

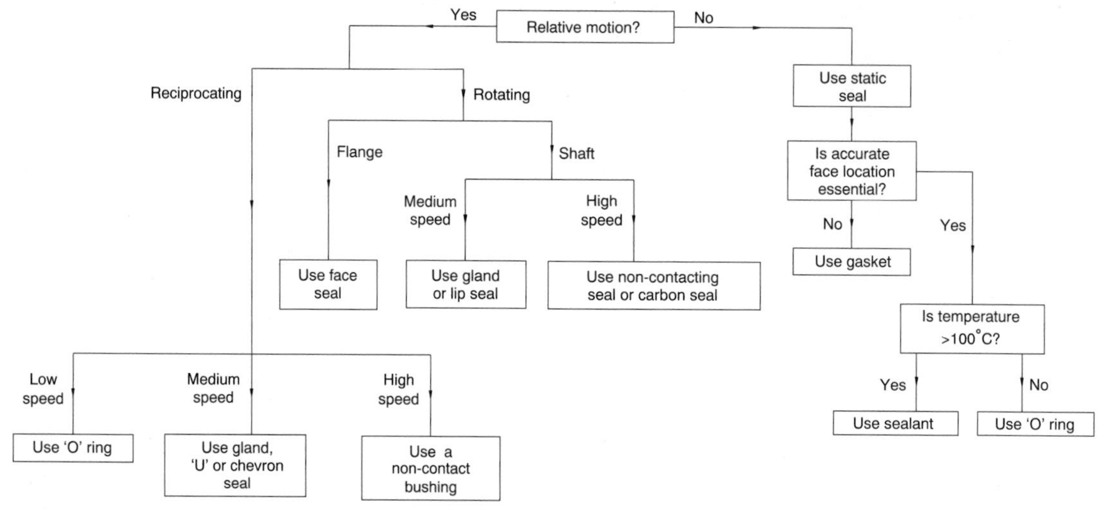

Fig. 5.2 Seal selection procedure. (Reproduced from Hamilton, 1994.)

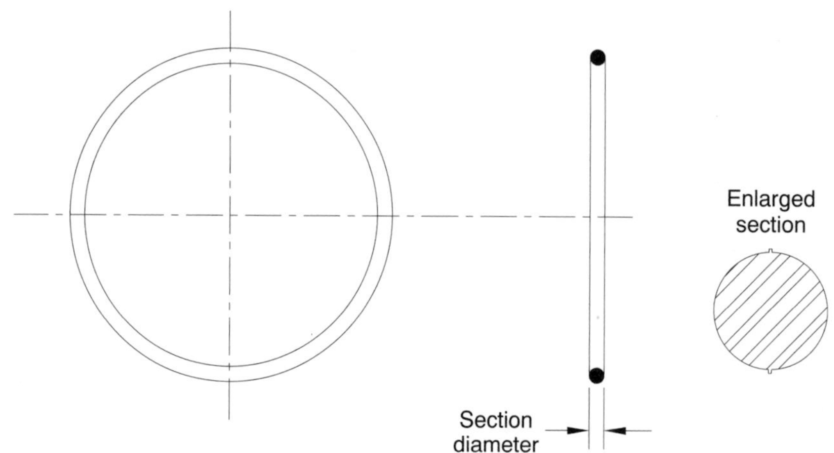

Fig. 5.3 General 'O' ring geometry.

enough to resist extrusion into clearances. Elastomeric seals and gaskets fulfil these criteria.

5.2.1 Elastomeric seal rings

The 'O' ring is a simple and versatile type of seal with a wide range of applications for both static and dynamic sealing. An 'O' ring seal is a moulded elastomeric ring 'nipped' in a cavity in which the seal is located. Figure 5.3 shows the specific geometry of an 'O' ring and Fig. 5.4 shows the principle of operation for an 'O' ring sealing against a fluid at various pressures.

Elastomeric seal rings require the seal material to have an interference fit with one of the mating parts of the assembly. 'O' rings are available in a wide range of sizes, with internal diameters from 3.1 mm to 249.1 mm and section diameters of 1.6, 2.4, 3.0, 4.1, 5.7 and 8.4 mm as defined in BS 4518. Table 5.1 shows the dimensions for a small number of the seal sizes available. A more extensive table is normally available in the form of a sales catalogue from manufacturers or in BS 4518. Figure 5.5 shows the groove dimensions which must be specified to house the 'O' ring seal and ensure the seal is nipped or compressed sufficiently to enable effective sealing.

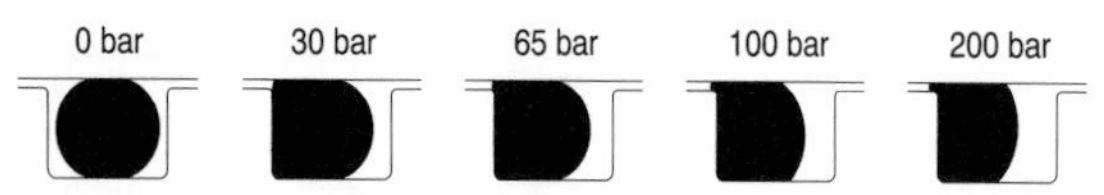

Fig. 5.4 Principle of operation of 'O' rings sealing a fluid against no, medium and high pressure.

Table 5.1 'O' ring seal dimensions (limited range tabulated only for illustration)

Reference number	Internal diameter (mm)	Section diameter (mm)	$d_{nominal}$ (mm) (Fig. 5.5(a))	$D_{nominal}$ (mm) (Fig. 5.5(a))	$D_{nominal}$ (mm) (Fig. 5.5(b))	$d_{nominal}$ (mm) (Fig. 5.5(b))	B (mm)	R (mm)
0031-16	3.1	1.6	3.5	5.8	6	3.7	2.3	0.5
0041-16	4.1	1.6	4.5	6.8	7	4.7	2.3	0.5
0051-16	5.1	1.6	5.5	7.8	8	5.7	2.3	0.5
0061-16	6.1	1.6	6.5	8.8	9	6.7	2.3	0.5
0071-16	7.1	1.6	7.5	9.8	10	7.7	2.3	0.5
0081-16	8.1	1.6	8.5	10.8	11	8.7	2.3	0.5
0091-16	9.1	1.6	9.5	11.8	12	9.7	2.3	0.5
0101-16	10.1	1.6	10.5	12.8	13	10.7	2.3	0.5
0111-16	11.1	1.6	11.5	13.8	14	11.7	2.3	0.5
0036-24	3.6	2.4	4	7.7	8	4.3	3.1	0.5
0046-24	4.6	2.4	5	8.7	9	5.3	3.1	0.5
0195-30	19.5	3.0	20	24.8	25	20.2	3.7	1.0
0443-57	44.3	5.7	45	54.7	55	45.3	6.4	1.0
1441-84	144.1	8.4	145	160	160	145	9.0	1.0
2491-84	249.1	8.4	250	265	265	250	9.0	1.0

Source: BS4518.

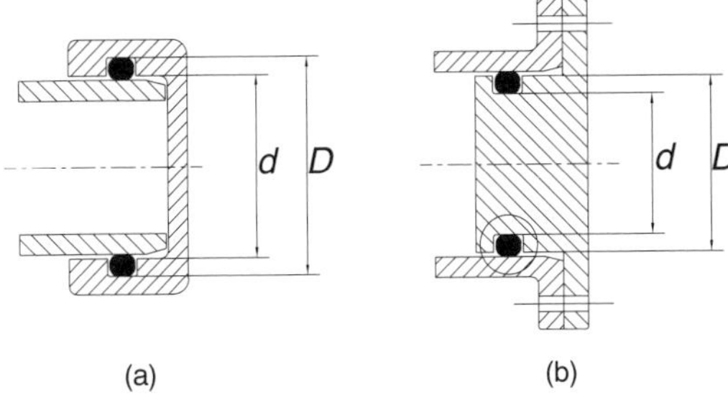

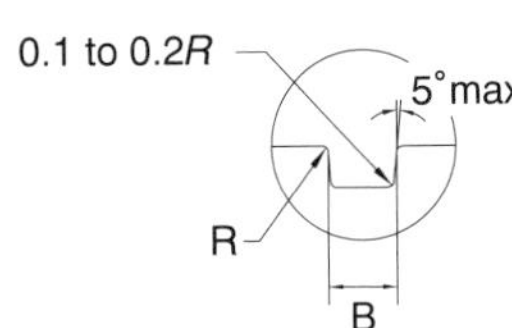

Fig. 5.5 'O' ring groove dimensions.

5.2.2 Gaskets

A gasket is a material or composite of materials clamped between two components with the purpose of preventing fluid flow. Figure 5.6 shows a typical application for a gasket seal. When first closed, a gasket seal is subject to compressive stresses produced by the assembly. Under working conditions, however, the compressive load may be relieved by the pressures generated within the assembly or machine. This must be accounted for in the detailed design or by use of a factor to allow for the relaxation of gasket compression. Typical gasket designs are illustrated in Fig. 5.7.

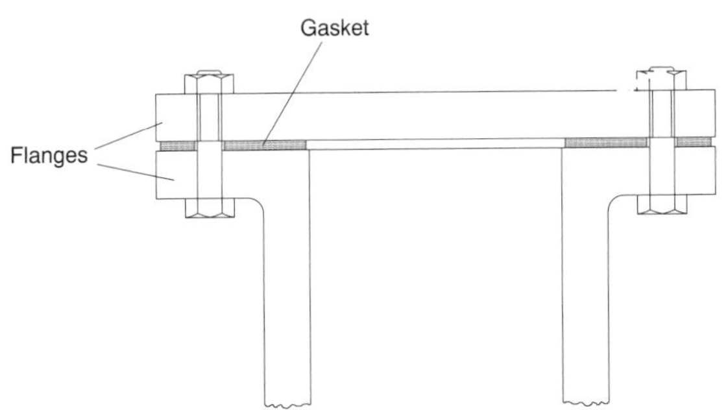

Fig. 5.6 Typical gasket application.

5.3 Dynamic seals

The term 'dynamic seal' is used to designate a device used to limit flow of fluid between surfaces which move relative to each other. The range of dynamic seals is extensive with devices for both rotary and reciprocating motion. The requirements of dynamic seals are often conflicting and require compromise. Effective sealing may require high contact pressure between a stationary component and a rotating component, but minimal wear is also desired for long seal life.

5.3.1 Seals for rotating machinery

The functions of seals on rotating shafts include retaining working fluids, retaining lubricants and excluding contaminants such as dirt and dust.

Type	Cross-section	Comment
Flat	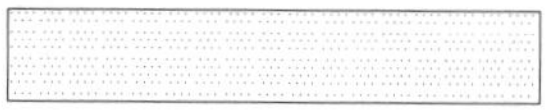	Available in a wide variety of materials. Easily formed into other shapes
Reinforced	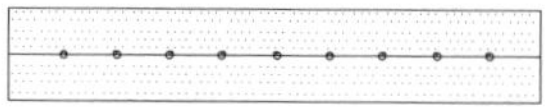	Fabric or metal reinforced. Improves torque retention and blowout resistance in comparison with flat types
Flat with rubber beads		Rubber beads located on flat or reinforced material. Gives high unit sealing pressure
Flat with metal grommet	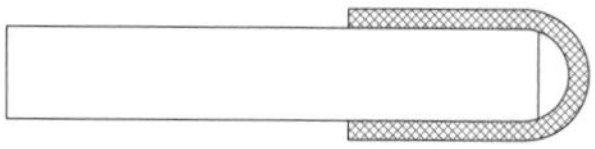	The metal grommet gives protection to the base material
Plain metal jacket	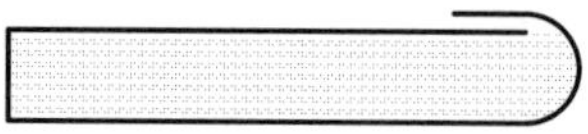	The metal jacket gives protection to the filler on one edge and across the surface
Corrugated or embossed	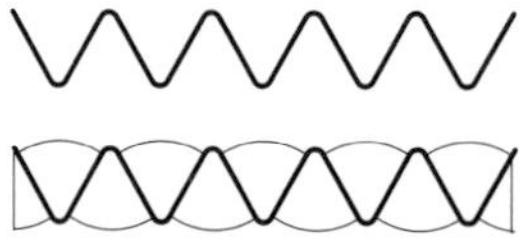	Corrugations provide increased sealing pressure capability
Profile	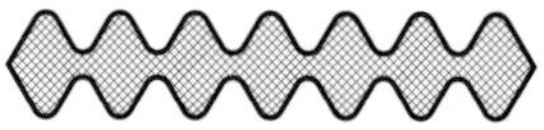	Multiple sealing surfaces
Spiral wound		Interleaving pattern of metal and filler

Fig. 5.7 Typical gasket designs. (Reproduced with alterations from Czernik, 1996.)

The selection of seal type depends on the shaft speed, working pressure and desired sealing effectiveness. Seals for rotary motion include 'O' rings, lip seals, face seals, sealing rings, compression packings and non-contacting seals such as bush and labyrinth seals.

'O' rings were covered in Section 5.2.1. Their application to rotating shafts is generally limited to use when the shaft speed is below 3.8 m/s and seal pressures below 14 bar.

The typical geometry for a radial lip seal, commonly known as an oil ring, is shown in Fig. 5.8. These seals are used to retain lubricants and exclude dirt and are well suited to moderate-speed and low-pressure applications. The outer case should be retained in the housing by an interference fit. The purpose of the garter spring is to maintain a uniform radial force on the shaft, ensuring contact between the elastomeric sealing ring and the rotating shaft.

A mechanical face seal consists of two sealing rings, one attached to the rotating member and one attached to the stationary component to form a sealing surface, usually perpendicular to the shaft axis, as illustrated in Fig. 5.9. During rotation the primary sealing ring attached to the shaft rubs with its seal face against the counterseal face of the stationary ring. The two interface contact areas function like bearings and are subject to frictional wear. Any leakage flow must pass across this interface. The rubbing contact is maintained by forces acting axially caused by hydraulics or mechanically, using, for example, a spring. The

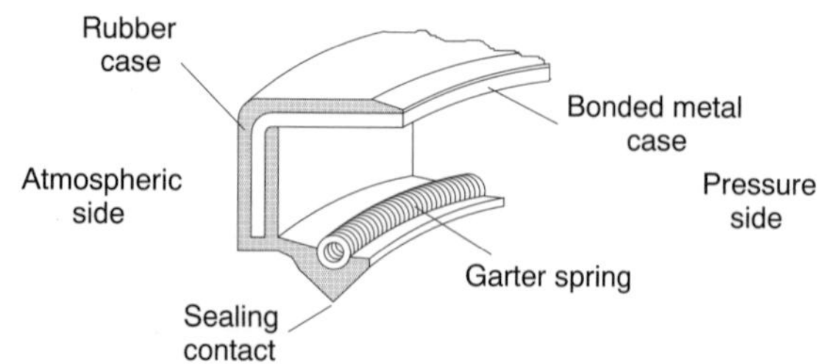

Fig. 5.8 Radial lip seal.

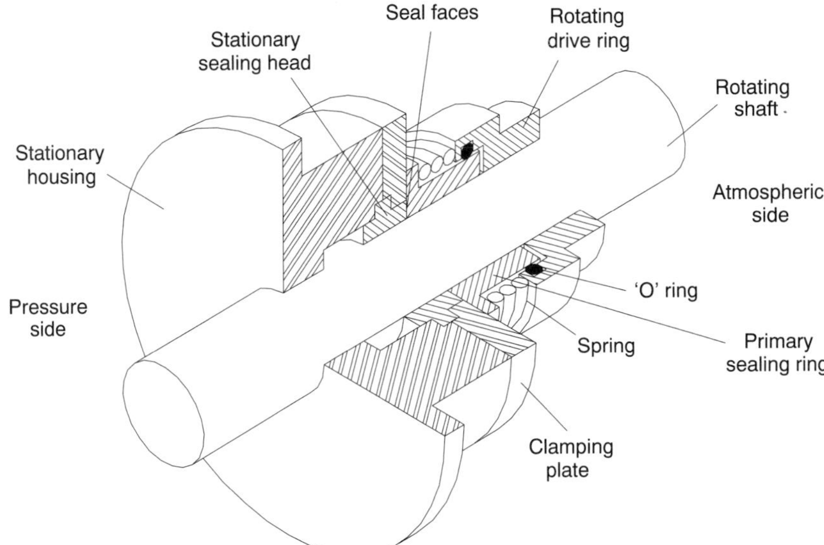

Fig. 5.9 A typical mechanical face seal.

rubbing action between the surfaces produces heat and wear. In order to minimise this, lubrication is used, which as well as limiting wear and heat build-up also serves to generate a fluid film which assists in producing a tight seal. The range of face seals is extensive, as reviewed by Summers-Smith (1992).

The term interstitial seal is used for seals which allow unrestricted relative motion between the stationary and moving components (i.e. no seal to shaft contact). Types include labyrinth and bush seals.

A labyrinth seal in its simplest form consists of a series of radial fins forming a restriction to an annular flow of fluid, as shown in Fig. 5.10. In order for the fluid to pass through the annular restriction it must accelerate. Just after the restriction the fluid will expand and decelerate with the formation of separation eddies in the cavity downstream of the fin, as illustrated schematically in Fig. 5.11. These turbulent eddies dissipate some of the energy of the flow reducing the pressure. This process will be repeated in subsequent cavities until the pressure reaches downstream conditions. Labyrinth seals are essentially a controlled clearance seal without rubbing contact. As there is no surface-to-surface contact, very high relative speeds are possible and the geometry can be arranged to limit leakage to tolerable levels. Leakage is, however, inevitable and can be quantified using the method outlined below. Care must be taken in the design of labyrinth seals. If the cavity following the radial fin is too small the flow will pass straight through without expanding and without the subsequential pressure drop. Despite careful labyrinth design some flow will

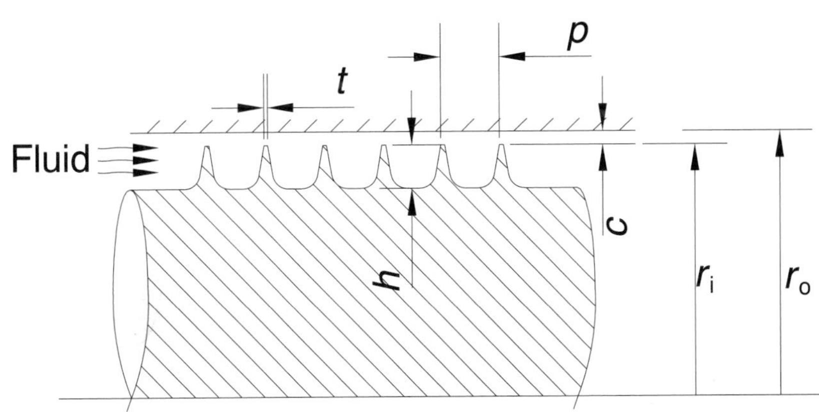

Fig. 5.10 Labyrinth seal geometry.

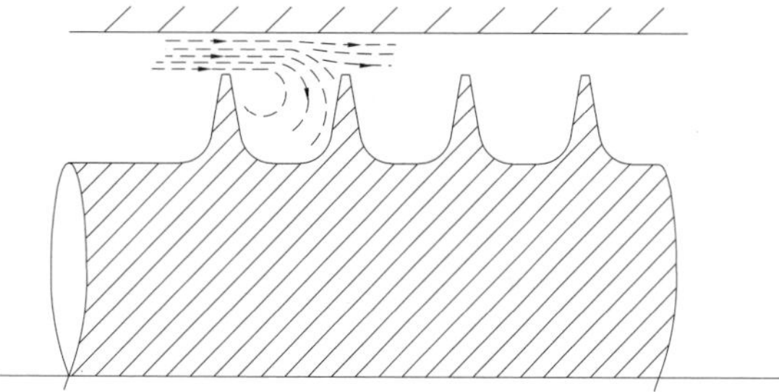

Fig. 5.11 Labyrinth seal flow.

inevitably be carried straight over from one fin to another. In order to reduce this effect, steps can be incorporated into labyrinth seal design and a wide range of labyrinth variants are possible, as illustrated in Fig. 5.12.

Flow through a labyrinth can be estimated using (Egli, 1935)

$$\dot{m} = A\alpha\gamma\varphi\sqrt{\rho_0 p_0} \tag{5.1}$$

where

$\dot{m}$ = mass flow rate (kg/s),
$A = \pi(r_o^2 - r_i^2)$ = area of the annular gap between the fin tips and the casing (m^2)

Fig. 5.12 Labyrinth seal designs to limit carryover flow (after Stocker, 1978).

α = flow coefficient
γ = carryover correction factor
φ = expansion ratio
p_0 = upstream pressure (Pa)
ρ_0 = density at the upstream conditions (kg/m^3).

The flow coefficient, α, is a function of the clearance to tip width ratio, but an average value of 0.71 can be used for $1.3 < c/t < 2.3$, where c = radial clearance (m) and t = thickness of fin (m).

The carryover correction factor, γ, varies as a function of the clearance to pitch ratio. As a crude approximation, γ can be taken as varying linearly for c/p values of 0 to 0.11 as listed below (p = pitch (m)):

$$\begin{aligned}
\gamma &= 1 + 11.1(c/p) \quad \text{for } n = 12\\
\gamma &= 1 + 10.2(c/p) \quad \text{for } n = 8\\
\gamma &= 1 + 8.82(c/p) \quad \text{for } n = 6\\
\gamma &= 1 + 6.73(c/p) \quad \text{for } n = 4\\
\gamma &= 1 + 5(c/p) \quad \text{for } n = 3\\
\gamma &= 1 + 3.27(c/p) \quad \text{for } n = 2
\end{aligned} \tag{5.2}$$

The expansion ratio, φ, is given by

$$\varphi = \sqrt{\frac{1 - (p_n/p_0)}{n + \ln(p_0/p_n)}} \tag{5.3}$$

where

p_n = downstream pressure following the nth labyrinth (Pa)
n = number of fins.

Although this method was developed some time ago the results are reasonably reliable. Many large companies, e.g. Rolls-Royce plc, do, however, have their own proprietary equations and methods for determining leakage rates through both straight and stepped labyrinths and, as an alternative, CFD (computational fluid dynamics) can also be used to model the seal flow. In general, fins are normally included on the rotor, although they can if necessary be incorporated on the stationary component if, for example, significant axial movements are anticipated. Six fins are normally found to be adequate and beyond six or seven fins there is little improvement. Typical values for the selection of values for the fin thickness, pitch and fin height are listed in Table 5.2 as a guideline. The flow rate can then be calculated using equation 5.1 for a given labyrinth radius and clearance.

Table 5.2 Guidelines for the selection of fin thickness, pitch and height

t (mm)	p (mm)	h (mm)
0.3–0.4	6–8	4–5
0.28–0.32	4–5	3–3.5
0.18–0.22	1.8–2.2	1.8–2.2

Example Determine the mass flow rate through a labyrinth seal on a 100 mm diameter shaft. The labyrinth consists of six fins, height 3.2 mm, pitch 4.5 mm, radial clearance 0.4 mm and tip width 0.3 mm. The pressure is being dropped from 4 bar absolute, 353 K, to atmospheric conditions (1.01 bar). Take the gas constant R as 287 J/kg per Kelvin.

Solution The outer radius of the annular gap is $(100/2) + 3.2 + 0.4 = 53.6$ mm. The inner radius of the annular gap is $(100/2) + 3.2 = 53.2$ mm. The annulus gap area is

$$A = \pi(r_o^2 - r_i^2)$$

$$= \pi[(53.6 \times 10^{-3})^2 - (53.2 \times 10^{-3})^2]$$

$$= 1.342 \times 10^{-4}\ \mathrm{m^2}$$

$\alpha = 0.71$; $\gamma = 1 + 8.82(c/p)$ for $n = 6$; $c = 0.4$ mm and $p = 4.5$ mm, so $\gamma = 1 + 8.82(0.4/4.5) = 1.784$.

$$\varphi = \sqrt{\frac{1 - (p_n/p_0)}{n + \ln(p_0/p_n)}}$$

$$= \sqrt{\frac{1 - (1.01 \times 10^5/4 \times 10^5)}{6 + \ln(4 \times 10^5/1.01 \times 10^5)}} = 0.3183$$

$p = \rho RT$, so the upstream density is given by

$$\rho_0 = p_0/RT_0$$

$$= 4 \times 10^5/(287 \times 353) = 3.948\ \mathrm{kg/m^3}$$

$$\dot{m} = 1.342 \times 10^{-4} \times 0.71 \times 1.784$$

$$\times 0.3183\sqrt{3.948 \times 4 \times 10^5}$$

$$= 0.0680\ \mathrm{kg/s}$$

Simple axial and radial bush seals are illustrated in Figs 5.13 and 5.14 and can be used for sealing both liquids and gases. Leakage through a concentric axial bush seal can be estimated using equation 5.4 for incompressible flow (flow where the density can be considered constant) and equation 5.5 for compressible flow. The leakage flow through a radial bush seal can be estimated by equation 5.6 for incompressible flow and equation 5.7 for compressible flow:

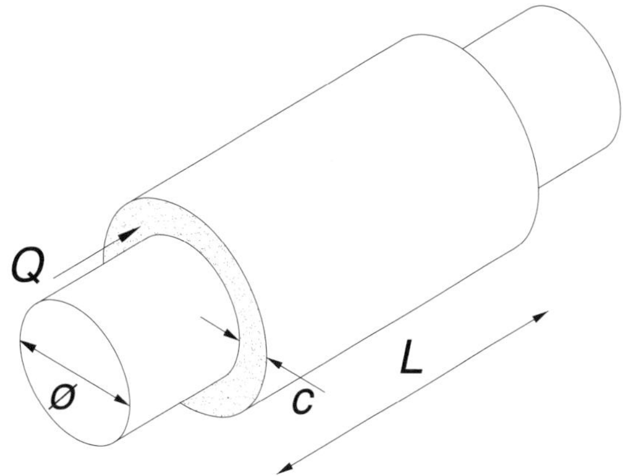

Fig. 5.13 Axial bush seal.

$$Q = \frac{\pi\phi c^3(p_0 - p_a)}{12\mu L} \tag{5.4}$$

$$Q = \frac{\pi\phi c^3(p_0^2 - p_a^2)}{24\mu L p_a} \tag{5.5}$$

where

Q = volumetric flow rate ($\mathrm{m^3/s}$)
ϕ = diameter of the shaft (m)
c = radial clearance (m)
p_0 = upstream pressure (Pa)
p_a = downstream pressure (Pa)
μ = absolute viscosity (Pa · s)
L = length of the axial bush seal (m).

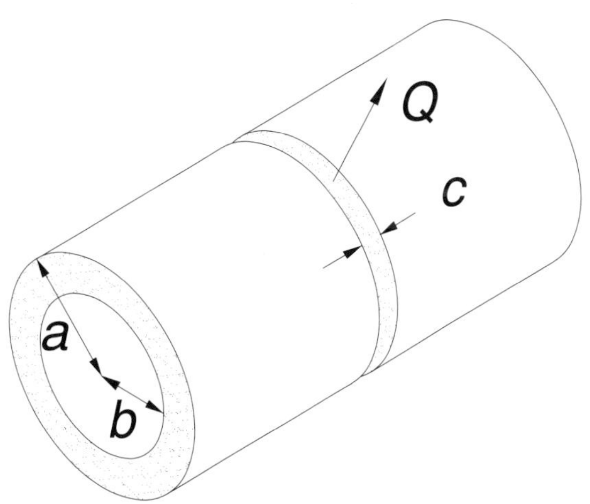

Fig. 5.14 Radial bush seal.

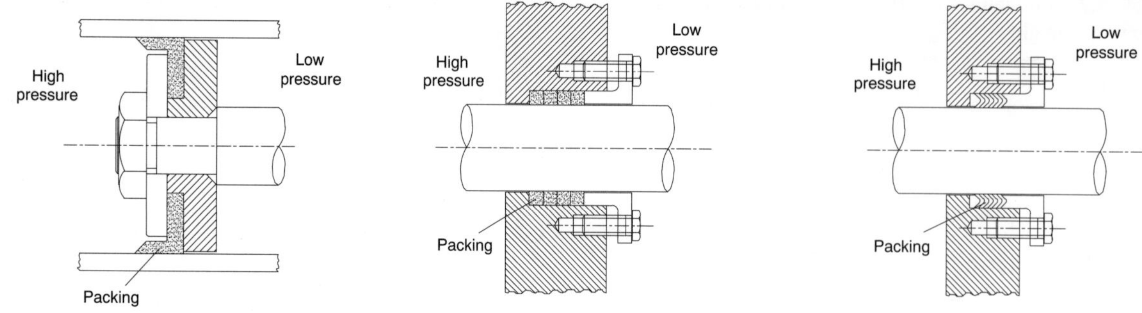

Fig. 5.15 Packing seals.

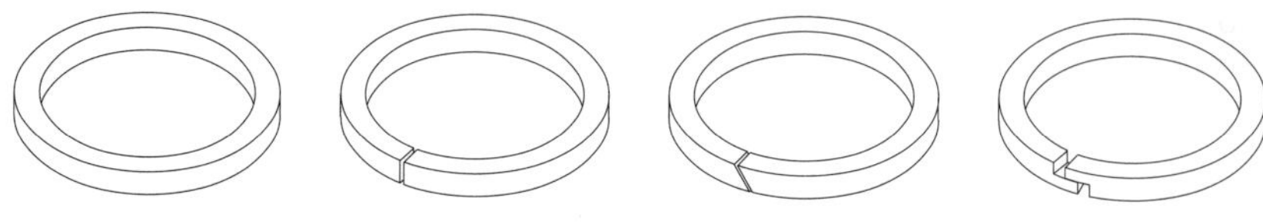

Fig. 5.16 Piston rings.

$$Q = \frac{\pi c^3 (p_0 - p_a)}{6\mu \ln(a/b)} \tag{5.6}$$

$$Q = \frac{\pi c^3 (p_0^2 - p_a^2)}{12\mu p_a} \tag{5.7}$$

where

c = axial clearance (m)
a = outer radius (m)
b = inner radius (m).

Example An axial bush seal consists of an annular gap with inner and outer radii of 50 mm and 50.5 mm respectively. The length of the seal is 40 mm. Determine the flow rate of oil through the seal if the pressures upstream and downstream of the seal are 7 bar and 5.5 bar respectively. The viscosity of the oil can be taken as 0.025 Pa·s.

Solution The radial clearance is $0.0505 - 0.05 = 0.0005$ m. $\phi = 0.1$ m, $L = 0.04$ m.

The volumetric flow rate is given by

$$Q = \frac{\pi \times 0.1 \times (0.5 \times 10^{-3})^3 (7 \times 10^5 - 5.5 \times 10^5)}{12 \times 0.025 \times 0.04}$$

$$= 4.909 \times 10^{-4}\ \mathrm{m^3/s}$$

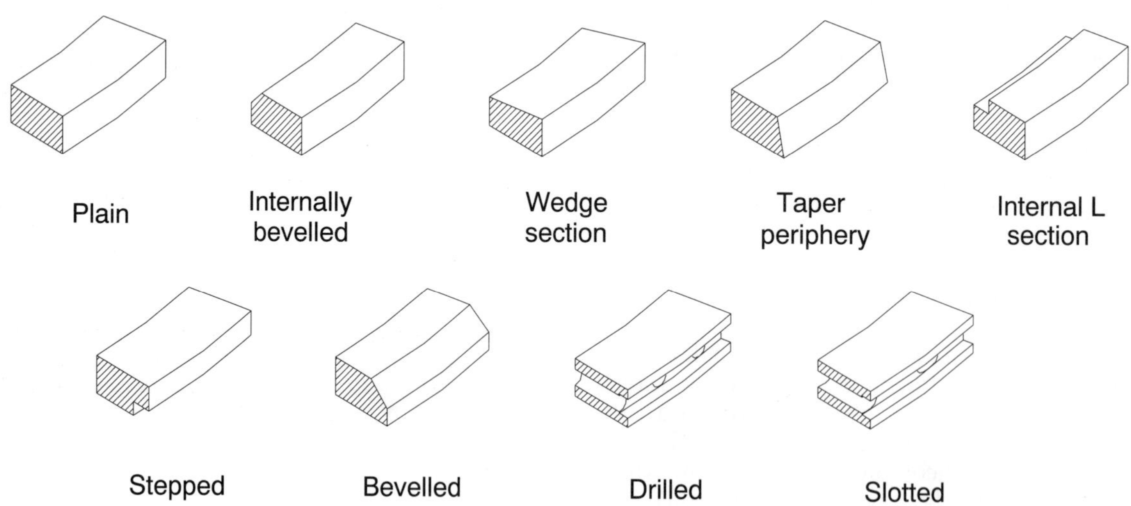

Fig. 5.17 Piston ring sections.

Table 5.3 Number of piston rings required to seal a given pressure

p_0 (bar)	Number of rings
<20	2
$20 < p_0 < 60$	3
$60 < p_0 < 100$	4
$100 < p_0 < 200$	5
>200	6+

5.3.2 Seals for reciprocating components

The seals principally used for reciprocating motion are packings and piston rings.

Packing seals are illustrated in Fig. 5.15. The seal essentially consists of a cup, V, U or X section of leather, solid rubber or fabric-reinforced rubber. The sealing principle is by direct contact with the reciprocating component. The contact pressure can be increased in the case of V packings by axial compression of the seals, although this obviously increases the friction on the shaft and wear rate. The principal uses of cup packings are as piston seals in hydraulic and pneumatic applications, U packings for piston rods and V packings for sealing piston rods or reciprocating shafts.

Piston rings are used to seal cylinders where the operating temperature is above the limit of elastomeric, fabric or polymeric materials. Piston rings are used in automotive cylinders for three purposes:

1. to seal the combustion chamber/cylinder head;
2. to transfer heat from the piston to the cylinder walls;
3. to control the flow of oil.

Piston rings are usually machined from a fine-grain alloy cast iron and must be split to allow for assembly over the piston. Conventional practice is to use three piston rings, with two compression rings sealing the high pressure and one to control the flow of oil. The range of piston rings available is extensive, as illustrated in Figs 5.16 and 5.17. For other applications the number of piston rings required can be determined using Table 5.3 assuming normal piston temperatures and running speeds.

Learning objectives checklist

Can you identify the different sealing devices? ☐

Can you select a seal type for rotating, reciprocating or static conditions? ☐

Can you calculate the leakage flow through a labyrinth seal? ☐

Can you calculate the leakage flow through a bush seal? ☐

References and sources of information

Books and papers

Buchter, H.H. 1979: *Industrial sealing technology*. Wiley.

Czernik, D.E. 1996: Gaskets. In Shigley, J.E. and Mischke, C.R. (eds), *Standard handbook of machine design*. McGraw Hill.

Egli, A. 1935: The leakage of steam through labyrinth seals. *Transactions of the ASME* **57**, 115–22, 445–6.

Engineering Sciences Data Unit 1980: ESDU 80012. Dynamic sealing of fluids 1: guide to selection of rotary seals.

Engineering Sciences Data Unit 1983: ESDU 83031. Dynamic sealing of fluids 2: guide to selection of reciprocating seals.

Hamilton, P. 1994: Seals. In Hurst, K. (ed.), *Rotary power transmission*. McGraw Hill.

Hopkins, R.B. 1996: Seals. In Shigley, J.E. and Mischke, C.R. (eds), *Standard handbook of machine design*. McGraw Hill.

Morse, W. 1969: *Seals*. Morgan-Grampian.

Neale, M.J. 1994: *Drives and seals*. Butterworth Heinemann.

Stocker, H.L. 1978: Determining and improving labyrinth seal performance in current and advanced high performance gas turbines. AGARD CP-237.

Stone, R. 1992: *Introduction to internal combustion engines*. Macmillan.

Summers-Smith, J.D. (ed.) 1992: *Mechanical seal practice for improved performance*. IMechE, MEP.

Warring, R.H. 1981: *Seals and sealing handbook*. Latty International.

Standards

BS 5341: Part 5: 1976. Piston rings up to 200 mm diameter for reciprocating internal combustion engines. Ring grooves. BSI.

BS 4518: 1982. Specification for metric dimensions of toroidal sealing rings ('O' rings) and their housings. BSI.

BS 2492: 1990. Specification for elastomeric seals for joints in pipework and pipelines. BSI.

BS 5341: Part 7: Section 7.4: 1992. Piston rings up to 200 mm diameter for reciprocating internal combustion engines. Designs, dimensions and designations for single piece rings. Specifications for oil control rings. BSI.

BS 7780: Part 1: 1994. Specification for rotary shaft lip type seals. Nominal dimensions and tolerances. BSI.

BS 7780: Part 2: 1994. Specification for rotary shaft lip type seals. Vocabulary. BSI.

Web sites

http://www.flexibox.com/

http://www.garlock-inc.com/

Seals worksheet

1. Determine an appropriate seal type for the following applications:
   (a) a high-speed, 100 mm diameter, rotating shaft for a gas turbine engine to limit the escape of air between the shaft and stationary housing;
   (b) to completely seal the back face of a rotodynamic pump rotating at 2970 rpm;
   (c) to seal the faces of two mating flanges.
2. Determine suitable groove dimensions for an 0101-16 'O' ring to seal against a solid cylinder.
3. Determine the flow rate of air through a labyrinth seal on a 150 mm diameter shaft. The labyrinth consists of eight fins, height 4.5 mm, pitch 7 mm, radial clearance 0.5 mm and tip width 0.35 mm. The absolute pressures upstream and downstream of the seal are 2 bar and 1.01 bar respectively. The temperature of the air upstream of the seal is approximately 323 K. Take the gas constant R as 287 J/kg per Kelvin.
4. The shaft for a small high-speed rotating machine has a diameter of 60 mm. The absolute pressure within the machine is 1.2×10^5 Pa and the operating temperature is approximately 318 K. Design a suitable labyrinth seal to limit the flow of air escaping from the machine to atmospheric conditions to 0.01 kg/s.
5. The shaft for a high-speed rotating machine has a diameter of 150 mm. The absolute pressure within the machine is 2.5×10^5 Pa and the operating temperature is approximately 443 K. Design a suitable labyrinth seal to limit the flow of air escaping from the machine to a chamber at an absolute pressure of 2.2 bar to 0.03 kg/s.
6. An axial bush seal consists of an annular gap with inner and outer radii of 0.025 and 0.0253 mm respectively. The length of the seal is 30 mm. Determine the flow rate of oil through the seal if the pressures upstream and downstream of the seal are 6 bar and 5.5 bar respectively. The viscosity of the oil can be taken as 0.02 Pa · s.
7. Determine the flow rate of air through an axial bush seal. The outer diameter of the inner cylinder is 100 mm and the radial clearance is 0.1 mm. The length of the seal is 50 mm. The pressures upstream and downstream of the seal are 1.4 bar and 1.01 bar respectively. The viscosity of air can be taken as 1.85×10^{-5} Pa · s.
8. A radial bush seal consists of a cylinder with an inner radius of 20 mm and an outer radius of 40 mm. The axial gap is 0.2 mm. Determine the flow rate of oil through the clearance if the pressure inside the cylinder is 6 bar and the pressure outside the cylinder is 5.2 bar. Take the viscosity of oil as 0.022 Pa · s.

Answers

Worked solutions to the above problems are given in Appendix B.

1. (a) Labyrinth seal. (b) Mechanical face seal. (c) Gasket.
2. $d = 10.5$ mm, $D = 12.8$ mm, $B = 2.3$ mm, $R = 0.5$ mm.
3. 0.048 kg/s.
4. No unique solution.
5. No unique solution.
6. 2.945×10^{-5} m^3/s.
7. 1.317×10^{-3} m^3/s.
8. 2.2×10^{-5} m^3/s.

6 Belt and chain drives

Belt and chain drives are used to transmit power from one rotational drive to another. A belt is a flexible power transmission element which runs tightly on a set of pulleys. A chain drive consists of a series of pin-connected links which run on a set of sprockets. The aims of this chapter are to introduce the various types of belt and chain drives and to develop design and selection procedures for flat, wedge and toothed belts and roller chains.

6.1 Introduction

Belt and chain drives consist of flexible elements running on either pulleys or sprockets, as illustrated in Fig. 6.1. The purpose of a belt or chain drive is to transmit power from one rotating shaft to another. The speed ratio between the driving and driven shaft is dependent on the ratio of the pulley or sprocket diameters, given by

$$V_{\text{pitch line}} = \omega_1 R_1 = \omega_2 R_2 \tag{6.1}$$

$$\text{Angular velocity ratio} = \frac{\omega_1}{\omega_2} = \frac{R_2}{R_1} \tag{6.2}$$

A belt drive transmits power between shafts by using a belt to connect pulleys on the shafts by means of frictional contact or mechanical interference. A chain consists of a series of links connected by pins. The chain is designed to mesh with corresponding teeth on sprockets located on both the driving and the driven shafts.

Power transmission between shafts can be achieved by a variety of means including belt, chain and gear drives and their use should be compared for suitability and optimisation for any given application. In addition to power transmission, constant speed ratio or synchronisation of the angular position of the driving and driven shaft may be critical to operation. This can be achieved by means of gears, chains or special toothed belts called synchronous or timing belts.

Both belt and chain drives can transmit power between shafts that are widely separated, giving designers greater scope for control over machine layout. In comparison to gears they do not require such precision in the location of centre distances. Both belt and chain drives are often cheaper than the equivalent gear drive. Belts and chains are generally complementary, covering a range of operational requirements.

In general, belt drives are used when rotational speeds are of the order of 10–60 m/s. At lower speeds the tension in the belt becomes too high for typical belt sections. At higher speeds centrifugal forces throw the belts off the pulleys, reducing the torque capacity, and dynamic phenomena reduce the effectiveness and life of the belt drive. Chain drives are typically used at lower speeds and consequently higher torques than belts. Recall that for a rotating machine, torque is proportional to power/ω. As the angular velocity reduces, for a given power, the torque increases.

Belt drives have numerous advantages over gear and chain drives, including easy installation, low

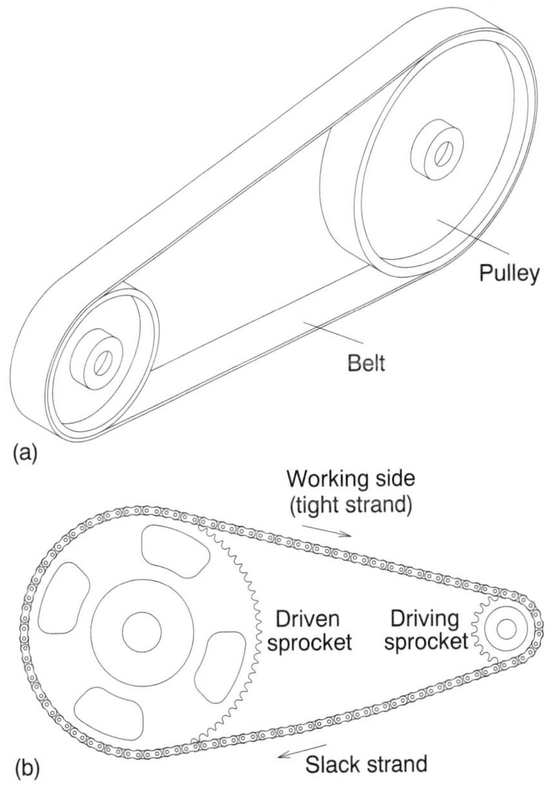

Fig. 6.1 (a) Belt drive; (b) chain drive.

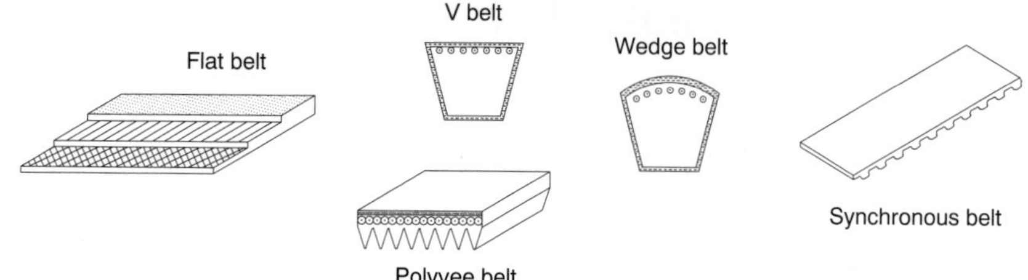

Fig. 6.2 Belt cross-sections.

maintenance, high reliability, adaptability to non-parallel drive and high transmission speeds. The principle disadvantages of belt drives are their limited power transmission capacity and limited speed ratio capability. Belt drives are less compact than either gear or chain drives and are susceptible to changes in environmental conditions such as contamination with lubricants. In addition, vibration and shock loading can damage belts.

Chains are usually more compact than belt drives for a given speed ratio and power capacity. Chain drives are generally more economical than the equivalent gear drive and are usually competitive with belt drives. Chains are inherently stronger than belt drives due to the use of steels in their manufacture, and can therefore support higher tension and transmit greater power. The disadvantages of chain drives are limited speed ratios and power transmission capability, and also safety issues. Chain drives can break and get thrown off the sprockets with large forces and high speeds. Guards should be provided for chain drives and some belt drives to prevent damage caused by a broken chain or belt, and also to prevent careless access to the chain or belt drive.

6.2 Belt drives

As mentioned previously, a belt drive consists of a flexible element which runs on a set of pulleys mounted on the shafts. Belt drives can be used to simply transmit power between one and another with the speed of the driving and driven shaft equal. In this case the pulley diameters would be equal. Alternatively, the driven shaft velocity can be decreased by using a bigger diameter pulley on the driven shaft than on the driving shaft. It is also possible to use belt drives to step up or increase the speed of a driven shaft, but this is a less common application.

There are various types of belt drive configurations, including flat, round, V, wedge and synchronous belt drives, each with their individual merits. The cross-sections of various belts are illustrated in Fig. 6.2. Most belts are manufactured from rubber or polymer-based materials.

Power is transmitted by means of friction in the case of flat, round, wedge and V belts, and by a combination of friction and positive mechanical interference in the case of synchronous belt drives. When the driven shaft rotates, friction between the pulley and the belt causes the belt to grip the pulley, increasing the tension in the side near to the point of first rotational contact. The tensile force exerts a tangential force, and associated torque, on the driven pulley. The opposite side of the belt is also under tension, but to a lesser extent, and is referred to as the slack side.

A frequent application of belt drives is to reduce the speed output from electric motors, which typically run at specific synchronous speeds that are high in comparison to the desired application drive speed. Because of their good 'twistability', belt drives are well suited to applications where the rotating shafts are in different planes. Some of the standard layouts are shown in Fig. 6.3. Belts are installed by moving the shafts closer together, slipping the belt over the pulleys and then moving the shafts back into their operating locations.

Flat belts have high strength, can be used for large speed ratios (>8:1), have a low pulley cost, give low noise levels and are good at absorbing torsional vibration. The belts are typically made from multiple plies, with each layer serving a special purpose. A typical three-ply belt consists of a friction ply made from synthetic rubber, polyurethane or chrome leather, a tension ply made from polyamide strips or polyester cord and an outer skin made from polyamide fabric, chrome leather or an elastomer. The corresponding pulleys are made from cast iron or polymer materials and are relatively smooth to limit wear. The driving force is limited by the friction between the belt and the pulley. Applications include manufacturing tools, saw mills, textile machinery, food processing machines, multiple spindle drives, pumps and compressors.

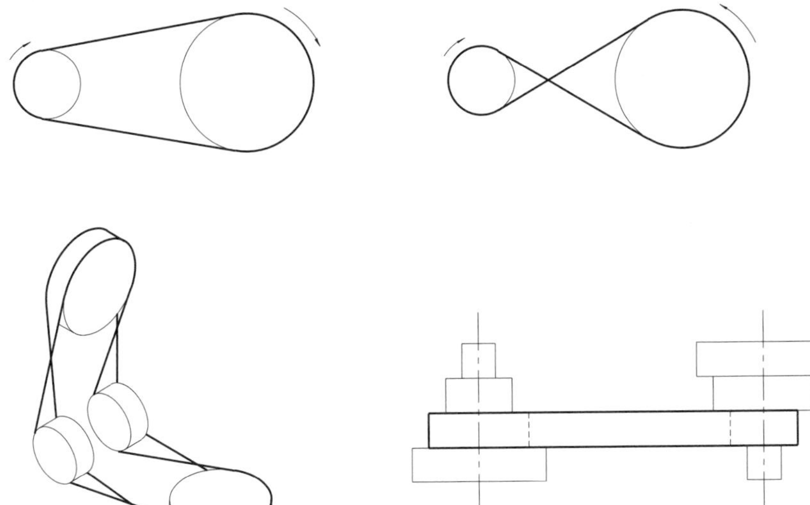

Fig. 6.3 Pulley configurations.

The most widely used type of belt in industrial and automotive applications is the V belt or wedge belt. It is familiar from its automotive application, where it is used to connect the crankshaft to accessory drives such as the alternator, water pump and cooling fan. It is also used for general engineering purposes, from domestic appliances to heavy duty rolling machines. The V or wedge shape causes the belt to wedge into the corresponding groove in the pulley, increasing friction and torque capacity. Multiple belts are commonly used so that a cheaper small cross-sectional area belt can be used to transmit more power. Note that long centre distances are not recommended for V or wedge belts.

Synchronous belts, also called timing belts, have teeth which mesh with corresponding teeth on the pulleys. This mechanical interference or positive contact between the pulleys and the belt provides angular synchronisation between the driving and the driven shafts and ensures a constant speed ratio. Synchronous belts combine the advantages of normal friction belt drives with the capability of synchronous drive. The meshing of the belt and pulleys is critical to their effective operation. The teeth of the advancing belt must mesh correctly with the corresponding grooves on the pulley wheels and remain in mesh throughout the arc of contact. To achieve this the pitch of the belts and the pulleys must correspond exactly. A disadvantage of synchronous belts can be the noise generated by compression of air between the teeth, especially at high speeds.

Table 6.1 lists the comparative merits of various belt drives.

6.2.1 Belt selection

Two approaches are presented here for the selection of a belt drive. Use can be made of the design procedures and accompanying charts provided by most belt drive manufacturers. The use of these charts is illustrated by considering a wedge belt drive. Alternatively, use can be made of fundamental relationships for the belt tensions

Table 6.1 Comparison of belt performance

Belt type	Optimum efficiency (%)	Maximum speed (m/s)	Minimum pulley diameter (mm)	Maximum speed ratio	Optimum tension ratio
Flat	98	70	40	20	2.5
V	80	30	67	7	5
Wedge	86	40	60	8	5
Synchronous	98	50	16	9	–

Reproduced from Hamilton (1994).

Table 6.2 Service factors

	Type of prime mover					
	Soft start,[a] duty (hours/day)			Heavy starts,[b] duty (hours/day)		
Type of driven machine	<10	10–16	>16	<10	10–16	>16
Light duty[c]	1.0	1.1	1.2	1.1	1.2	1.3
Medium duty[d]	1.1	1.2	1.3	1.2	1.3	1.4
Heavy duty[e]	1.2	1.3	1.4	1.4	1.5	1.6
Extra heavy duty[f]	1.3	1.4	1.5	1.5	1.6	1.8

[a] Electric motors: AC – star delta start, DC – shunt wound. IC engines with four or more cylinders. All prime movers fitted with centrifugal clutches, dry or fluid couplings or electronic soft starts.
[b] Electric motors: AC – direct on line start, DC series and compound wound. IC engines with less than four cylinders. Prime movers not fitted with soft start devices.
[c] Agitators (uniform density), blowers, exhausters and fans (up to 7.5 kW), centrifugal compressors, rotodynamic pumps, uniformly loaded belt conveyors etc.
[d] Agitators (variable density), blowers, exhausters and fans (over 7.5 kW), rotary compressors and pumps (other than centrifugal), non-uniformly loaded conveyors, generators, machine tools, printing machinery, sawmill machinery etc.
[e] Brick machinery, bucket elevators, reciprocating compressors and pumps, heavy duty conveyors, hoists, pulverisers, punches, presses, quarry plant, textile machinery etc.
[f] Crushers etc.
After Fenner Power Transmission UK.

and torque transmission and a belt selection based on its maximum permissible tensile stress. These equations are illustrated by an example using a flat belt drive.

The typical design and selection procedure for use in conjunction with power-speed rating charts supplied by commercial belt companies is outlined below.

1. Define the operating conditions: these include the nominal power to be transmitted, the rotational speeds of the shafts, and any space, layout or other constraints such as environmental conditions.
2. Determine the service factor: service factors are used to down-rate the power transmission capability, listed by belt suppliers, to account for the differences between practical applications and test conditions. Table 6.2 lists typical values for service factors.
3. Calculate the design power: this is the product of the nominal power and the service factor.
4. Select the belt type: to assist in the selection of which type of belt drive to use; the procedure given in Fig. 6.4 can be used as a guideline.
5. Select a belt: using a manufacturer's rating chart, select a specific belt for the design power and speed.
6. Select the pulley diameters: pulleys are normally available in standard sizes. Choose the smallest sizes available so that the speed ratio is acceptable.
7. Set the centre distance: this is dependent on the application. As a general guideline the centre

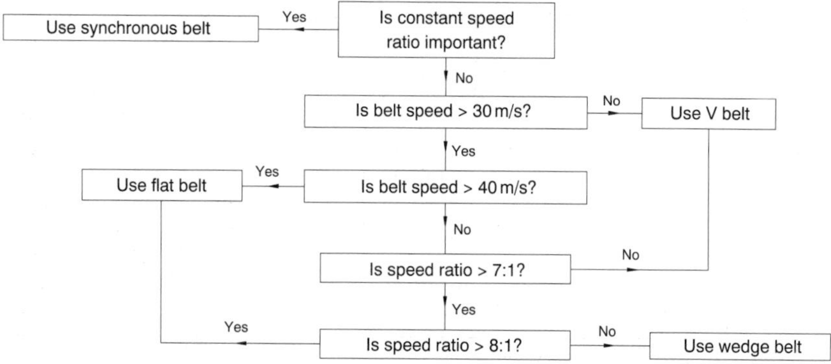

Fig. 6.4 Procedure for the selection of belt type. (Reproduced from Hamilton, 1994.)

distance should be greater than the diameter of the larger pulley.

8. Determine the belt length: note that belts are usually manufactured in standard lengths, so some iteration around the design parameters may be necessary to arrive at a satisfactory compromise.
9. Apply power correction factors: these are used to compensate for speed ratio and the belt geometry, and are provided within belt manufacturers' design guides.
10. Determine the allowable power per belt (or per belt width for flat belts): this is a function of the belt dimensions and is available from manufacturers' design guides.
11. Determine the number of belts: the number of belts is given by dividing the design power by the allowable power per belt and rounding up to the nearest integer.

Figure 6.5 shows a design chart for power rating versus speed and Tables 6.3–6.6 list values for the minimum pulley diameter, centre distance and power ratings for wedge belts. The use of these in conjunction with the procedure given is illustrated by the following example.

Example Select a wedge belt and determine the pulley diameters for a reciprocating compressor driven by a 28 kW two-cylinder diesel engine. The engine speed is 1500 rpm and the compressor speed is 950 rpm. The proposed distance between the engine and compressor shaft centres is approximately 1.5 m. The system is expected to be used for less than 10 hours per day.

Solution The speed ratio is

$$\frac{1500}{950} = 1.58$$

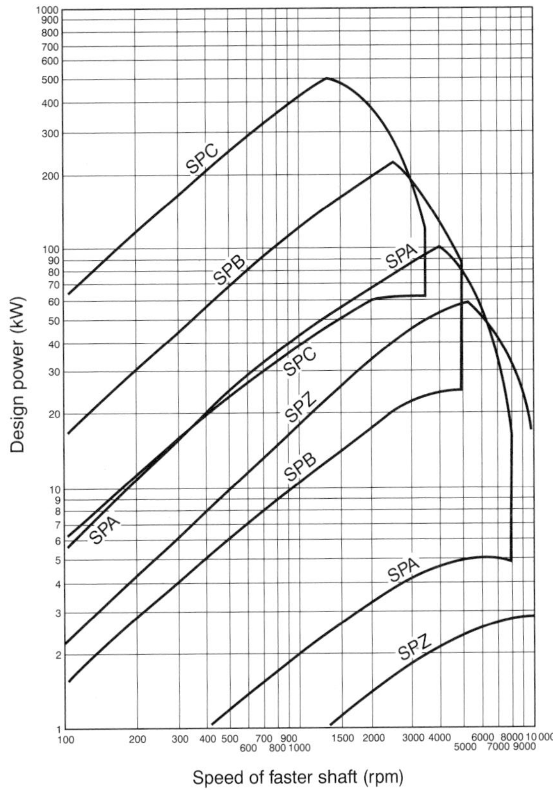

Fig. 6.5 Selection chart for wedge belts. (Courtesy of Fenner Drives UK.)

From Table 6.2 for 'heavy start', 'heavy duty' and less than 10 hours operation per day, the service factor is 1.4:

Design power = 28 × 1.4 = 39.2 kW

From Fig. 6.5, the combination of power equal to 39.2 kW and the speed equal to 1500 rpm is found to be within the range suitable for SPB belt drives.

Table 6.3 Minimum recommended pulley diameters for wedge belt drives

Speed of faster shaft (rpm)	Minimum pulley diameters (mm) for design power (kW) of																			
	<1	3.0	4.0	5.0	7.5	10.0	15.0	20.0	25	30	40	50	60	75	90	110	130	150	200	250
500	56	90	100	112	125	140	180	200	212	236	250	280	280	315	375	400	450	475	500	560
600	56	85	90	100	112	125	140	180	200	212	224	250	265	280	300	335	375	400	475	500
720	56	80	85	90	100	106	132	150	160	170	200	236	250	265	280	300	335	375	450	500
960	56	75	80	85	95	100	112	132	150	180	180	200	224	250	280	280	300	335	400	450
1200	56	71	80	80	95	95	106	118	132	150	160	180	200	236	236	250	265	300	335	355
1440	56	63	75	80	85	85	100	112	125	140	160	170	190	212	236	236	250	280	315	335
1800	56	63	71	75	80	85	95	106	112	125	150	160	170	190	212	224	236	265	300	335
2880	56	60	67	67	80	80	85	90	100	112	125	140	160	170	180	212	224	236	–	–

Courtesy of Fenner Power Transmission UK.

Table 6.4 Centre distances for selected SPB wedge belts

	Pitch diameter of pulleys		Power per belt (kW)		Combined arc and belt length correction factor															
					0.85		0.90		0.95		1.00			1.05			1.10		1.15	
					Belt length (mm)															
Speed ratio	Driver	Driven	1440 rpm	960 rpm	1250	1400	1800	2000	2240	2500	2800	3150	3550	4000	4500	5000	5600	6300	7100	8000
1.27	315	400	24.56	17.91	–	–	–	436	557	687	837	1013	1213	1438	1688	1938	2238	2588	2988	3438
1.27	118	150	6.37	4.58	414	489	689	789	–	–	–	–	–	–	–	–	–	–	–	–
1.28	125	160	7.24	5.18	401	476	676	776	–	–	–	–	–	–	–	–	–	–	–	–
1.29	140	180	7.69	5.54	373	448	648	748	869	998	1148	1342	1524	1749	1999	2249	2549	2899	3299	3749
1.29	132	170	8.10	5.78	387	462	663	763	–	–	–	–	–	–	–	–	–	–	–	–
1.53	118	180	6.51	4.67	390	465	665	765	–	–	–	–	–	–	–	–	–	–	–	–
1.56	180	280	12.03	8.61	259	335	536	637	757	887	1038	1213	1413	1638	1888	2138	2438	2788	3188	3638
1.56	160	250	9.95	7.13	300	375	576	676	797	927	1077	1252	1452	1677	1927	2178	2478	2828	3228	3678
1.57	150	236	8.89	6.37	319	394	595	696	816	946	1096	1271	1471	1696	1946	2196	2496	2847	3247	3697
1.57	200	315	14.05	10.05	–	290	492	593	713	844	994	1169	1369	1594	1845	2095	2395	2745	3145	3595
1.58	224	355	16.54	11.86	–	–	440	541	662	793	943	1118	1319	1544	1794	2044	2344	2694	3095	3545
1.59	315	500	24.81	18.09	–	–	–	–	471	603	754	930	1131	1357	1607	1858	2158	2508	2908	3359
1.60	125	200	7.49	5.35	368	443	644	744	–	–	–	–	–	–	–	–	–	–	–	–
1.60	140	224	7.95	5.71	336	412	613	713	833	963	1113	1288	1489	1714	1964	2214	2514	2864	3264	3714
1.60	250	400	19.02	13.68	–	–	382	484	605	736	886	1062	1262	1488	1738	1988	2288	2368	3039	3489
1.87	190	355	13.17	9.42	–	–	465	566	687	818	968	1144	1344	1570	1820	2070	2371	2721	3121	3571
1.89	212	400	15.37	11.00	–	–	409	511	633	764	915	1090	1291	1516	1767	2017	2317	2668	3068	3518
1.89	125	236	7.49	5.35	337	413	614	714	–	–	–	–	–	–	–	–	–	–	–	–
1.89	112	212	5.88	4.23	367	443	644	744	–	–	–	–	–	–	–	–	–	–	–	–
1.89	132	250	8.36	5.95	320	396	597	697	–	–	–	–	–	–	–	–	–	–	–	–
2.09	170	355	11.21	8.02	–	–	479	580	702	833	983	1159	1360	1585	1835	2086	2386	2736	3136	3586
2.10	150	315	9.11	6.53	–	324	528	629	750	881	1031	1207	1407	1633	1883	2133	2433	2784	3184	3634
2.11	190	400	13.26	9.48	–	–	424	526	648	780	931	1107	1307	1533	1784	2034	2334	2685	3085	3535
2.11	112	236	5.97	4.30	346	422	624	724	–	–	–	–	–	–	–	–	–	–	–	–
2.12	118	250	6.72	4.81	429	406	607	708	–	–	–	–	–	–	–	–	–	–	–	–

2.23	224	500	16.64	11.92	–	–	–	408	534	667	820	997	1198	1425	1676	1926	2227	2578	2978	3429
2.24	125	280	7.59	5.42	297	374	577	677	–	–	–	–	–	–	–	–	–	–	–	–
2.25	140	315	8.04	5.78	252	331	535	637	758	888	1039	1214	1415	1640	1891	2141	2441	2791	3191	3642
2.25	280	630	21.87	15.81	–	–	–	–	364	505	662	842	1046	1273	1525	1777	2078	2429	2830	3281
2.35	170	400	11.21	8.02	–	–	437	540	663	794	945	1121	1322	1548	1799	2049	2350	2700	3100	3550
2.54	315	800	24.81	18.15	–	–	–	–	–	–	–	654	865	1097	1353	1606	1909	2261	2663	3115
2.63	190	500	13.26	9.48	–	–	–	430	557	691	844	1021	1223	1450	1701	1952	2253	2603	3004	3455
2.67	150	400	9.11	6.53	–	–	451	554	677	808	960	1136	1337	1563	1814	2064	2365	2715	3116	3566
2.67	118	315	6.72	4.81	267	346	551	652	–	–	–	–	–	–	–	–	–	–	–	–
2.67	236	630	17.79	12.77	–	–	–	–	390	533	692	873	1077	1305	1557	1809	2111	2462	2863	3314
3.39	236	800	17.86	12.82	–	–	–	–	–	–	–	705	918	1152	1408	1662	1966	2319	2722	3174
3.50	180	630	12.31	8.80	–	–	–	–	425	569	729	911	1116	1345	1598	1850	2152	2504	2905	3356
3.57	112	400	6.04	4.34	–	–	476	580	–	–	–	–	–	–	–	–	–	–	–	–
3.57	140	500	8.11	5.82	–	–	351	462	590	725	879	1057	1259	1486	1738	1989	2290	2641	3042	3493
3.57	224	800	16.71	11.97	–	–	–	–	–	–	–	713	926	1160	1416	1671	1975	2328	2731	3183
3.57	280	1000	21.94	15.85	–	–	–	–	–	–	–	–	673	925	1190	1450	1758	2114	2519	2973
3.71	170	630	11.28	8.06	–	–	–	–	431	576	736	918	1123	1352	1605	1857	2159	2511	2913	3364
3.77	212	800	15.53	11.11	–	–	–	–	–	–	–	720	934	1168	1425	1679	1983	2337	2739	3192
3.79	132	500	8.52	6.06	–	–	356	467	–	–	–	–	–	–	–	–	–	–	–	–
3.94	160	630	10.23	7.32	–	–	–	–	437	582	742	925	1130	1359	1612	1865	2167	2519	2920	3371
4.72	212	1000	15.53	11.11	–	–	–	–	–	–	–	–	714	968	1235	1496	1805	2162	2568	3022
5.00	160	800	10.23	7.32	–	–	–	–	–	–	554	753	968	1203	1461	1716	2021	2374	2778	3230
5.00	200	1000	14.34	10.25	–	–	–	–	–	–	–	–	722	976	1243	1504	1813	2171	2576	3031
5.26	190	1000	13.33	9.53	–	–	–	–	–	–	–	–	728	982	1250	1511	1820	2178	2584	3038
5.33	150	800	9.18	6.57	–	–	–	–	–	–	–	759	975	1210	1468	1723	2028	2382	2785	3238

Courtesy of Fenner Power Transmission UK. Note: This is only a partial selection from a typical catalogue relevant to the worked examples and worksheet questions.

Table 6.5 Power ratings for SPB wedge belts

Speed of faster shaft (rpm)	Rated power (kW) per belt for small pulley pitch diameter (mm) of												
	140	150	160	170	180	190	200	212	224	236	250	280	315
100	0.73	0.82	0.92	1.01	1.10	1.20	1.29	1.40	1.51	1.62	1.74	2.01	2.33
200	1.33	1.51	1.69	1.87	2.05	2.22	2.40	2.61	2.82	3.02	3.26	3.78	4.37
300	1.89	2.15	2.41	2.67	2.93	3.18	3.44	3.74	4.04	4.35	4.70	5.44	6.30
400	2.42	2.76	3.09	3.43	3.77	4.10	4.43	4.83	5.22	5.61	6.07	7.04	8.15
500	2.92	3.33	3.75	4.16	4.57	4.98	5.39	5.87	6.36	6.84	7.39	8.58	9.94
600	3.40	3.89	4.38	4.87	5.35	5.83	6.31	6.89	7.45	8.02	8.67	10.06	11.66
700	3.86	4.43	4.99	5.55	6.11	6.66	7.21	7.87	8.52	9.17	9.92	11.50	13.32
720	3.95	4.53	5.11	5.69	6.26	6.82	7.39	8.06	8.73	9.39	10.16	11.79	13.65
800	4.31	4.95	5.59	6.22	6.84	7.47	8.08	8.82	9.55	10.28	11.12	12.90	14.93
900	4.75	5.46	6.16	6.86	7.56	8.25	8.93	9.75	10.56	11.36	12.29	14.25	16.47
960	5.00	5.75	6.50	7.24	7.98	8.71	9.43	10.29	11.15	11.99	12.97	15.03	17.37
1000	5.17	5.95	6.72	7.49	8.25	9.01	9.76	10.65	11.53	12.41	13.42	15.55	17.96
1100	5.58	6.42	7.27	8.10	8.93	9.75	10.56	11.52	12.48	13.43	14.52	16.80	19.39
1200	5.97	6.89	7.79	8.69	9.58	10.46	11.34	12.37	13.40	14.41	15.57	18.01	20.75
1300	6.36	7.34	8.31	9.27	10.22	11.16	12.09	13.19	14.28	15.36	16.59	19.17	22.05
1400	6.73	7.77	8.81	9.83	10.84	11.84	12.82	13.99	15.14	16.27	17.57	20.28	23.28
1440	6.88	7.95	9.00	10.05	11.08	12.10	13.11	14.30	15.47	16.63	17.96	20.70	23.75
1500	7.09	8.20	9.29	10.37	11.44	12.49	13.53	14.76	15.97	17.15	18.51	21.33	24.43
1600	7.44	8.61	9.76	10.90	12.02	13.12	14.21	15.50	16.76	18.00	19.41	22.33	25.51
1700	7.78	9.01	10.21	11.40	12.58	13.73	14.87	16.21	17.52	18.81	20.27	23.27	26.51
1800	8.11	9.39	10.65	11.90	13.12	14.32	15.50	16.89	18.25	19.58	21.08	24.15	27.43
1900	8.43	9.76	11.08	12.37	13.64	14.88	16.11	17.54	18.94	20.31	21.85	24.97	28.27
2000	8.73	10.12	11.48	12.82	14.14	15.43	16.69	18.16	19.60	20.99	22.57	25.72	29.01
2100	9.02	10.46	11.88	13.26	14.62	15.94	17.24	18.75	20.22	21.64	23.23	26.41	29.67
2200	9.31	10.79	12.25	13.68	15.07	16.44	17.76	19.31	20.80	22.24	23.85	27.03	30.22
2300	9.57	11.11	12.61	14.08	15.51	16.90	18.26	19.83	21.35	22.80	24.42	27.57	30.68
2400	9.83	11.41	12.95	14.46	15.92	17.34	18.72	20.32	21.85	23.31	24.93	28.05	31.04
2500	10.08	11.70	13.28	14.82	16.31	17.76	19.16	20.77	22.31	23.78	25.38	28.44	–
2600	10.31	11.97	13.59	15.16	16.68	18.14	19.56	21.19	22.73	24.19	25.78	28.76	–
2700	10.53	12.23	13.88	15.47	17.02	18.50	19.93	21.56	23.11	24.56	26.12	28.99	–
2800	10.73	12.47	14.15	15.77	17.33	18.83	20.27	21.90	23.44	24.87	26.40	–	–
2880	10.89	12.65	14.35	15.99	17.57	19.07	20.51	22.14	23.67	25.08	26.57	–	–
2900	10.93	12.69	14.40	16.04	17.62	19.13	20.57	22.20	23.72	25.12	26.61	–	–
3000	11.10	12.90	14.63	16.30	17.89	19.40	20.84	22.46	23.96	25.33	26.76	–	–

Courtesy of Fenner Power Transmission UK.

From Table 6.3 for a design power of 39.2 kW and driving shaft speed of 1500 rpm, an approximate minimum pulley diameter of 160 mm is suitable.

From Table 6.4, the actual pulley diameters can be selected by tracing down the left-hand side to the speed ratio and reading across for the pitch diameters of the driving and driven pulleys. The minimum diameter from Table 6.3 can be used as a guideline when there is a choice.

From Table 6.4 for a speed ratio of 1.58, suitable pitch diameters are $D_1 = 224$ mm and $D_2 = 355$ mm. The nearest centre distance listed to the 1.5 m desired is 1.544 m, and tracing the column upwards defines the belt length as 4000 mm and by tracing the shading upwards the arc correction factor is 1.05.

From Table 6.5, the rated power per belt for $n_1 = 1500$ rpm and $D_1 = 224$ mm is 15.97 kW.

From Table 6.6, the additional power per belt accounting for the speed ratio is 1.11 kW.

The corrected value for the power per belt is $(15.97 + 1.11) \times 1.05 = 17.93$ kW.

The total number of belts is therefore

$$\frac{39.2}{17.93} = 2.19$$

As a partial belt cannot be used, round this number up to the nearest integer. Thus three SPB4000 belts should be used, running on

Table 6.6 Additional power increment per SPB belt

Speed of faster shaft (rpm)	Additional power (kW) per belt for speed ratio of									
	1.00 to 1.01	1.02 to 1.05	1.06 to 1.11	1.12 to 1.18	1.19 to 1.26	1.27 to 1.38	1.39 to 1.57	1.58 to 1.94	1.95 to 3.38	3.39 and over
100	0.00	0.01	0.02	0.04	0.04	0.06	0.07	0.07	0.08	0.08
200	0.00	0.01	0.04	0.07	0.09	0.11	0.13	0.15	0.16	0.17
300	0.00	0.02	0.06	0.10	0.14	0.17	0.20	0.22	0.24	0.25
400	0.00	0.03	0.07	0.13	0.19	0.22	0.26	0.29	0.32	0.34
500	0.00	0.04	0.09	0.17	0.23	0.28	0.33	0.37	0.40	0.43
600	0.00	0.04	0.12	0.20	0.28	0.34	0.40	0.45	0.48	0.51
700	0.00	0.05	0.13	0.24	0.33	0.39	0.46	0.52	0.57	0.59
720	0.00	0.05	0.14	0.25	0.33	0.41	0.48	0.54	0.59	0.62
800	0.00	0.06	0.16	0.28	0.37	0.45	0.53	0.60	0.65	0.69
900	0.00	0.07	0.18	0.31	0.42	0.51	0.60	0.66	0.72	0.77
960	0.00	0.07	0.19	0.32	0.44	0.54	0.62	0.70	0.77	0.81
1000	0.00	0.07	0.19	0.34	0.46	0.56	0.66	0.74	0.81	0.86
1100	0.00	0.08	0.22	0.37	0.51	0.62	0.72	0.81	0.89	0.94
1200	0.00	0.09	0.23	0.41	0.56	0.68	0.79	0.89	0.97	1.03
1300	0.00	0.09	0.25	0.44	0.60	0.73	0.86	0.96	1.05	1.11
1400	0.00	0.10	0.28	0.48	0.65	0.79	0.93	1.04	1.13	1.20
1440	0.00	0.10	0.28	0.48	0.66	0.79	0.94	1.06	1.15	1.21
1500	0.00	0.10	0.29	0.51	0.69	0.84	0.99	1.11	1.21	1.28
1600	0.00	0.11	0.31	0.54	0.75	0.90	1.05	1.19	1.29	1.37
1700	0.00	0.12	0.34	0.58	0.79	0.95	1.12	1.26	1.37	1.45
1800	0.00	0.13	0.35	0.61	0.84	1.01	1.19	1.34	1.45	1.54
1900	0.00	0.13	0.37	0.65	0.88	1.07	1.25	1.41	1.54	1.63
2000	0.00	0.14	0.39	0.68	0.93	1.13	1.32	1.48	1.62	1.71
2100	0.00	0.15	0.41	0.72	0.98	1.18	1.39	1.56	1.69	1.79
2200	0.00	0.16	0.43	0.75	1.02	1.24	1.45	1.63	1.78	1.88
2300	0.00	0.16	0.45	0.78	1.07	1.29	1.51	1.71	1.86	1.97
2400	0.00	0.17	0.47	0.82	1.11	1.35	1.58	1.78	1.94	2.05
2500	0.00	0.18	0.49	0.85	1.16	1.41	1.65	1.86	2.02	2.14
2600	0.00	0.19	0.51	0.89	1.21	1.46	1.72	1.92	2.10	2.22
2700	0.00	0.19	0.53	0.92	1.25	1.52	1.78	1.99	2.18	2.31
2800	0.00	0.20	0.54	0.95	1.29	1.57	1.84	2.07	2.26	2.39
2880	0.00	0.20	0.56	0.97	1.32	1.60	1.88	2.11	2.31	2.44
2900	0.00	0.21	0.57	0.99	1.34	1.63	1.91	2.15	2.34	2.48
3000	0.00	0.22	0.59	1.02	1.39	1.69	1.98	2.23	2.42	2.57

Courtesy of Fenner Power Transmission UK.

sprockets of 224 mm and 315 mm pitch diameter with a centre distance of 1.544 m.

For the simple belt drive configuration shown in Fig. 6.6 the angles of contact between the belt and the pulleys are given by

$$\theta_d = \pi - 2\sin^{-1}\frac{D-d}{2C} \tag{6.3}$$

$$\theta_D = \pi + 2\sin^{-1}\frac{D-d}{2C} \tag{6.4}$$

where

d = diameter of the small pulley (m)
D = diameter of the large pulley (m)
C = distance between the pulley centres (m)
θ_d = angle of contact between the belt and the small pulley (rad)
θ_D = angle of contact between the belt and the large pulley (rad).

The length of the belt can be obtained by summing the arc lengths of contact and the spanned distances and is given by

$$L = \sqrt{4C^2 - (D-d)^2} + \tfrac{1}{2}(D\theta_D + d\theta_d) \tag{6.5}$$

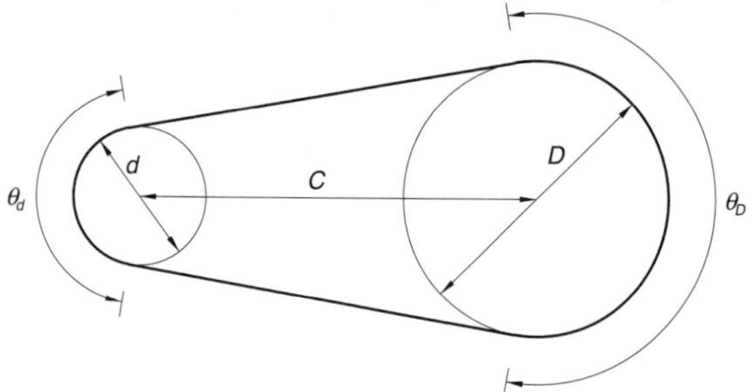

Fig. 6.6 Belt drive geometry definition.

The power transmitted by a belt drive is given by

$$\text{Power} = (F_1 - F_2)V \tag{6.6}$$

where

F_1 = belt tension in the tight side (N)
F_2 = belt tension in the slack side (N)
V = belt speed (m/s).

The torque is given by

$$\text{Torque} = (F_1 - F_2)r \tag{6.7}$$

Assuming that the friction is uniform throughout the arc of contact and ignoring centrifugal effects the ratio of the tensions in the belts can be modelled by Eytlewein's formula:

$$\frac{F_1}{F_2} = e^{\mu\theta} \tag{6.8}$$

where

μ = coefficient of friction
θ = angle of contact (rad), usually taken as the angle for the smaller pulley.

The centrifugal forces acting on the belt along the arcs of contact reduce the surface pressure. The centrifugal force is given by

$$F_c = \rho V^2 A = mV^2 \tag{6.9}$$

where

ρ = density of the belt material (kg/m^3)
A = cross-sectional area of the belt (m^2)
m = mass per unit length of the belt (kg/m).

The centrifugal force acts on both the tight and the slack sides of the belt and Eytlewein's formula can be modified to model the effect:

$$\frac{F_1 - F_c}{F_2 - F_c} = e^{\mu\theta} \tag{6.10}$$

The maximum allowable tension, $F_{1,\max}$, in the tight side of a belt depends on the allowable stress of the belt material, $\sigma_{\max}$:

$$F_{1,\max} = \sigma_{\max} A \tag{6.11}$$

The required cross-sectional area for a belt drive can be found from

$$A = \frac{F_1 - F_2}{\sigma_1 - \sigma_2} \tag{6.12}$$

Example A fan is belt driven by an electric motor running at 1500 rpm. The pulley diameters for the fan and motor are 500 mm and 355 mm respectively. A flat belt has been selected with a width of 100 mm, thickness of 3.5 mm, coefficient of friction of 0.8, density of 1100 kg/m^3 and permissible stress of 11 MN/m^2. The centre distance is 1500 mm. Determine the power capacity of the belt.

Solution The arcs of contact for the driving and driven pulleys are

$$\theta_d = \pi - 2\sin^{-1}\frac{D-d}{2C}$$

$$= \pi - 2\sin^{-1}\frac{500-355}{2 \times 1500} = 3.045\ \text{rad}$$

$$\theta_D = \pi + 2\sin^{-1}\frac{D-d}{2C}$$

$$= \pi + 2\sin^{-1}\frac{500-355}{2 \times 1500} = 3.238\ \text{rad}$$

The maximum tension in the tight side is given as a function of the maximum permissible stress in the belt by

$$F_1 = \sigma_{\max} A = 11 \times 10^6 (3.5 \times 10^{-3} \times 100 \times 10^{-3})$$

$$= 3850\ \text{N}$$

The belt velocity is

$$V = 1500 \times \frac{2\pi}{60} \times \frac{0.355}{2} = 27.88\ \text{m/s}$$

The mass per unit length is

$$m = \rho A = 1100 \times 0.1 \times 0.0035 = 0.385\ \text{kg/m}$$

The centrifugal load is given by

$$F_c = 0.385(27.88)^2 = 299.3\ \text{N}$$

Using equation 6.10:

$$\frac{F_1 - F_c}{F_2 - F_c} = e^{\mu\theta} = \frac{3850 - 299.3}{F_2 - 299.3} = e^{0.8 \times 3.045}$$

which gives

$$F_2 = 610.0\ \text{N}$$

Table 6.7 Guideline values for the maximum permissible stress for high-performance flat belts

Multi-ply structure			Maximum permissible stress (MN/m^2)
Friction surface coating	Core	Top surface	
Elastomer	Polyamide sheet	Polyamide fabric	8.3–19.3
Elastomer	Polyamide sheet	Elastomer	6.6–13.7
Chrome leather	Polyamide sheet	None	6.3–11.4
Chrome leather	Polyamide sheet	Polyamide fabric	5.7–14.7
Chrome leather	Polyamide sheet	Chrome leather	4–8
Elastomer	Polyester cord	Elastomer	$\leq$21.8
Chrome leather	Polyester cord	Polyamide fabric	5.2–12
Chrome leather	Polyester cord	Chrome leather	3.1–8

The power capacity is given by

$$(F_1 - F_2)V = (3850 - 610)27.88 = 90.3\,\text{kW}$$

6.3 Chain drives

A chain is a power transmission device consisting of a series of pin-connected links, as illustrated in Fig. 6.7. The chain transmits power between two rotating shafts by meshing with toothed sprockets, as shown in Fig. 6.8.

Chain drives are usually manufactured using high strength steel and for this reason are capable of transmitting high torque. Chain drives are complementary and competitive with belt drives, serving the function of transmitting a wide range of powers for shaft speeds up to 6000 rpm. At higher speeds the cyclic impact between the chain links and the sprocket teeth, high noise and difficulties in providing lubrication, limit the application of chain drives. Table 6.8 shows a comparison of chain, belt and gear attributes. Chain drives are principally used for power transmission, conveyors and for supporting or lifting loads. Applications range from motorcycle and bicycle transmissions, automotive camshaft drives, machine tools and aerospace drives such as the thruster nozzles for the Harrier, to conveyors and packaging machinery. Efficiencies of up to 98.9 per cent and, where necessary, ratios of up to 9:1 and power transmission of several hundred kilowatts can be achieved. Chain drives are typically used with reduction ratios of up to 3:1, giving high efficiency, and ratios of up to 5:1, giving reasonable efficiency.

The range of chain drives is extensive, as illustrated in Fig. 6.9. The most common type is the roller chain, which is used for high power transmission and conveyor applications. The selection and specification of this type of chain is introduced in Section 6.3.1. The roller chain is made up of a series of links. Each chain consists of side plates, pins, bushes and rollers, as shown in Fig. 6.10. The chain runs on toothed sprockets, and as the teeth of the sprockets engage with the rollers, a

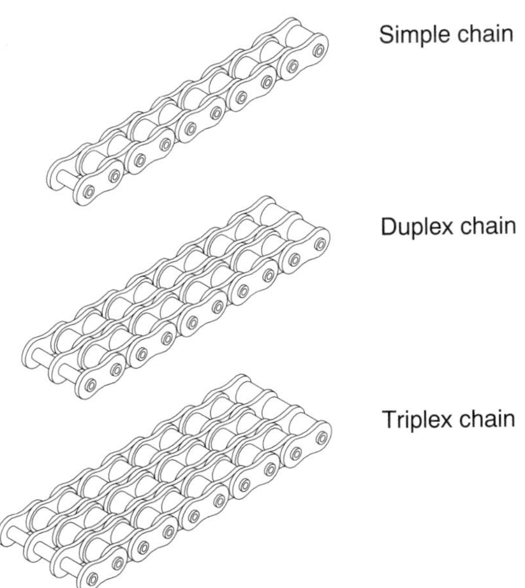

Fig. 6.7 Roller chain.

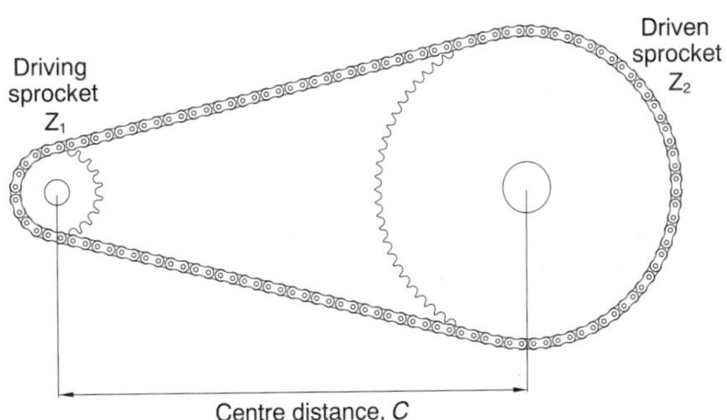

Fig. 6.8 Simple chain drive.

Table 6.8 Comparison of chain, belt and gear performance

Feature	Chain drive	Belt drive	Gear drive
Efficiency	A	A	A
Positive drive	A	A[a]	A
Large centre distance	A	A	C
Wear resistance	A	B	A
Multiple drives	A	A	C
Heat resistance	A	C	B
Chemical resistance	A	C	B
Oil resistance	A	C	A
Power range	A	B	A
Speed range	C	A	A
Ease of maintenance	A	B	C
Environment	A	C	A

A = excellent, B = good, C = poor.
[a] Using a synchronous belt drive.

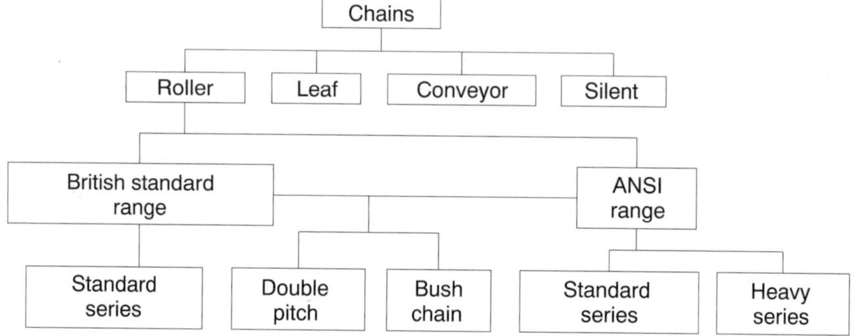

Fig. 6.9 Chain types.

rolling motion occurs and the chain articulates on to the sprocket. The roller chain is classified by its pitch. This is the distance between the corresponding points of adjacent links. The roller chain is available in stainless steel and nylon (for hygienic applications), as well as a series of other materials for specialist applications.

The conveyor chain is specially designed for use in materials handling and conveyor equipment and is characterised by its long pitch, large roller diameters and high tensile strength. The applications of the conveyor chain demand that it is capable of pulling significant loads, usually along a straight line at relatively low speeds. The sides of conveyor chains often incorporate special features to aid connection to conveyor components, as illustrated in Fig. 6.11.

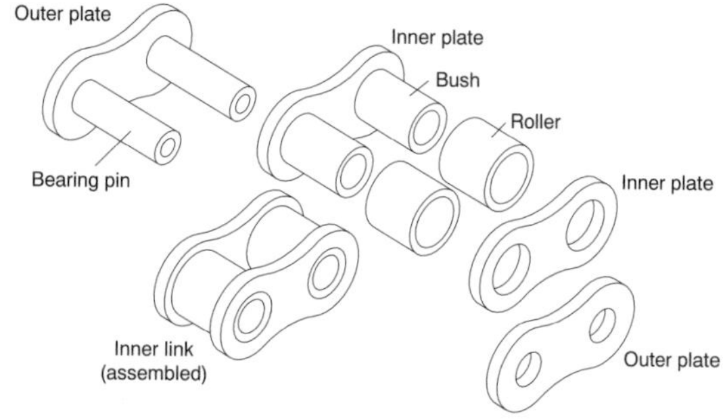

Fig. 6.10 Roller chain components.

The leaf chain consists of a series of pin-connected side plates, as illustrated in Fig. 6.12. It is generally used for load balancing applications and is essentially a special form of transmission chain.

The silent chain (also called the inverted tooth chain) has the teeth formed in the link plates, as shown in Fig. 6.13. The chain consists of alternately mounted links so that the chain can articulate onto the mating sprocket teeth. Silent chains can operate at higher speeds than comparatively sized roller chains and, as the name suggests, more quietly.

The direction of rotation relative to the chain layout is important to the effective operation of

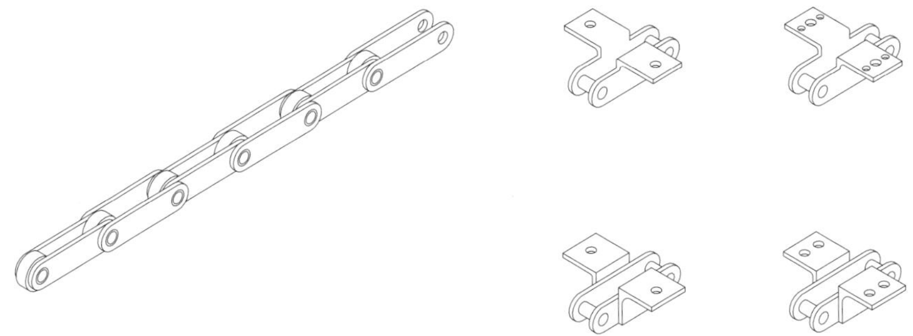

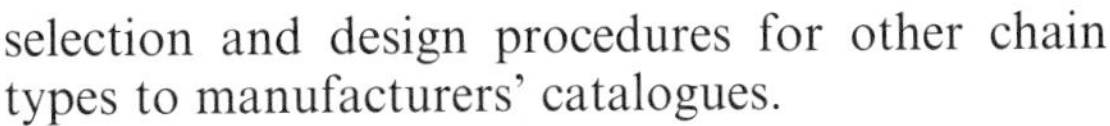

Fig. 6.11 Conveyor chain.

the chain drive. Figure 6.14 shows recommended and acceptable practice for chain drive layouts. The idler sprockets shown are used to engage the chain in the slack region for chain drives with large centre distances or when the drive is vertical.

Chain design is based on ensuring that the power transmission capacity is within limits for three modes of failure: fatigue, impact and galling. Chains are designed so that the maximum tensile stress is below the fatigue endurance limit for finite life of the material. Failure would nevertheless eventually occur but it would be due to wear, not fatigue. In service, failures due to wear can be eliminated by inspection and replacement intervals. When the chain rollers mesh with the sprocket teeth an impact occurs and a Hertz contact stress occurs, similar to that found for gear meshing. The power rating charts for chain drives limit the selection of the drive, so that these modes of failure should not occur assuming proper installation, operation and lubrication.

Details of the selection procedure for standard roller chains is outlined in Section 6.3.1. For details of chain geometry the reader is referred to the standards listed in the references, and for details of selection and design procedures for other chain types to manufacturers' catalogues.

6.3.1 Roller chain selection

Once the use of a chain drive has been shown to be preferable to other forms of drive the type of chain to be used can be selected from the range available, as illustrated in Fig. 6.9. The next step is the design of the chain drive layout and selection of the standard components available from chain manufacturers. The method outlined here is for roller chains. Procedures for the selection of other chain types can be found in manufacturers' catalogues.

The method is based upon the use of power rating charts for the chain drive, which ensure 15 000 hours operation assuming proper installation, operation and lubrication. The steps for the method are itemised below.

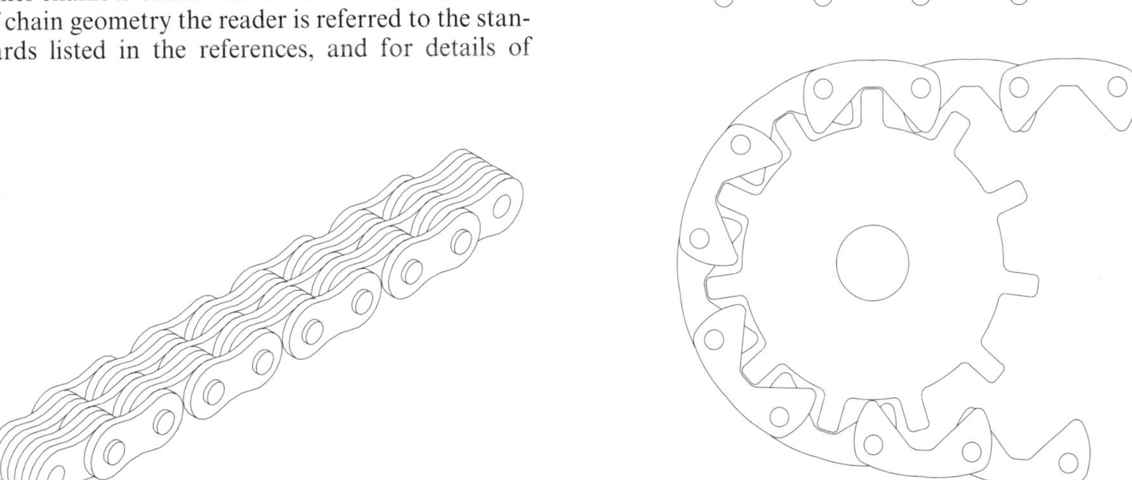

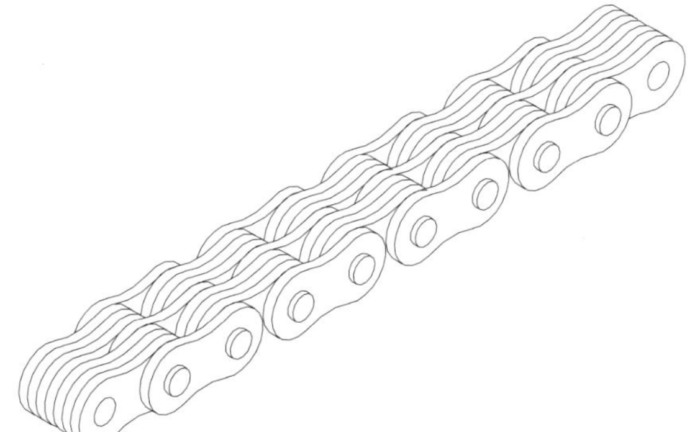

Fig. 6.12 Leaf chain.

Fig. 6.13 Silent chain.

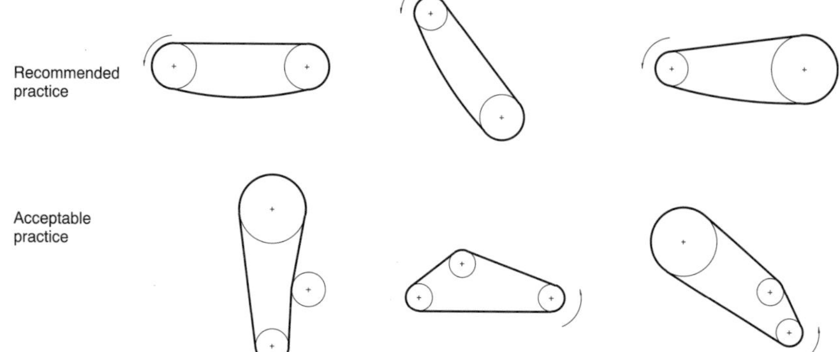

Fig. 6.14 Typical chain drive layouts.

1. Determine the power to be transmitted.
2. Determine the speeds of the driving and driven shafts.
3. Determine the characteristics of the driving and driven shaft, e.g. type of running, whether smooth or shock loadings etc.
4. Set the approximate centre distance: this should normally be in the range of 30 to 50 times the chain pitch.
5. Select the speed ratio: this is dependent on the standard sizes available, as listed in Table 6.9. Ideally sprockets should have a minimum of 19 teeth. For high-speed drives subjected to transient loads the minimum number of teeth rises to 25. Note that the maximum number of teeth should not exceed 114.
6. Establish the application and tooth factors: Table 6.10 gives values for the application factor f_1. The tooth factor is given by $f_1 = 19/N_1$ assuming that the selection rating charts are based on a 19-tooth sprocket.
7. Calculate the selection power: selection power $=$ power $\times f_1 \times f_2$.
8. Select the chain drive pitch: use power speed rating charts as supplied by chain manufacturers (see Fig. 6.15). The smallest pitch of a simple chain should be used as this normally gives the most economical drive. If the power requirement at a given speed is beyond the capacity of a single strand of chain then the use of a multistrand chain, such as duplex (two strands), triplex (three strands) and up to decuplex (ten strands) for the ANSI range, permits higher power to be transmitted and can be considered.
9. Calculate the chain length: equation 6.13 gives the chain length as a function of the number of pitches. Note that the value for the length should be rounded up to the nearest even integer.
10. Calculate the exact centre distance: this can be calculated using equation 6.14.
11. Specify the lubrication method.

The chain length, in pitches, is given by

$$L = \frac{N_1 + N_2}{2} + \frac{2C}{p} + \left(\frac{N_2 - N_1}{2\pi}\right)^2 \frac{p}{C} \tag{6.13}$$

where

$L =$ number of pitches

Table 6.9 Chain reduction ratios as a function of the standard sprockets available

	Number of teeth in drive sprocket, N_1					
Number of teeth in driven sprocket, N_2	15	17	19	21	23	25
25	–	–	–	–	–	1.00
38	2.53	2.23	2.00	1.80	1.65	1.52
57	3.80	3.35	3.00	2.71	2.48	2.28
76	5.07	4.47	4.00	3.62	3.30	3.04
95	6.33	5.59	5.00	4.52	4.13	3.80
114	7.60	6.70	6.00	5.43	4.96	4.56

Reproduced from Renold (1996).

Table 6.10 Application factor

Driven machine characteristics	Driver characteristics		
	Smooth running[a]	Slight shocks[b]	Heavy shocks[c]
Smooth running[d]	1	1.1	1.3
Moderate shocks[e]	1.4	1.5	1.7
Heavy shocks[f]	1.8	1.9	2.1

[a] Electric motors, IC engines with hydraulic coupling etc.
[b] IC engines with more than six cylinders, electric motors with frequent starts etc.
[c] IC engines with less than six cylinders etc.
[d] Fans, pumps, compressors, printing machines, uniformly loaded conveyors etc.
[e] Concrete mixing machines, non-uniformly loaded conveyors, mixers etc.
[f] Planars, presses, drilling rigs etc.
Reproduced from Renold (1996).

N_1 = number of teeth in the driving sprocket
N_2 = number of teeth in the driven sprocket
C = centre distance (m)
p = chain pitch (m).

The exact centre distance is given by

$$C = \frac{p}{8}\left[2L - N_2 - N_1 + \sqrt{(2L - N_2 - N_1)^2 - \frac{\pi}{3.88}(N_2 - N_1)^2}\right] \tag{6.14}$$

Roller chain drives have been standardised according to International, British, DIN and American Standards. The standard pitches available in ISO 606 are listed in Table 6.11.

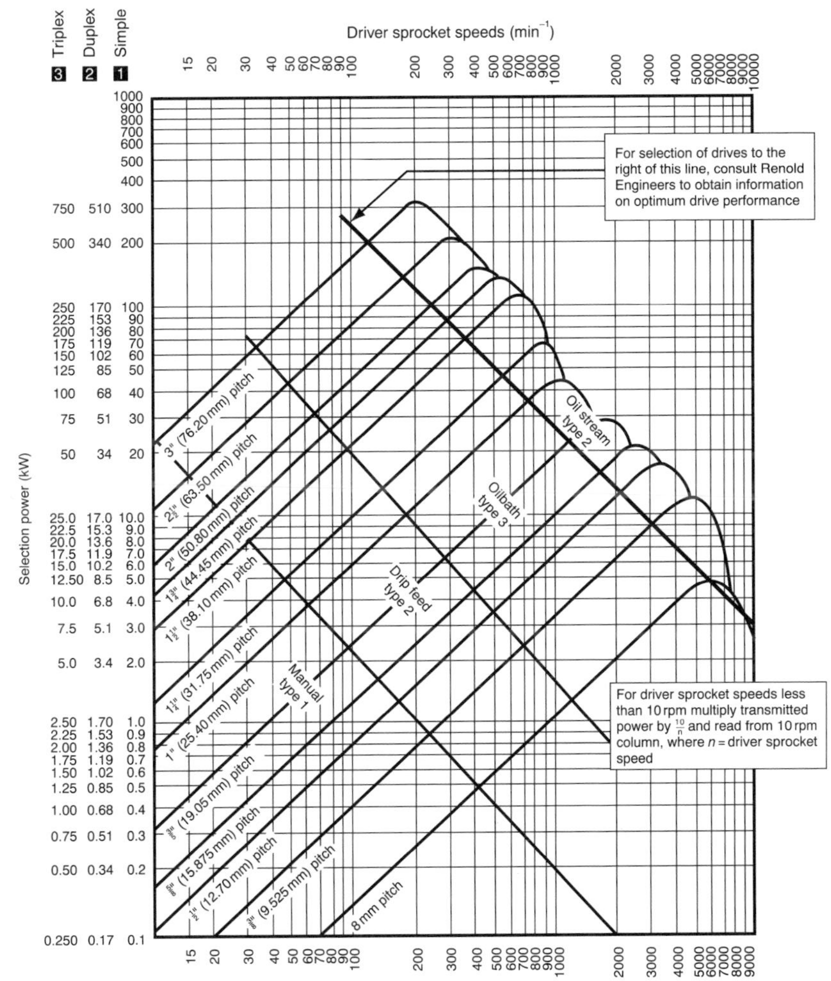

Fig. 6.15 British Standard chain drives – rating chart using 19T driver sprocket. (Courtesy of Renold Chain.)

Table 6.11 Selected standard pitches

ISO 606, BS 228, DIN 8187 code	ANSI code	Pitch (mm)
05B		8.0
06B	35	9.525
08B	40	12.7
10B	50	15.875
12B	60	19.05
16B	80	25.4
20B	100	31.75
24B	120	38.1
28B	140	44.45
32B	160	50.8
36A		57.15
40B	200	63.5
48B		76.2
56B		88.9
64B		101.6
72B		114.3

Chain drives should be protected against dirt and moisture (tell this to a mountain biker!). Lubrication should be provided using a non-detergent mineral-based oil. For the majority of applications, multigrade SAE20/50 is suitable. There are five principal types of lubrication: manual application, drip feed, bath, stream (*see* Fig. 6.16) and dry lubrication. Grease lubrication is not recommended but can be used for chain speeds of less than 4 m/s. In order to ensure the grease penetrates the working parts of the chain, the grease should be heated until liquid and the chain immersed until the air has been displaced. This process should be repeated at regular service intervals. For dry lubrication, solid lubricant is contained in a volatile carrier fluid. When applied to the chain the carrier transports the lubricant into the chain and then evaporates, leaving the chain lubricated but dry to touch. Applications for dry lubrication include food processing, dusty environments and fabric handling.

The pitch diameters for the driving and driven sprockets are given by

$$D_1 = \frac{N_1 p}{\pi} \qquad D_2 = \frac{N_2 p}{\pi} \tag{6.15}$$

and the angle of contact (in radians) between the chain and the sprockets by

$$\begin{aligned} \theta_1 &= \pi - \frac{2\pi(N_1 - N_2)}{2pC} \\ \theta_2 &= \pi + \frac{2\pi(N_1 - N_2)}{2pC} \end{aligned} \tag{6.16}$$

Note that the maximum angle of wrap recommended for the small sprocket is 120°.

The chain tension is given by

$$\text{Chain tension} = \frac{\text{Power}}{N_1 \omega_1 p/2\pi} \tag{6.17}$$

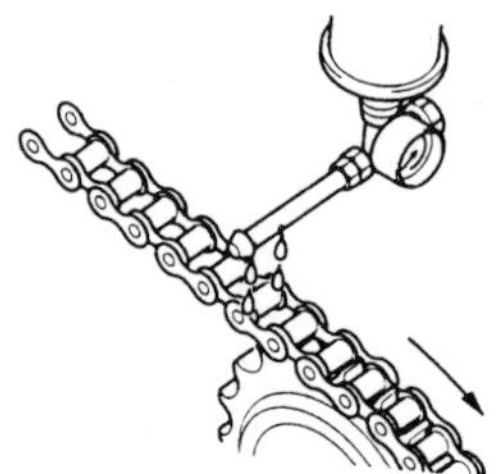

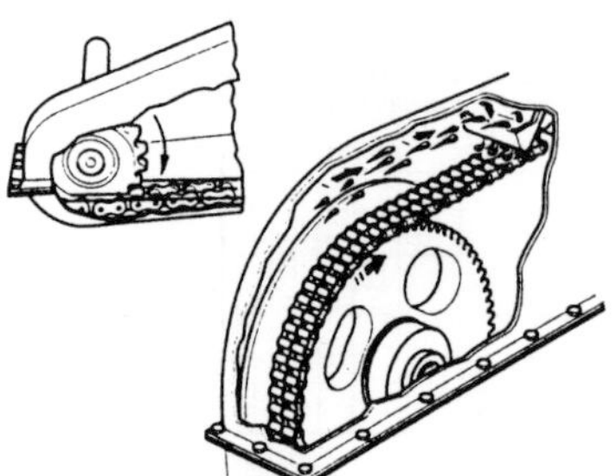

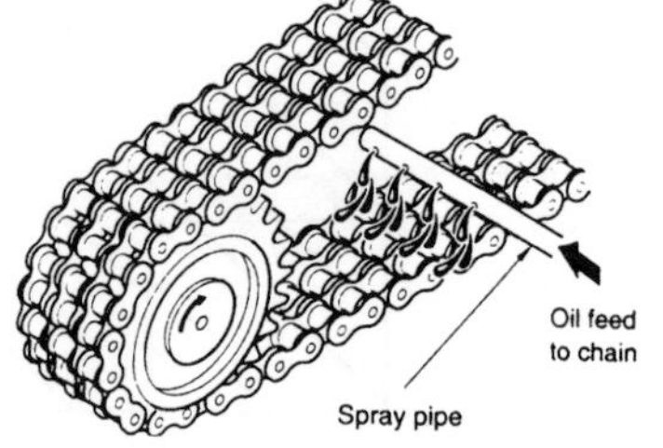

Fig. 6.16 Chain lubrication methods. (Courtesy of Renold Chain.)

Table 6.12 Selection of sprocket materials

Sprocket	Smooth running	Moderate shocks	Heavy shocks
Up to 29 teeth	080M40 or 070M55	080M40 or 070M55 hardened and tempered or case hardened mild steel	080M40 or 070M55 hardened and tempered or case hardened mild steel
Over 30 teeth	Cast iron	Mild steel	080M40 or 070M55 hardened and tempered or case hardened mild steel

Adapted from Renold (1996).

Standard sprockets can be purchased. The choice of sprocket material depends on the number of teeth and the operating conditions, as shown in Table 6.12.

Example A chain drive is required for a gear pump operating at 400 rpm, driven by a 5.5 kW electric motor running at 1440 rpm. The centre distance between the motor and pump shafts is approximately 470 mm.

Solution The desired reduction ratio is

$$\frac{1440}{400} = 3.6$$

The nearest ratio available (*see* Table 6.9) using standard-sized sprockets is 3.62. This requires a driving sprocket of 21 teeth and a driven sprocket of 76 teeth.

The application factor from Table 6.10 is $f_1 = 1.0$.

The tooth factor is

$$f_2 = \frac{19}{N_1} = \frac{19}{21} = 0.905$$

The selection power is

$$5.5 \times 1.0 \times 0.905 = 4.98\,\text{kW}$$

Using the BS/ISO selection chart (Fig. 6.15), a 12.7 mm pitch simple BS chain drive is suitable.

The chain length is given by

$$L = \frac{N_1 + N_2}{2} + \frac{2C}{p} + \left(\frac{N_2 - N_1}{2\pi}\right)^2 \frac{p}{C}$$

$$= \frac{21 + 76}{2} + \frac{2 \times 470}{12.7} + \left(\frac{76 - 21}{2\pi}\right)^2 \frac{12.7}{470}$$

$$= 124.6\,\text{pitches}$$

Rounding up to the nearest even integer gives $L = 126$ pitches.

The exact centre distance can now be calculated using

$$C = \frac{p}{8}\left[2L - N_2 - N_1 + \sqrt{(2L - N_2 - N_1)^2 - \frac{\pi}{3.88}(N_2 - N_1)^2}\right]$$

$$C = \frac{12.7}{8}\left[2 \times 126 - 76 - 21 + \sqrt{(2 \times 126 - 76 - 21)^2 - \frac{\pi}{3.88}(76 - 21)^2}\right]$$

$$= 479.2\,\text{mm}$$

From the rating chart (Fig. 6.15), the required lubrication type is oil bath. Use of an SAE20/50 multigrade lubricant would likely suffice in the absence of more detailed knowledge concerning the operating conditions.

Learning objectives checklist

Can you select a particular wedge belt for a given speed and power requirement? ☐

Can you determine the width of flat belt for a given speed and power requirement? ☐

Can you select an appropriate chain drive for a given speed and power requirement? ☐

Can you differentiate whether to use a chain, belt or gear drive for a given application? ☐

References and sources of information

Books and papers

Fenner Power Transmission UK. *Belt drives design manual.*

Hamilton, P. 1994: Belt drives. In Hurst, K. (ed.), *Rotary power transmission design*. McGraw Hill.

Renold Power Transmission Ltd 1996: *The designer guide*, 2nd edition.

Standards

BS 5801: 1979. Specification for flat top chains and associated chain wheels for conveyors. BSI.

BS 2969: 1980. Specification for high tensile steel chains (round link) for chain conveyors and coal ploughs. BSI.

BS AU 150b: 1984. Specification for automotive V belts and pulleys. BSI.
BS 2947: 1985. Specification for steel roller chains, types S and C, attachments and chain wheels for agricultural and similar machinery. BSI.
BS AU 218: 1987. Specification for automotive synchronous belt drives. BSI.
BS 4548: 1987. Specification for synchronous belt drives for industrial applications. BSI.
BS 4116: Part 1: 1992. Conveyor chains, their attachments and associated chain wheels. Specification for chains (metric series). BSI.
BS 7615: 1992. Specification for motor cycle chains. BSI.
BS 5594: 1993. Specification for leaf chains, clevises and sheaves. BSI.
BS 228: 1994. Specification for short pitch transmission precision roller chains and chain wheels. BSI.
BS 3790: 1995. Specification for endless wedge belt drives and endless V belt drives. BSI.
BS ISO 10823: 1996. Guidance on the selection of roller chain drives. BSI.
ISO 4348 (1978). Flat top chains and associated chain wheels for conveyors. ISO.
ISO 610 (1980). High tensile steel chains for chain conveyors and coal ploughs. ISO.
ISO 487 (1984). Steel roller chains, attachments and chain wheels for agricultural and similar machinery. ISO.
ISO 1275 (1984). Extended pitch precision roller chains and chain wheels. ISO.
ISO 5294 (1989) Synchronous belt drives – Pulleys. ISO.
ISO 5296-1 (1989) Sychronous belt drives – Belts – Part 1: Pitch codes MXL, XL, L, H, XH and XXH – Metric and inch dimensions. ISO.
ISO 155 (1989) Belt drives – Pulleys – Limiting values for adjustment of centres. ISO.
ISO 22 (1991) Belt drives – Flat transmission belts and corresponding pulleys – Dimensions and tolerances. ISO.
ISO 4184 (1992). Belt drives – Classical and narrow V belts – Lengths in datum system. ISO.
ISO 9633 (1992). Cycle chains – characteristics and test methods. ISO.
ISO 606 (1994). Specification for short pitch transmission precision roller chains and chain wheels. ISO.
ISO 4183 (1995). Belt drives – Classical and narrow V belts – Grooved pulleys (system based on datum width). ISO.
ISO 5292 (1995). Belt drives – V belts and V ribbed belts – Calculation of power ratings. ISO.

Web sites

http://www.fennerdrives.com/precision/
http://www.industry.net/siegling.america
http://www.powertransmission.com/index.htm

Belt and chain drives worksheet

1. A four-cylinder diesel engine running at 2000 rpm, developing 45 kW, is being used to drive a medium duty agricultural machine running at 890 rpm. The distance between the pulley centres is approximately 80 cm. The expected use is less than 10 hours per day. Using a wedge belt drive, select suitable pulley diameters and determine the type and number of belts required.
2. A wedge belt drive is required to transmit 18.5 kW from an electric motor running at 1455 rpm, to a uniformly loaded conveyor running at 400 rpm. The desired centre distance is 1.4 m and expected use is 15 hours a day. Select a suitable belt, or belts, and determine the pulley diameters.
3. A wedge belt drive is required for an electric motor driven crusher. The electric motor speed and power are 720 rpm and 11 kW respectively. The desired crusher speed is 140 rpm. Limitations on space within the crushing machine restrict the use of the maximum pulley diameter to 90 cm and the pulley centre distance to less than 1 m. The device will be operated for approximately 12 hours a day. Select a suitable belt drive and pulley diameters for this application.
4. A flat belt is required to transmit 22 kW from a 250 mm diameter pulley running at 1450 rpm, to a 355 mm diameter pulley. The coefficient of friction can be taken as 0.7, the density of the belt is 1100 kg/m^3 and the maximum permissible stress is 7 MPa. The distance between the shaft centres is 1.8 m. The proposed belt is 3.5 mm thick. Calculate the width required.
5. A flat belt drive is required for a surface grinder to transmit 5 kW from a 100 mm diameter pulley running at 1500 rpm, to a 250 mm diameter pulley. The coefficient of friction can be taken as 0.75, the density of the belt as 1100 kg/m^3 and the maximum permissible stress as 9 MPa. The distance between the shaft centres is 0.6 m. The proposed belt is 3.5 mm thick. Determine the belt width required.
6. A flat belt drive is required for a piston pump. The pump is driven by a 37 kW electric motor, running at 1470 rpm, with a pulley of 250 mm diameter. The pump pulley is 700 mm in diameter. The coefficient of friction can be taken as 0.8, the density of the belt as 1100 kg/m^3 and the maximum permissible stress as 6 MPa. The distance between the shaft centres is 1.2 m. The proposed belt is 2.9 mm thick. Determine the belt width required.
7. The application of an existing machine has changed such that the desired transmission power has increased to 120 kW. The existing

design consists of a driving pulley running at 3000 rpm, with a pulley diameter of 200 mm and the driven pulley of 250 m diameter. The belt is 150 mm wide, 4.2 mm thick and has a maximum permissible stress of 6.6 MPa. The belt density and coefficient of friction are 1100 kg/m^3 and 0.75 respectively. The centre distance is 1.4 m. Is the belt drive suitable?

8. An extractor fan is belt driven by a 75 kW electric motor running at 2946 rpm. The pulley diameters on the fan drive and the motor drive are 300 and 200 mm respectively. The centre distance is approximately 1.4 m. Using a flat belt with a coefficient of friction of 0.8, density 1100 kg/m^3 and maximum permissible stress of 6 MPa, 4.3 mm thick, determine the belt width required.
9. Specify a suitable chain drive for a gear pump operating at 400 rpm, driven by a 18.5 kW electric motor running at 725 rpm. The centre distance between the motor and pump shafts is approximately 470 mm.
10. Specify a suitable chain drive for a packaging machine operating at 75 rpm, driven by a 2.2 kW electric motor running at 710 rpm. The maximum permissible centre distance between the motor and pump shafts is approximately 1 m.
11. Specify a suitable chain drive for a gear pump operating at 400 rpm, driven by a 30 kW electric motor running at 728 rpm. The centre distance between the motor and pump shafts is approximately 1 m.
12. Specify a suitable chain drive for a gear pump operating at 400 rpm, driven by a 3 kW electric motor running at 2820 rpm. The centre distance between the motor and pump shafts is approximately 300 mm.
13. An agricultural application which runs for up to 16 hours per day requires a drive system to step down an 18.5 kW electric motor running at 725 rpm to 30 rpm. Determine a suitable drive system.

Answers

Worked solutions to the above problems are given in Appendix B.

1. $D_1 = 280$ mm, $D_2 = 630$ mm, $C = 662$ mm, SPB2800, 3 belts.
2. $D_1 = 224$ mm, $D_2 = 800$ mm, $C = 1.416$ m, SPB4500, 2 belts.
3. $D_1 = 160$ mm, $D_2 = 800$ mm, $C = 968$ mm, SPB3550, 3 belts.
4. $w = 57$ mm.
5. $w = 23$ mm.
6. $w = 133$ mm.
7. Power capacity = 98.5 kW. No.
8. 130 mm.
9. $N_1 = 21$, $N_2 = 38$, simple chain, $p = 25.4$ mm, $C = 484.1$ mm.
10. No unique solution.
11. $N_1 = 21$, $N_2 = 38$, simple chain, $p = 25.4$ mm, $C = 1020$ mm.
12. $N_1 = 19$, $N_2 = 114$, simple chain, $p = 9.525$ mm, $C = 306.7$ mm.
13. No unique solution. The application probably calls for a double reduction unit consisting of belt drive for the first reduction and a chain drive for the second reduction.

7 Clutches and brakes

A clutch is a device which permits the smooth, gradual connection of two shafts rotating at different speeds. A brake enables the controlled dissipation of energy to slow down, stop or control the speed of a system. This chapter describes the basic principles of frictional clutches and brakes and outlines design and selection procedures for disc clutches and disc and drum brakes.

7.1 Introduction

When a rotating machine is started it must be accelerated from rest to the desired speed. A clutch is a device used to connect or disconnect a driven component from a prime mover such as an engine or motor. A familiar application is the use of a clutch between the engine's crankshaft and the gearbox in automotive applications. The need for the clutch arises from the relatively high torque requirement to get a vehicle moving and the low torque output from an internal combustion engine at low levels of rotational speed. The disconnection of the engine from the drive enables the engine to speed up unloaded to about 1000 rpm where it is generating sufficient torque to drive the transmission. The clutch can then be engaged, allowing power to be transmitted to the gearbox, transmission shafts and wheels. A brake is a device used to reduce or control the speed of a system or bring it to rest. Typical applications of a clutch and brake are illustrated in Fig. 7.1. Clutches and brakes are similar devices providing frictional, magnetic or mechanical connection between two components. If one component rotates and the other is fixed to a non-rotating plane of reference, the device will function as a brake and if both rotate then as a clutch.

Whenever the speed or direction of motion of a body is changed there is force exerted on the body. If the body is rotating, a torque must be applied to the system to speed it up or slow it down. If the speed changes, so does the energy, either by addition or absorption. The acceleration, α, of a rotating machine is given by $\alpha = T/I$, where T is the torque and I is the mass moment of inertia. The mass moment of inertia can often be approximated by considering an assembly to be made up of a series of cylinders ($I_{\text{cylinder}} = \frac{1}{2}\rho\pi L(r_o^4 - r_i^4)$) and discs ($I_{\text{disc}} = \frac{1}{2}\rho\pi L r_o^4$) and summing the individual values for the disc and cylinder mass moments of inertia. The desired level of acceleration will depend on the application and once the acceleration has been set and I determined, the approximate value of torque required to accelerate or brake the load can be estimated. This can be used as the principal starting point for the design or selection of the clutch or brake geometry.

Torque is equal to the ratio of power and angular velocity. In other words, torque is inversely proportional to angular velocity. This implies that it is usually advisable to locate the clutch or brake on the highest speed shaft in the system so that the required torque is a minimum. Size, cost and response time are all lower when the torque is lower. The disadvantage is that the speed differential between the components can

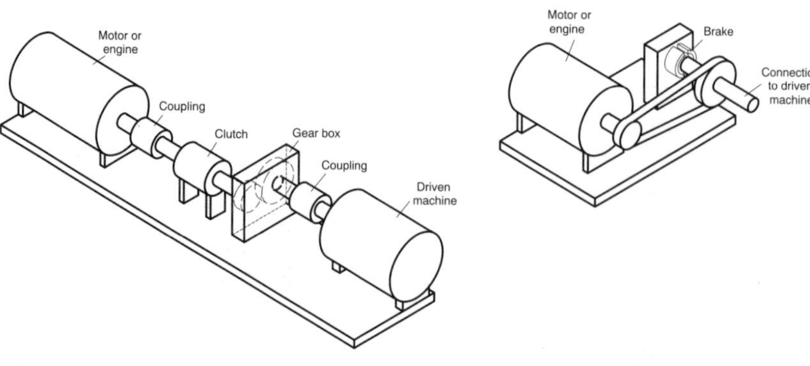

Fig. 7.1 Typical applications of clutches and brakes.

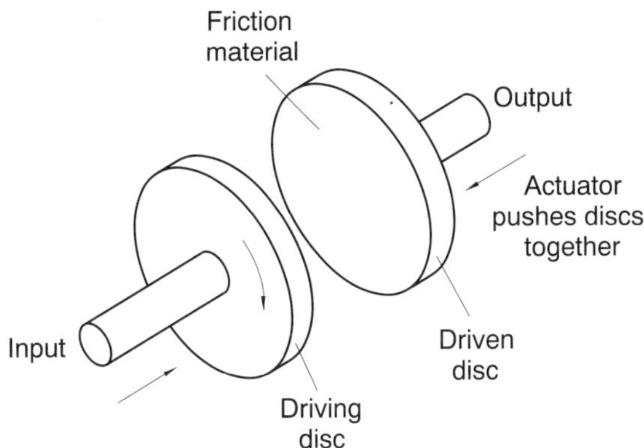

Fig. 7.2 Idealised friction disc clutch or brake.

result in increased slipping and associated frictional heating, potentially causing overheating problems.

Friction type clutches and brakes are the most common. Two or more surfaces are pushed together with a normal force to generate a friction torque (*see* Fig. 7.2). Normally, at least one of the surfaces is metal and the other a high-friction material referred to as the lining. The frictional contact can occur radially, as on a cylindrical arrangement, or axially as in a disc arrangement. The function of a frictional clutch or brake surface material is to develop a substantial friction force when a normal force is applied. Ideally, a material with a high coefficient of friction, constant properties, good resistance to wear and chemical compatibility is required. Clutches and brakes transfer or dissipate significant quantities of energy and their design must enable the absorption and transfer of this heat without damage to the component parts of the surroundings.

With the exception of high-volume automotive clutches and brakes, engineers rarely need to design a clutch or a brake from scratch. Clutch and brake assemblies can be purchased from specialist suppliers (listed in business directories (Section 9.5)) and the engineer's task is to define the torque and speed requirements, the loading characteristics and the system inertias, and to select an appropriately sized clutch or brake and the lining materials.

7.2 Clutches

The function of a clutch is to permit the connection and disconnection of two shafts, either when both are stationary or when there is a difference in the relative rotational speeds of the shafts. Clutch connection can be achieved by a number of techniques from direct mechanical friction, electromagnetic coupling, hydraulic or pneumatic means, or by some combination. There are various types of clutches, as outlined in Fig. 7.3. The devices considered here are of the friction type. Clutches must be designed principally to satisfy four requirements:

1. The necessary actuation force should not be excessive.
2. The coefficient of friction should be constant.
3. The energy converted to heat must be dissipated.
4. Wear must be limited to provide reasonable clutch life.

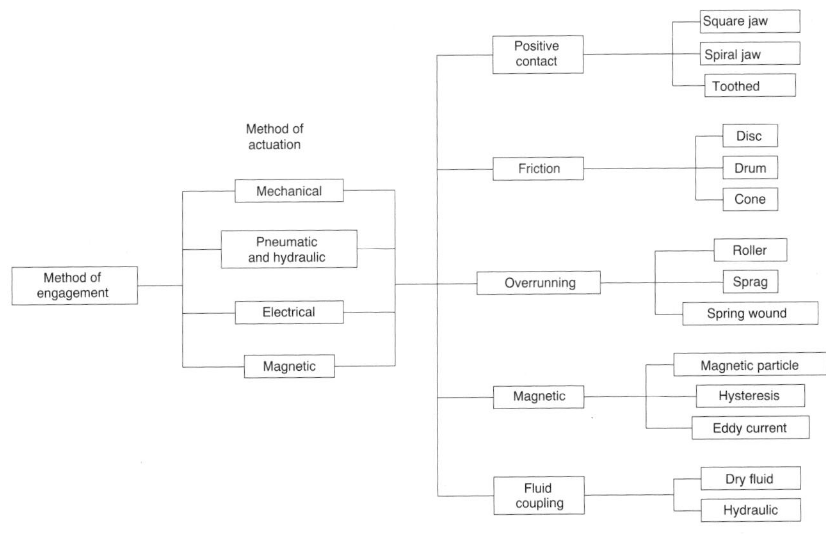

Fig. 7.3 Clutch classification. (Reproduced with adaptations from Hindhede *et al.*, 1983.)

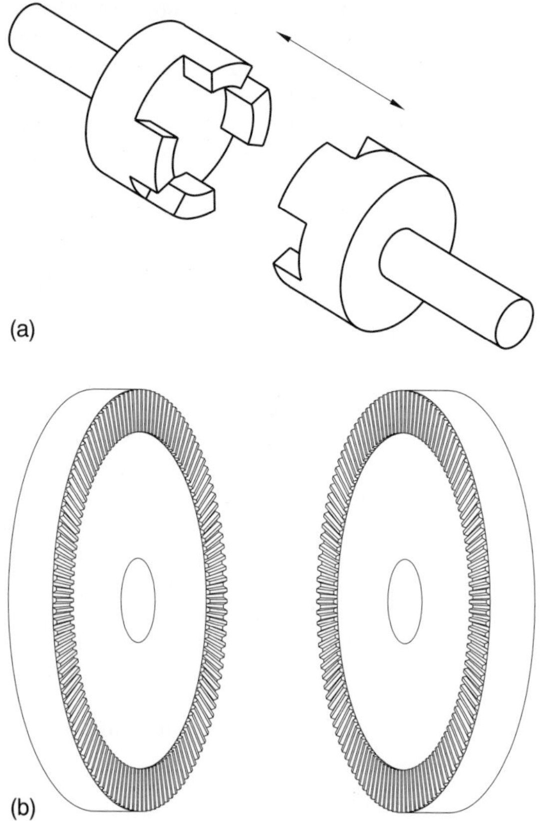

Fig. 7.4 Positive contact clutches: (a) square jaw; (b) multiple serration.

Alternatively, the objective in clutch design can be stated as maximisation of a maintainable friction coefficient and minimisation of wear. Correct clutch design and selection is critical. A clutch that is too small for an application will slip and overheat. A clutch that is too large will have a high inertia and may overload the drive.

Positive contact clutches have teeth or serrations which provide mechanical interference between mating components. Figure 7.4 shows a square jaw and a multiple serration positive contact clutch. This principle can be combined with frictional surfaces as in an automotive synchromesh clutch. As helical gears cannot be shifted in and out of mesh easily, the pragmatic approach is to keep the gears engaged in mesh and allow the gear to rotate freely on the shaft when no power is required, but provide positive location of the gear on the shaft when necessary.

Friction clutches consist of two surfaces, or two sets of surfaces, which can be forced into frictional contact. A frictional clutch allows gradual engagement between two shafts. The frictional surfaces can be forced together by springs, hydraulic pistons or magnetically. Various forms exist such as disc, cone and radial clutches. The design of the primary geometry for disc clutches is described in Section 7.2.1.

Overrunning clutches operate automatically based on the relative velocity of the mating components. They allow relative motion in one direction

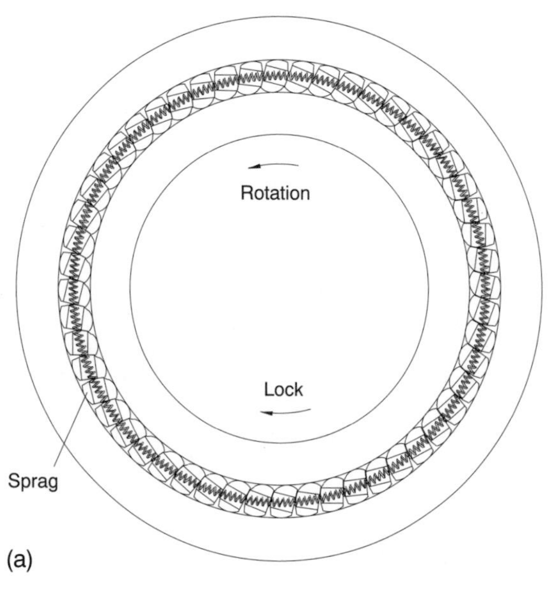

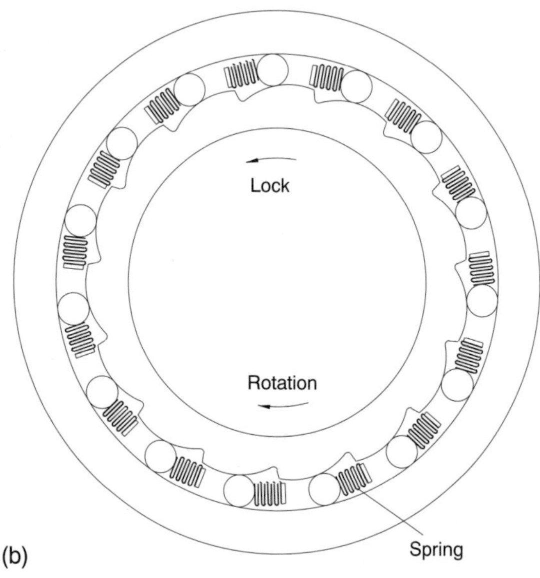

Fig. 7.5 Overrunning clutches: (a) sprag; (b) roller.

only. If the rotation attempts to reverse, the constituent components of the clutch grab the shaft and lock up. Applications include back stops, indexing and freewheeling, as on a bicycle when the wheel speed is greater than the drive sprocket. The range of overrunning clutches includes sprag, roller and spring wound clutches. Figure 7.5(a) illustrates the principal components of a sprag clutch which consist of an inner and outer race, like a bearing, but instead of balls the space between the races contains specially shaped elements called 'sprags' which permit rotation in one direction only. If the clutch motion is reversed, the sprags jam and lock the clutch. Another form of one-way clutch is the roller clutch (Fig. 7.5(b)) which consists of balls or rollers located in wedge-shaped chambers between inner and outer races. Once again rotation is possible in one direction, but when this is reversed the rollers wedge themselves between the races, jamming the inner and outer races together.

Magnetic clutches use a magnetic field to couple the rotating components together. Magnetic clutches are smooth, quiet and have a long life, as there is no direct mechanical contact and hence no wear except at the bearings.

Fluid couplings transmit torques through a fluid such as oil. A fluid coupling always transmits some torque because of the swirling nature of the fluid contained within the device. For this reason it is necessary to maintain some braking force in automatic transmission cars, which utilise fluid coupling torque converters in the automatic gearbox.

Centrifugal clutches engage automatically when the shaft speed exceeds some critical value. Friction elements are forced radially outwards and engage against the inner radius of a mating cylindrical drum. Common applications of centrifugal clutches include chain saws, overload releases and go-karts.

The selection of clutch type and configuration depends on the application. Table 7.1 can be used as the first step in determining the type of clutch to be used. Clutches are rarely designed from scratch. Either an existing design is available and is being modified for a new application or a clutch can be bought in from a specialist manufacturer. In the latter case the type, size and the materials for the clutch lining must be specified. This requires determination of the system characteristics such as speed, torque, loading characteristic (e.g. shock loads) and operating temperatures. Many of these factors have been lumped into a multiplier called a service factor. A lining material is typically tested under steady conditions using an electric motor drive. The torque capacity obtained from this test is then de-rated by the service factor according to the particular application, to take account of vibrations and loading conditions. Table 7.2 gives an indication of the typical values for service factors.

7.2.1 Design of disc clutches

Disc clutches can consist of single or multiple discs, as illustrated in Figs 7.6 and 7.7. Generally multiple disc clutches enable greater torque capacity but are

Table 7.1 Clutch selection criteria

Type of clutch	Characteristics	Typical applications
Sprag	One-way clutch. Profiled elements jam against the outer edge to provide drive. High torque capacity	One-way operation, e.g. backstop for hoists
Roller	One-way clutch. Rollers ride up ramps and drive by wedging into place	One-way operation
Cone clutch	Embodies the mechanical principle of the wedge which reduces the axial force required to transmit a given torque	Contractor's plant, feed drives for machine tools
Single disc clutch	Used when diameter is not restricted. Simple construction	Automobile drives
Multiple disc clutch	The power transmitted can be increased by using more plates, allowing a reduction in diameter	Machine tool head stocks, motorcycles
Centrifugal clutch	Automatic engagement at a critical speed	Electric motor drives, industrial diesel drives
Magnetic	Compact. Low wear	Machine tool gearboxes, numerical control machine tools

Reproduced with adaptations from Neale (1994).

Table 7.2 Service factors

Description of general system	Typical driven system	Type of driver: Small electric motors, turbines	IC engines (4–6 cylinders), medium to large electric motors	IC engines (2–3 cylinders)	Single-cylinder engines
Steady power source, steady load, no shock or overload	Belt drive, small generators, centrifugal pumps, fans, machine tools	1.5	1.7	1.9	2.2
Steady power source with some irregularity of load up to 1.5 times nominal power	Light machinery for wood, metal and textiles, conveyor belts	1.8	2.0	2.4	2.7
	Larger conveyor belts, larger machines, reciprocating pumps	2.0	2.2	2.4	2.7
Frequent start–stops, overloads, cycling, high inertia starts, high power, pulsating power source	Presses, punches, piston pumps, cranes, hoists	2.5	2.7	2.9	3.2
	Stone crushers, roll mills, heavy mixers, single cylinder compressors	3.0	3.2	3.4	3.7

Source: Vedamuttu (1994).

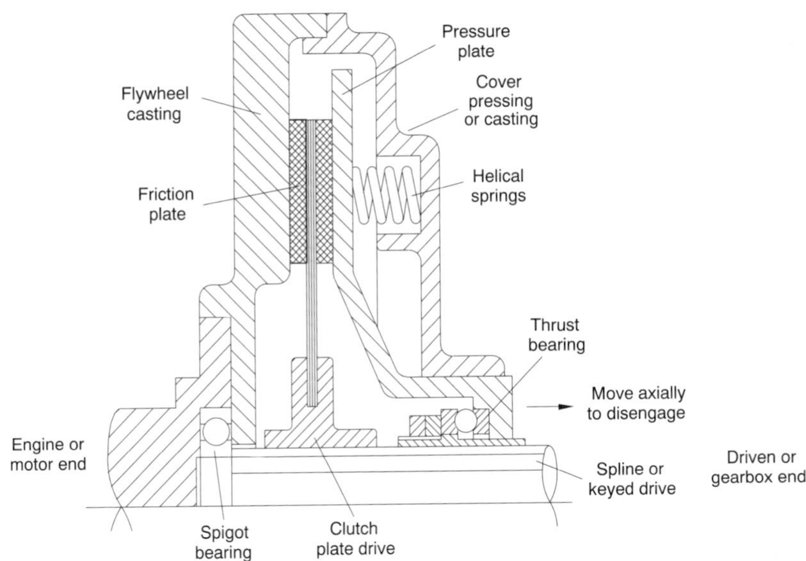

Fig. 7.6 Automotive disc clutch. (Reproduced with adaptations from Neale, 1994.)

harder to cool. Frictional clutches can be run dry or wet using oil. Typical coefficients of friction are 0.07 for a wet clutch and 0.45 for a dry clutch. While running a clutch wet in oil reduces the coefficient of friction it enhances heat transfer and the potential for cooling of the components. The expedient solution to the reduction of the friction coefficient is to use more discs, hence the use of multiple disc clutches.

Two basic assumptions are used in the development of procedures for disc clutch design, based upon a uniform rate of wear at the mating surfaces or a uniform pressure distribution between the mating surfaces. The equations for both of these methods are outlined in this section.

The assumption of a uniform pressure distribution at the interface between mating surfaces is valid for an unworn accurately manufactured clutch with rigid outer discs.

The area of an elemental annular ring on a disc clutch is $\delta A = 2\pi r \delta r$. Now $F = pA$, where p is the assumed uniform interface pressure, so

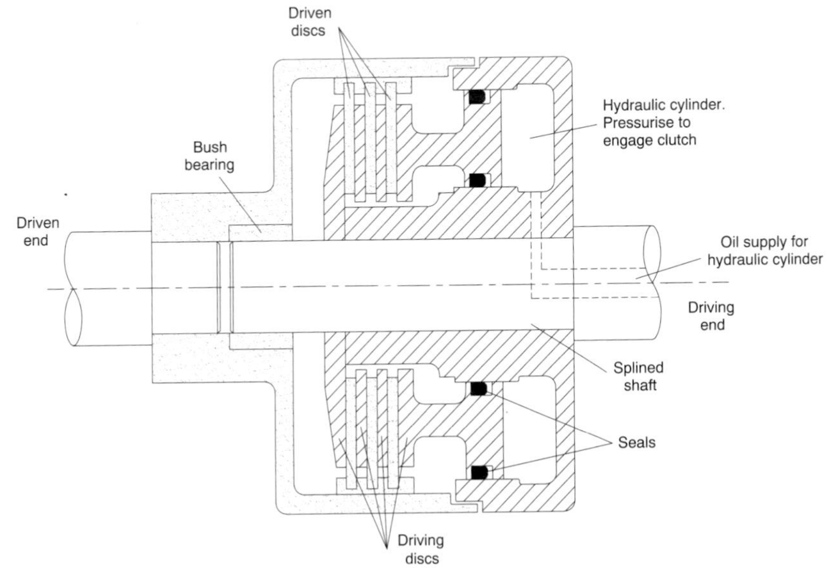

Fig. 7.7 Multiple disc clutch. (Reproduced with adaptations from Juvinall and Marshek, 1991.)

Table 7.3 Typical values for dynamic friction coefficients, permissible contact pressures and temperature limits

Material	μ_{dry}	μ_{oil}	p_{max} (MN/m^2)	T (°C)
Moulded compounds	0.25–0.45	0.06–0.10	1.035–2.07	200–260
Woven materials	0.25–0.45	0.08–0.10	0.345–0.69	200–260
Sintered metal	0.15–0.45	0.05–0.08	1.035–2.07	230–680
Cork	0.30–0.50	0.15–0.25	0.055–0.1	80
Wood	0.20–0.45	0.12–0.16	0.345–0.62	90
Cast iron	0.15–0.25	0.03–0.06	0.69–1.725	260
Paper based	–	0.10–0.17		
Graphite/resin	–	0.10–0.14		

$\delta F = 2\pi r p \delta r$. For the disc, the normal force acting on the entire face is

$$F = \int_{r_i}^{r_o} 2\pi r p dr = 2\pi p \left[\frac{r^2}{2}\right]_{r_i}^{r_o} \tag{7.1}$$

$$F = \pi p (r_o^2 - r_i^2) \tag{7.2}$$

Note that F is also the necessary force required to clamp the clutch discs together.

The friction torque, δT, that can be developed on an elemental ring is the product of the elemental normal force, given by $\mu \delta F$, and the radius:

$$\delta T = r\mu\delta F = 2\mu\pi r^2 p \delta r \tag{7.3}$$

where μ is the coefficient of friction which models the less than ideal frictional contact which occurs between two surfaces.

The total torque is given by integration of equation 7.3 between the limits of the annular ring, r_i and r_o:

$$T = \int_{r_i}^{r_o} 2\mu\pi r^2 p dr$$

$$= 2\pi p\mu \left[\frac{r^3}{3}\right]_{r_i}^{r_o} = \frac{2}{3}\pi p\mu (r_o^3 - r_i^3) \tag{7.4}$$

This equation represents the torque capacity of a clutch with a single frictional interface. In practice, clutches use an even number of frictional surfaces, as illustrated in Figs 7.6 and 7.7.

For a clutch with N faces, the torque capacity is given by

$$T = \tfrac{2}{3}\pi p\mu N (r_o^3 - r_i^3) \tag{7.5}$$

Using equations 7.2 and 7.5 and substituting for the pressure, p, gives an equation for the torque capacity as a function of the axial clamping force:

$$T = \frac{2}{3}\mu F N \left(\frac{r_o^3 - r_i^3}{r_o^2 - r_i^2}\right) \tag{7.6}$$

The equations assuming uniform wear are developed below. The wear rate is assumed to be proportional to the product of the pressure and velocity. So

$$pr\omega = \text{constant}$$

where ω is the angular velocity (rad/s).

For a constant angular velocity the maximum pressure will occur at the smallest radius:

$$p_{max} r_i \omega = \text{constant}$$

Eliminating the angular velocity and constant from the above two equations gives a relationship for the pressure as a function of the radius:

$$p = p_{max} \frac{r_i}{r}$$

The value for the maximum permissible pressure is dependent on the clutch lining material; typical values are listed in Table 7.3.

The elemental axial force on an elemental annular ring is given by

$$\delta F = 2\pi p r \delta r \tag{7.7}$$

integrating to give the total axial force:

$$F = \int_{r_i}^{r_o} 2\pi p r dr$$

$$= \int_{r_i}^{r_o} 2\pi p_{max} \frac{r_i}{r} r dr = 2\pi p_{max} r_i (r_o - r_i) \tag{7.8}$$

The elemental torque is given by

$$\delta T = \mu r \delta F \tag{7.9}$$

So

$$T = \int_{r_i}^{r_o} 2\mu\pi p_{max} r_i r dr = p_{max}\mu\pi r_i (r_o^2 - r_i^2) \tag{7.10}$$

Rearranging equation 7.8 gives

$$p_{max} = \frac{F}{2\pi r_i (r_o - r_i)} \tag{7.11}$$

Substituting equation 7.11 into equation 7.10 gives

$$T = \frac{\mu F}{2}\left(\frac{r_o^2 - r_i^2}{r_o - r_i}\right) = \frac{\mu F}{2}(r_o + r_i) \qquad (7.12)$$

For N frictional surfaces:

$$T = \int_{r_i}^{r_o} 2\mu\pi N p_{max} r_i r dr$$

$$= p_{max}\mu\pi N r_i (r_o^2 - r_i^2) \qquad (7.13)$$

which gives

$$T = \frac{\mu N F}{2}(r_o + r_i) \qquad (7.14)$$

By differentiating equation 7.10 with respect to r_i and equating the result to zero, the maximum torque for any outer radius r_o is found to occur when $r_i = \sqrt{1/3}r_o$. This useful formula can be used to set the inner radius if the outer radius is constrained to a particular value.

Note that equation 7.14 indicates a lower torque capacity than the uniform pressure assumption. This is because the higher initial wear at the outer diameter shifts the centre of pressure towards the inner radius.

Clutches are usually designed based on uniform wear. The uniform wear assumption gives a lower torque capacity clutch than the uniform pressure assumption. The preliminary design procedure for disc clutch design requires the determination of the torque and speed, specification of space limitations, selection of materials, i.e. the coefficient of friction and the maximum permissible pressure, and the selection of principal radii, r_o and r_i. Common practice is to set the value of r_i between $0.45r_o$ and $0.8r_o$. This procedure for determining the initial geometry is itemised below.

1. Determine the service factor.
2. Determine the required torque capacity, $T = \text{power}/\omega$.
3. Determine the coefficient of friction μ.
4. Determine the outer radius r_o.
5. Find the inner radius r_i.
6. Find the axial actuation force required.

The material used for clutch plates is typically grey cast iron or steel. The friction surface will consist of a lined material which may be moulded, woven, sintered or solid. Moulded linings consist of a polymeric resin used to bind powdered fibrous material and brass and zinc chips. Table 7.3 lists typical values for the performance of friction linings.

Example A clutch is required for transmission of power between a four-cylinder internal combustion engine and a small machine. Determine the radial dimensions for a single dry disc clutch with a moulded lining which should transmit 5 kW at 1800 rpm. Base the design on the uniform wear assumption.

Solution From Table 7.2 a service factor of 2 should be used. The design will therefore be undertaken using a power of $2 \times 5\,\text{kW} = 10\,\text{kW}$.

The torque is given by

$$T = \frac{\text{Power}}{\omega} = \frac{10\,000}{1800 \times (2\pi/60)} = 53\,\text{N}\cdot\text{m}$$

From Table 7.3, taking midrange values for the friction coefficient and the maximum permissible pressure for moulded linings gives $\mu = 0.35$ and $p_{max} = 1.55\,\text{MN/m}^2$.

Taking $r_i = \sqrt{1/3}r_o$ and substituting for r_i in equation 7.10 gives

$$T = \pi\mu\sqrt{1/3}r_o p_{max}[r_o^2 - (\sqrt{1/3}r_o)^2]$$

$$= \pi\mu\sqrt{1/3}p_{max}(r_o^3 - \tfrac{1}{3}r_o^3)$$

$$= \pi\mu\sqrt{4/27}p_{max}r_o^3$$

$$r_o = \left(\frac{T}{\pi\mu p_{max}\sqrt{4/27}}\right)^{1/3}$$

$$= \left(\frac{53.05}{\pi 0.35 \times 1.55 \times 10^6\sqrt{4/27}}\right)^{1/3}$$

$$= 0.04324\,\text{m}$$

$$r_i = \sqrt{1/3}r_o = 0.02497\,\text{m}$$

$$F = 2\pi r_i p_{max}(r_o - r_i)$$

$$= 2\pi \times 0.02497 \times 1.55 \times 10^6(0.04324 - 0.02497)$$

$$= 4443\,\text{N}$$

So the clutch consists of a disc of inner and outer radius 25 mm and 43 mm respectively, with a moulded lining having a coefficient of friction value of 0.35, a maximum permissible contact pressure of 1.55 MPa and an actuating force of 4.4 kN.

Example A multiple disc clutch, running in oil, is required for a motorcycle with a three-cylinder engine. The power demand is 75 kW at 8500 rpm. The preliminary design layout indicates that the maximum diameter of the clutch discs should not exceed 100 mm. In addition, previous designs have indicated that a moulded lining with a coefficient of friction of 0.068 in oil and a maximum permissible pressure of 1.2 MPa is reliable. Within these specifications determine the radii for the discs, the number of discs required and the clamping force.

Solution The torque is given by

$$T = \frac{\text{Service factor} \times \text{Power}}{\omega}$$

$$= \frac{3.4 \times 75\,000}{8500 \times (2\pi/60)} = 286.5\,\text{N}\cdot\text{m}$$

Select the outer radius to be the largest possible, i.e. $r_o = 50$ mm. Using $r_i = \sqrt{1/3}r_o$, $r_i = 28.87$ mm. From equation 7.13, the number of frictional surfaces, N, can be determined:

$$N = \frac{T}{\pi p_{max} r_i \mu (r_o^2 - r_i^2)}$$

$$= \frac{286.5}{\pi 1.2 \times 10^6 \times 0.02887 \times 0.068(0.05^2 - 0.02887^2)}$$

$$= 23.23$$

This must be an even number, so the number of frictional surfaces is taken as $N = 24$. This requires 13 driving discs and 12 driven discs to implement.

Using equation 7.14, the clamping force can be calculated:

$$F = \frac{2T}{\mu N (r_o + r_i)}$$

$$= \frac{2 \times 286.5}{0.068 \times 24(0.05 + 0.02887)} = 4452\,\text{N}$$

7.3 Brakes

The basic function of a brake is to absorb kinetic energy and dissipate it in the form of heat. An idea of the magnitude of energy that must be dissipated can be obtained from considering the familiar example of a car undergoing an emergency stop in 7 s from 60 mph (96 km/h). If the car's mass is 1400 kg, and assuming that 65 per cent of the car's weight is loaded on to the front axles during rapid braking, then the load on the front axle is

$$1400 \times 9.81 \times 0.65 = 8927\,\text{N}$$

This will be shared between two brakes, so the energy that must be absorbed by one brake is

$$E = \tfrac{1}{2} m (V_i^2 - V_f^2)$$

$$= \frac{1}{2} \times \left(\frac{8927}{9.81} \times 0.5\right) \times \left[\left(\frac{96 \times 10^3}{3600}\right)^2 - 0^2\right]$$

$$= 161.8\,\text{kJ}$$

If the car brakes uniformly in 7 s, then the heat that must be dissipated is $161.8 \times 10^3/7 = 23.1$ kW. From your experience of heat transfer from, say, domestic heaters, you will recognise that this is a significant quantity of heat to transfer away from the relatively compact components which make up brake assemblies.

Convective heat transfer can be modelled by Fourier's equation:

$$Q = hA\Delta T = hA(T_s - T_f) \qquad (7.15)$$

where

Q = heat (W)
h = heat transfer coefficient (W/m^2 per K)
T_s = temperature of the surface (K or °C)
T_f = temperature of the surrounding fluid (K or °C)
A = surface area (m^2).

This equation indicates that the ability of a brake to dissipate the heat generated increases as the surface area increases, or as the heat transfer coefficient rises. For air, the heat transfer coefficient is usually dependent on the local flow velocity and on the geometry. A method often used for disc brakes to increase both the surface area and the local flow is

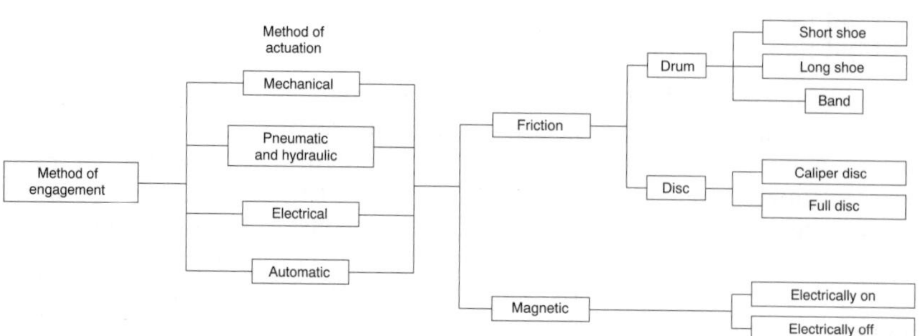

Fig. 7.8 Brake classification.

to machine multiple axial or radial holes in the disc (this also reduces the mass and inertia).

There are numerous brake types, as shown in Fig. 7.8. The selection and configuration of a brake depends on the requirements. Table 7.4 gives an indication of brake operation against various criteria. The brake factor listed in Table 7.4 is the ratio of frictional braking force generated to the actuating force applied.

Brakes can be designed so that once engaged the actuating force applied is assisted by the braking torque. This kind of brake is called a self-energising brake and is useful for braking large loads. Great care must be exercised in brake design. It is possible, and sometimes desirable, to design a brake which once engaged will grab and lock up (called self-locking action).

A critical aspect of all brakes are the materials used for frictional contact. Normally one component will comprise a steel or cast iron disc or drum; this is brought into frictional contact against a geometrically similar component with a brake lining made up of one of the materials listed in Table 7.3.

Section 7.3.1 gives details about the configuration design of disc brakes and Section 7.3.2 introduces the design of drum brakes.

7.3.1 Disc brakes

Disc brakes are familiar from automotive applications, where they are used extensively for car and motorcycle wheels. These typically consist of a cast iron disc, bolted to the wheel hub. This is sandwiched between two pads actuated by pistons, supported in a caliper mounted on the stub shaft (*see* Fig. 7.9). When the brake pedal is pressed, hydraulically pressurised fluid is forced into the cylinders, pushing the opposing pistons and brake pads into frictional contact with the disc. The advantages of this form of braking are steady braking, easy ventilation, balancing thrust loads and design simplicity. There is no self-energising action so the braking action is proportional to the applied force. The use of a discrete pad allows the disc to cool as it rotates, enabling heat transfer between the cooler disc and the hot brake pad. As the pads either side of the disc are pushed on to the disc with equal forces, the net thrust load on the disc cancels.

With reference to Fig. 7.10, the torque capacity per pad is given by

$$T = \mu F r_e \tag{7.16}$$

where r_e is an effective radius.

The actuating force assuming constant pressure is given by

$$F = p_{av}\theta\frac{r_o^2 - r_i^2}{2} \tag{7.17}$$

or, assuming uniform wear, by

$$F = p_{max}\theta r_i(r_o - r_i) \tag{7.18}$$

where θ (in radians) is the included angle of the pad, r_i is the inner radius of the pad and r_o is the outer radius of the pad.

The relationship between the average and the maximum pressure for the uniform wear assumption is given by

$$\frac{p_{av}}{p_{max}} = \frac{2r_i/r_o}{1 + (r_i/r_o)} \tag{7.19}$$

For an annular disc brake the effective radius is given by equation 7.20 assuming constant pressure, and equation 7.21 assuming uniform wear:

$$r_e = \frac{2(r_o^3 - r_i^3)}{3(r_o^2 - r_i^2)} \tag{7.20}$$

$$r_e = \frac{r_i + r_o}{2} \tag{7.21}$$

For circular pads the effective radius is given by $r_e = r\delta$, where values for δ are given in Table 7.5 as a function of the ratio of the pad radius and the radial location, R/r. The actuating force for circular pads can be calculated using

$$F = \pi R^2 p_{av} \tag{7.22}$$

Example A caliper brake is required for the front wheels of a sports car with a braking capacity of 820 N·m for each brake. Preliminary design estimates have set the brake geometry as $r_i = 100$ mm, $r_o = 160$ mm and $\theta = 45°$. A pad with a coefficient of friction of 0.35 has been selected. Determine the required actuating force and the average and maximum contact pressures.

Solution The torque capacity per pad = 820/2 = 410 N·m.

The effective radius is

$$r_e = \frac{0.1 + 0.16}{2} = 0.13\,\text{m}$$

The actuating force is given by

$$F = \frac{T}{\mu r_e} = \frac{410}{0.35 \times 0.13} = 9.011\,\text{kN}$$

Table 7.4 Comparative table of brake performance

Type of brake	Maximum operating temperature	Brake factor	Stability	Dryness	Dust and dirt	Typical applications
Differential band brake	Low	High	Low	Unstable but still effective	Good	Winches, hoists, excavators, tractors
External drum brake (leading trailing edge)	Low	Medium	Medium	Unstable if humid, poor if wet	Good	Mills, elevators, winders
Internal drum brake (leading trailing edge)	Higher than external brake	Medium	Medium	Unstable if humid, ineffective if wet	Good if sealed	Vehicles (rear axles on passenger cars)
Internal drum brake (two leading shoes)	Higher than external brake	High	Low	Unstable if humid, ineffective if wet	Good if sealed	Vehicles (rear axles on passenger cars)
Internal drum brake (duo-servo)	Low	High	Low	Unstable if humid, ineffective if wet	Good if sealed	Vehicles (rear axles on passenger cars)
Caliper disc brake	High	Low	High	Good	Poor	Vehicles and industrial machinery
Full disc brake	High	Low	High	Good	Poor	Machine tools and other industrial machinery

Reproduced from Neale (1994).

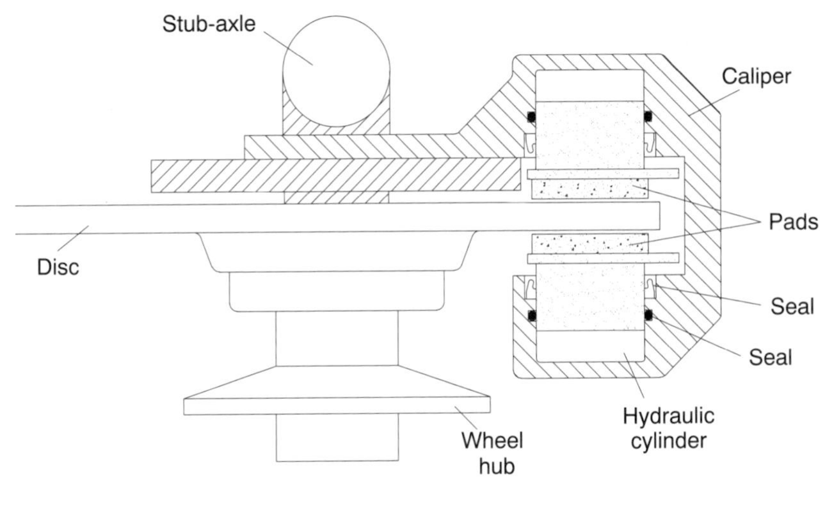

Fig. 7.9 Automotive disc brake. (Reproduced with adaptations from Heisler, 1985.)

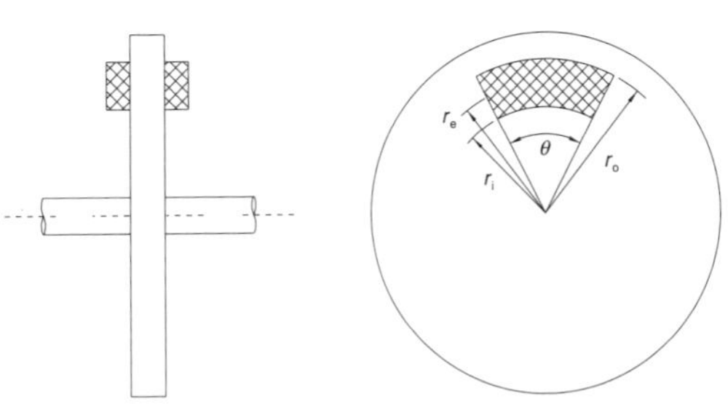

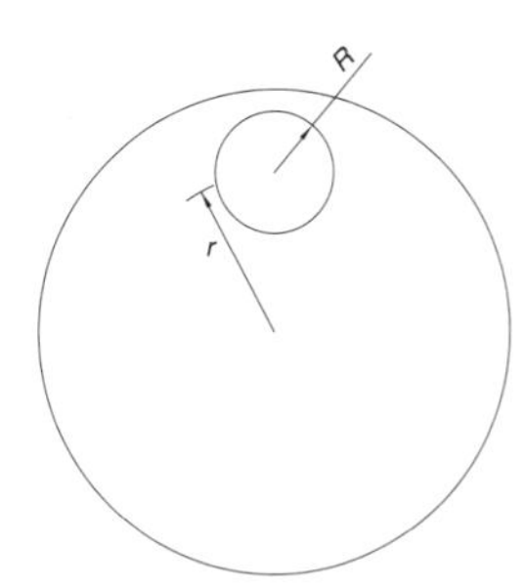

Fig. 7.10 Caliper disc brake.

The maximum contact pressure is given by

$$p_{\max} = \frac{F}{\theta r_i(r_o - r_i)}$$

$$= \frac{9.011 \times 10^3}{45 \times (2\pi/360) \times 0.1 \times (0.16 - 0.1)}$$

$$= 1.912\,\mathrm{MN/m^2}$$

The average pressure is given by

$$p_{av} = p_{\max} \frac{2r_i/r_o}{1 + (r_i/r_o)} = 1.471\,\mathrm{MN/m^2}$$

Table 7.5 Circular pad disc brake design values

R/r	$\delta = R_e/r$	$p_{\max}/p_{av}$
0	1.000	1.000
0.1	0.983	1.093
0.2	0.969	1.212
0.3	0.957	1.367
0.4	0.947	1.578
0.5	0.938	1.875

Source: Fazekas (1972).

Full disc brakes, consisting of a complete annular ring pad, are principally used for industrial machinery. The disc clutch equations developed in Section 7.2.1 are applicable to their design. The disc configuration can be designed to function as either a clutch or a brake (a clutch–brake combination) to transmit a load or control its speed.

7.3.2 Drum brakes

Drum brakes apply friction to the external or internal circumference of a cylinder. A drum brake consists of the brake shoe, which has the friction material bonded to it, and the brake drum. For braking, the shoe is forced against the drum, developing the friction torque. Drum brakes can be divided into two groups depending on whether the brake shoe is external or internal to the drum. A further classification can be made in terms of the length of the brake shoe: short, long or complete band.

Short-shoe internal brakes are used for centrifugal brakes which engage at a particular critical speed. Long-shoe internal drum brakes are used principally in automotive applications. Drum brakes (or clutches) can be designed to be self-energising. Once engaged the friction force increases the normal force non-linearly, increasing the friction torque, as in a positive feedback loop. This can be advantageous in braking large loads but makes control much more difficult. One problem associated with some drum brakes is stability. If the brake has been designed so that the braking torque is not sensitive to small changes in the coefficient of friction, which would occur if the brake is worn or wet, then the brake is said to be stable. If a small change in the coefficient of friction causes a significant change to the braking torque, the brake is unstable, and will tend to grab if the friction coefficient rises or the braking torque will drop noticeably if the friction coefficient reduces.

Short-shoe external drum brakes The schematic of a short-shoe external drum brake is given in Fig. 7.11. If the included angle of contact between the brake shoe and the brake drum is less than 45°, the force between the shoe and the drum is relatively uniform and can be modelled by a single concentrated load F_n at the centre of the contact area. If the maximum permissible pressure is p_{max}, the force F_n can be estimated by

$$F_n = p_{max} r \theta w \tag{7.23}$$

where

w = width of the brake shoe (m)
θ = angle of contact between the brake shoe and the lining (rad).

The frictional force, F_f, is given by

$$F_f = \mu F_n \tag{7.24}$$

where μ is the coefficient of friction.

The torque on the brake drum is

$$T = F_f r = \mu F_n r \tag{7.25}$$

Summing moments, for the shoe arm, about the pivot gives

$$\sum M_{pivot} = aF_a - bF_n + cF_f = 0$$
$$F_a = \frac{bF_n - cF_f}{a} = F_n \frac{b - \mu c}{a} \tag{7.26}$$

Resolving forces gives the reactions at the pivot:

$$R_x = -F_f \tag{7.27}$$
$$R_y = F_a - F_n \tag{7.28}$$

Note that for the configuration and direction of rotation shown in Fig. 7.11, the friction moment $\mu F_n c$ adds or combines with the actuating moment aF_a. Once the actuating force is applied the friction generated at the shoe acts to increase the braking torque. This kind of braking action is called self-energising. If the brake direction is reversed, the friction moment term $\mu F_n c$ becomes negative and the applied load F_a must be maintained to generate braking torque. This combination is called self de-energising. From equation 7.26, note that if the brake is self-energising and if $\mu c > b$, then the force required to actuate the brake is zero or negative and the brake action is called self-locking. If the shoe touches the drum it

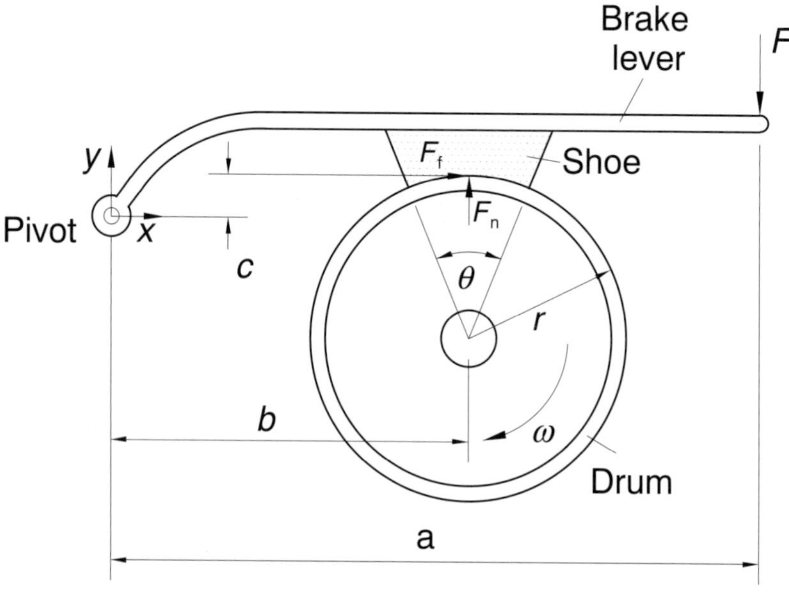

Fig. 7.11 Short-shoe external drum brake.

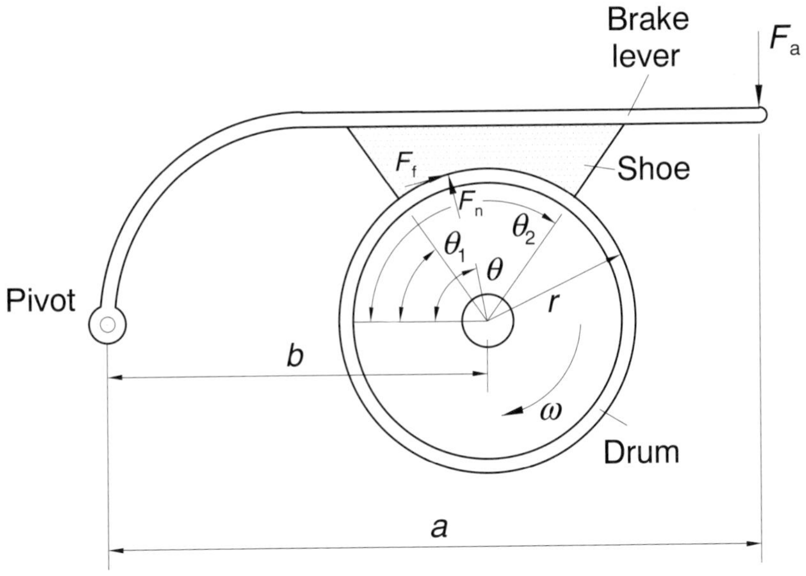

Fig. 7.12 Long-shoe external drum brake.

will grab and lock. This is usually undesirable, exceptions being hoist stops or overrunning clutch type applications.

Long-shoe external drum brakes If the included angle of contact between the brake shoe and the drum is greater than 45°, then the pressure between the shoe and the brake lining cannot be regarded as uniform, and the approximations made for the short-shoe brake analysis are inadequate. Most drum brakes use contact angles greater than 90°. Brake shoes are not rigid and the local deflection of the shoe affects the pressure distribution. Detailed shoe analysis is possible using finite element software. For initial synthesis/specification of brake geometry a simpler analysis suffices.

For a single block brake (*see* Fig. 7.12) the force exerted on the drum by the brake shoe must be supported by the bearings. To balance this load and provide a compact braking arrangement two opposing brake shoes are usually used in a caliper arrangement, as shown in Fig. 7.13.

The following equations can be used (with reference to Fig. 7.12) to determine the performance of a long-shoe brake.

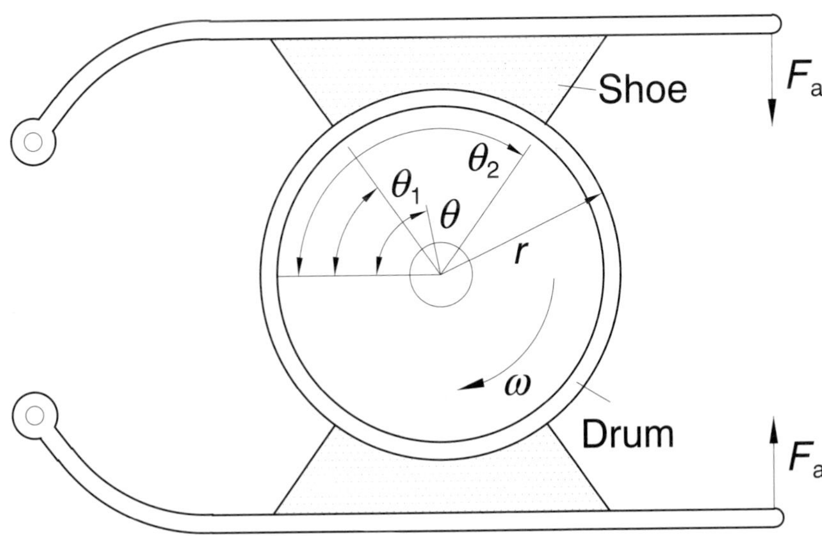

Fig. 7.13 Double long-shoe external drum brake.

The braking torque T is given by

$$T = \mu w r^2 \frac{p_{max}}{(\sin\theta)_{max}}(\cos\theta_1 - \cos\theta_2) \quad (7.29)$$

where

μ = coefficient of friction
w = width of the brake shoe (m)
r = radius of drum (m)
p_{max} = maximum allowable pressure for the lining material (N/m^2)
θ = angular location (rad)
$(\sin\theta)_{max}$ = maximum value of $\sin\theta$
θ_1 = centre angle from the shoe pivot to the heel of the lining (rad)
θ_2 = centre angle from the shoe pivot to the toe of the lining (rad).

This is based on the assumption that the local pressure p at an angular location θ is related to the maximum pressure, p_{max}, by

$$p = \frac{p_{max}\sin\theta}{(\sin\theta)_{max}} \quad (7.30)$$

The local pressure p will be a maximum when $\theta = 90°$. If $\theta_2 < 90°$ then the pressure will be a maximum at the toe, θ_2. The relationship given in equation 7.30 assumes that there is no deflection at the shoe or the drum, no wear on the drum, and that the shoe wear is proportional to the frictional work and hence the local pressure. Note that if $\theta = 0$ the pressure is zero. This indicates that frictional material near the pivot or heel of the brake does not contribute significantly to the braking action. For this reason, common practice is to leave out the frictional material near the heel and start it at an angle typically between $\theta_1 = 10°$ and $\theta_1 = 30°$.

With the direction of rotation shown in Fig. 7.12 (i.e. the brake is self-energising), the magnitude of the actuation force is given by

$$F_a = \frac{M_n - M_f}{a} \quad (7.31)$$

where

M_n = moment of the normal forces (N·m)
M_f = moment due to the frictional forces (N·m)
a = orthogonal distance between the brake pivot and the line of action of the applied force (m).

The normal and frictional moments can be determined using respectively:

$$M_n = \frac{wrbp_{max}}{(\sin\theta)_{max}}\left[\frac{1}{2}(\theta_2 - \theta_1) - \frac{1}{4}(\sin 2\theta_2 - \sin 2\theta_1)\right] \quad (7.32)$$

$$M_f = \frac{\mu w r p_{max}}{(\sin\theta)_{max}} \times \left[r(\cos\theta_1 - \cos\theta_2) + \frac{b}{4}(\cos 2\theta_2 - \cos 2\theta_1)\right] \quad (7.33)$$

If the geometry and materials are selected such that $M_f = M_n$, then the actuation force becomes zero. Such a brake would be self-locking. The slightest contact between the shoe and drum would bring the two surfaces into contact and the brake would snatch, giving rapid braking. Alternatively, values for the brake geometry and materials can be selected to give different levels of self-energisation, depending on the relative magnitudes of M_n and M_f.

If the direction of rotation for the drum shown in Fig. 7.11 is reversed, the brake becomes self-de-energising and the actuation force is given by

$$F_a = \frac{M_n + M_f}{a} \quad (7.34)$$

The pivot reactions can be determined by resolving the horizontal and vertical forces. For a self-energising brake they are given by

$$R_x = \frac{wrp_{max}}{(\sin\theta)_{max}}\left[[0.5(\sin^2\theta_2 - \sin^2\theta_1)] - \mu \times \left(\frac{\theta_2}{2} - \frac{\theta_1}{2} - \frac{1}{4}(\sin 2\theta_2 - \sin 2\theta_1)\right)\right] - F_x \quad (7.35)$$

$$R_y = \frac{wrp_{max}}{(\sin\theta)_{max}}\left[[0.5(\sin^2\theta_2 - \sin^2\theta_1)] + \mu \times \left(\frac{\theta_2}{2} - \frac{\theta_1}{2} - \frac{1}{4}(\sin 2\theta_2 - \sin 2\theta_1)\right)\right] + F_y \quad (7.36)$$

and for a self-de-energising brake by

$$R_x = \frac{wrp_{max}}{(\sin\theta)_{max}}\left[[0.5(\sin^2\theta_2 - \sin^2\theta_1)] + \mu \times \left(\frac{\theta_2}{2} - \frac{\theta_1}{2} - \frac{1}{4}(\sin 2\theta_2 - \sin 2\theta_1)\right)\right] - F_x \quad (7.37)$$

$$R_y = \frac{wrp_{max}}{(\sin\theta)_{max}}\left[[0.5(\sin^2\theta_2 - \sin^2\theta_1)] - \mu \times \left(\frac{\theta_2}{2} - \frac{\theta_1}{2} - \frac{1}{4}(\sin 2\theta_2 - \sin 2\theta_1)\right)\right] + F_y \quad (7.38)$$

Example Design a long-shoe drum brake to produce a friction torque of 75 N·m, to stop a drum rotating at 140 rpm. Initial design calculations have indicated that a shoe lining with $\mu = 0.25$, and using a value of $p_{max} = 0.5 \times 10^6$ N/m^2 in the design, will give suitable life.

Solution First propose trial values for the brake geometry, say $r = 0.1$ m, $b = 0.2$ m, $a = 0.3$ m, $\theta_1 = 30°, \theta_2 = 150°$. Using equation 7.29 and solving for the width of the shoe,

$$w = \frac{T(\sin\theta)_{max}}{\mu r^2 p_{max}(\cos\theta_1 - \cos\theta_2)}$$

$$= \frac{75 \sin 90}{0.25 \times 0.1^2 \times 0.5 \times 10^6(\cos 30 - \cos 150)}$$

$$= 0.0346 \text{ m}$$

Select the width to be 35 mm as this is a standard size. The actual maximum pressure experienced will be

$$p_{max} = 0.5 \times 10^6 \frac{0.0346}{0.035} = 494\,900 \text{ N/m}^2$$

From equation 7.32 the moment of the normal force with respect to the shoe pivot is

$$M_n = \frac{0.035 \times 0.1 \times 0.2 \times 0.4949 \times 10^6}{\sin 90} \times \left[\frac{1}{2}\left(120 \times \frac{2\pi}{360}\right) - \frac{1}{4}(\sin 300 - \sin 60)\right]$$

$$= 512.8 \text{ N·m}$$

From equation 7.33 the moment of the frictional forces with respect to the shoe pivot is

$$M_f = \frac{0.25 \times 0.035 \times 0.1 \times 0.4949 \times 10^6}{\sin 90} \times \left[0.1(\cos 30 - \cos 150) + \frac{0.2}{4}(\cos 300 - \cos 60)\right]$$

$$= 75 \text{ N·m}$$

From equation 7.31 the actuation force is

$$F_a = \frac{M_n - M_f}{a} = \frac{512.8 - 75}{0.3} = 1459 \text{ N}$$

For the double long-shoe external drum brake illustrated in Fig. 7.13, the top shoe is self-energising and the frictional moment reduces the actuation load. The bottom shoe, however, is self-de-energising and its frictional moment acts to reduce the maximum pressure which occurs on the bottom brake shoe. The normal and frictional moments for a self-energising and self-de-energising brake are related by

$$M'_n = \frac{M_n p'_{max}}{p_{max}} \tag{7.39}$$

$$M'_f = \frac{M_f p'_{max}}{p_{max}} \tag{7.40}$$

where

M'_n = moment of the normal forces for the self-de-energising brake (N·m)
M'_f = moment due to the frictional forces for the self-de-energising brake (N·m)
p'_{max} = the maximum pressure on the self-de-energising brake (N/m^2).

Example For the double long-shoe external drum brake illustrated in Fig. 7.14, determine the limiting force on the lever such that the maximum pressure on the brake lining does not exceed 1.4 MPa, and determine the torque capacity of the brake. The face width of the shoes is 30 mm and the coefficient of friction between the shoes and the drum can be taken as 0.28.

Solution First it is necessary to calculate values for θ_1 and θ_2 as these are not indicated directly on the diagram:

$$\theta_1 = 20° - \tan^{-1}\left(\frac{20}{120}\right) = 10.54°$$

$$\theta_2 = 20° + 130° - \tan^{-1}\left(\frac{20}{120}\right) = 140.5°$$

The maximum value of $\sin\theta$ would be $\sin 90 = 1$.

The distance between the pivot and the drum centre is

$$b = \sqrt{0.02^2 + 0.12^2} = 0.1217 \text{ m}$$

The normal moment is given by

$$M_n = \frac{wrbp_{max}}{(\sin\theta)_{max}}\left[\frac{1}{2}(\theta_2 - \theta_1) - \frac{1}{4}(\sin 2\theta_2 - \sin 2\theta_1)\right]$$

$$= \frac{0.03 \times 0.1 \times 0.1217 \times 1.4 \times 10^6}{\sin 90} \times \left[\frac{1}{2}\left((140.5 - 10.54) \times \frac{2\pi}{360}\right) - \frac{1}{4}(\sin 281 - \sin 21.08)\right]$$

$$= 751.1 \text{ N·m}$$

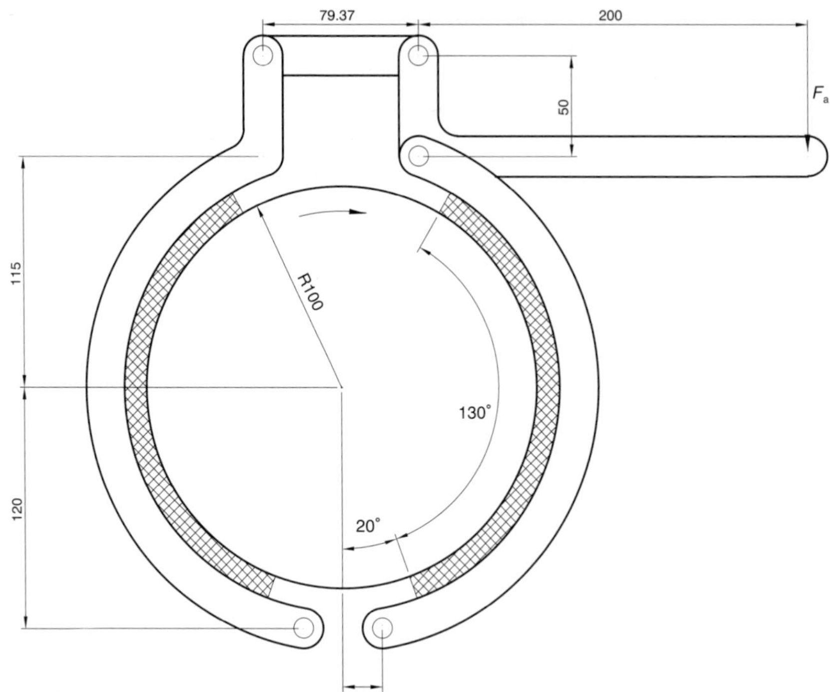

Fig. 7.14 Double long-shoe external drum brake.

$$M_f = \frac{\mu w r p_{max}}{(\sin\theta)_{max}} \times \left[r(\cos\theta_1 - \cos\theta_2) + \frac{b}{4}(\cos 2\theta_1 - \cos 2\theta_1) \right]$$

$$= \frac{0.28 \times 0.03 \times 0.1 \times 1.4 \times 10^6}{\sin 90} \times \left[0.1(\cos 10.54 - \cos 140.5) + \frac{0.1217}{4}(\cos 281 - \cos 21.08) \right]$$

$$= 179.8\,\text{N} \cdot \text{m}$$

The orthogonal distance between the actuation force and the pivot is $a = 0.12 + 0.115 + 0.05 = 0.285$ m.

The actuation load on the left hand shoe is given by

$$F_{a\text{ left shoe}} = \frac{M_n - M_f}{a} = \frac{751.1 - 179.8}{0.285} = 2004\,\text{N}$$

The torque contribution from the left hand shoe is given by

$$T_{\text{left shoe}} = \mu w r^2 \frac{p_{max}}{(\sin\theta)_{max}} (\cos\theta_1 - \cos\theta_2)$$

$$= 0.28 \times 0.03 \times 0.1^2 \times 1.4 \times 10^6 \times (\cos 10.54 - \cos 140.5) = 206.4\,\text{N} \cdot \text{m}$$

The actuation force on the right hand shoe can be determined by considering each member of the lever mechanism as a free body (*see* Fig. 7.15):

$$F - A_V + B_V = 0$$

$$A_H = B_H$$

$$B_H = C_H, \; A_H = C_H$$

$0.2F = 0.05B_H$, $F = B_H/4$. For $B_H = 2204$ N,

$$F = 2004/4 = 501\,\text{N}$$

So the limiting lever force is $F = 501$ N:

$$C_V = 0, B_V = 0$$

The actuating force for the right hand lever is the resultant of F and B_H. The resultant angle is given by $\tan^{-1}(0.05/0.2) = 14.04°$:

$$F_{a\text{ right shoe}} = \frac{2004}{\cos 14.04} = 2065\,\text{N}$$

The orthogonal distance between the actuation force vector and the pivot is given by

$$a = (0.235 - 0.01969 \tan 14.04) \times \cos 14.04 = 0.2232\,\text{m}$$

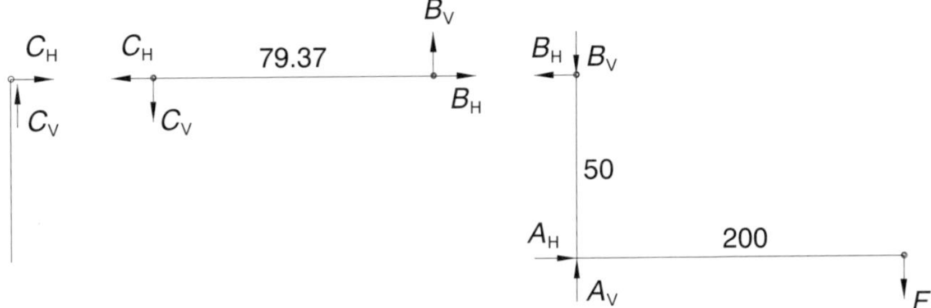

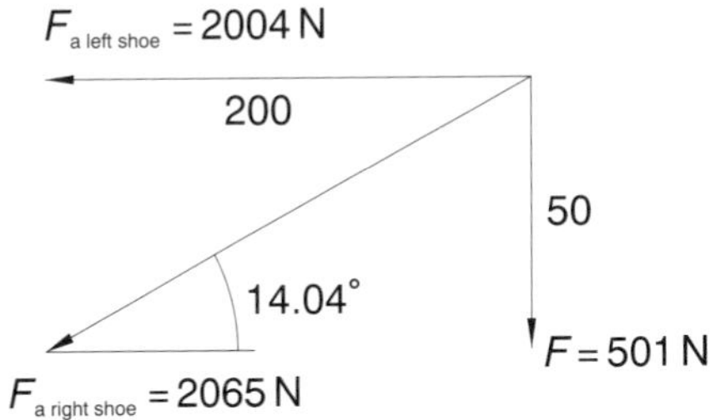

Fig. 7.15 Free body diagrams.

The normal and frictional moments for the right hand shoe can be determined using equations 7.39 and 7.40:

$$M'_n = \frac{M_n p'_{max}}{p_{max}} = \frac{751.1 p'_{max}}{1.4 \times 10^6}$$

$$M'_f = \frac{M_f p'_{max}}{p_{max}} = \frac{179.8 p'_{max}}{1.4 \times 10^6}$$

For the right hand shoe the maximum pressure can be determined from

$$F_{a\ right\ shoe} = \frac{M'_n + M'_f}{a} = 2065$$

$$= \frac{751.1 p'_{max} - 179.8 p'_{max}}{1.4 \times 10^6 \times 0.2232}$$

$$p'_{max} = 1.130 \times 10^6\ \mathrm{N/m^2}$$

The torque contribution from the right hand shoe is

$$T_{right\ shoe} = \mu w r^2 \frac{p'_{max}}{(\sin\theta)_{max}} (\cos\theta_1 - \cos\theta_2)$$

$$= 0.28 \times 0.03 \times 0.1^2 \times 1.13 \times 10^6$$

$$\times (\cos 10.54 - \cos 140.5)$$

$$= 166.6\ \mathrm{N \cdot m}$$

The total torque is given by

$$T_{total} = T_{left\ shoe} + T_{right\ shoe}$$

$$= 206.4 + 166.6 = 373\ \mathrm{N \cdot m}$$

Long-shoe internal drum brakes Most drum brakes use internal shoes which expand against the inner radius of the drum. Long-shoe internal drum brakes are principally used in automotive applications. An automotive drum brake typically comprises two brake shoes and linings supported on a back plate bolted to the axle casing. The shoes are pivoted at one end on anchor pins or abutments fixed on to the back plate (*see* Fig. 7.16). The brake can be actuated by a double hydraulic piston expander, forcing the free ends of the brake apart so that the non-rotating shoes

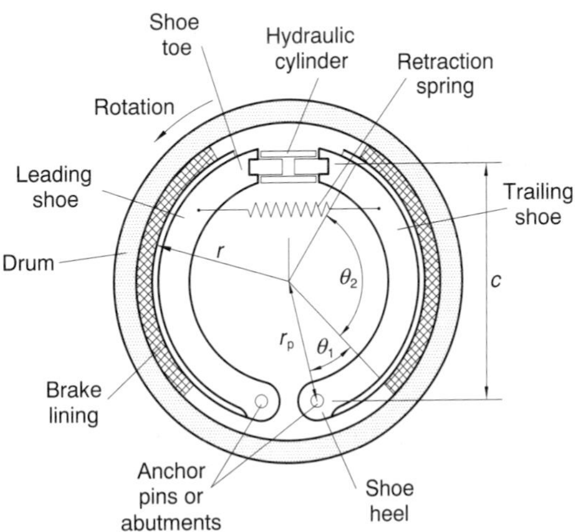

Fig. 7.16 Double long-shoe internal drum brake.

come into frictional contact with the rotating brake drum. A leading and trailing shoe layout consists of a pair of shoes pivoted at a common anchor point, as shown in Fig. 7.16. The leading shoe is identified as the shoe whose expander piston moves in the direction of rotation of the drum. The frictional drag between the shoe and the drum will tend to assist the expander piston in forcing the shoe against the drum; this action is referred to as self-energising or the self-servo action of the shoe. The trailing shoe is the one whose expander piston moves in the direction opposite to the rotation of the drum. The frictional force opposes the expander and hence a trailing brake shoe provides less braking torque than an equivalent leading shoe actuated by the same force. The equations developed for external long-shoe drum brakes are also valid for internal long-shoe drum brakes.

Example Determine the actuating force and the braking capacity for the double internal long-shoe brake, illustrated in Fig. 7.17. The lining is sintered metal with a coefficient of friction of 0.32, and the maximum lining pressure is 1.2 MPa. The drum radius is 68 mm and the shoe width is 25 mm.

Solution

$b = \sqrt{0.015^2 + 0.055^2} = 0.05701\,\text{m}$

As the brake lining angles relative to the pivot, brake axis line, are not explicitly shown on the diagram, they must be calculated:

$\theta_1 = 4.745°, \theta_2 = 124.7°$

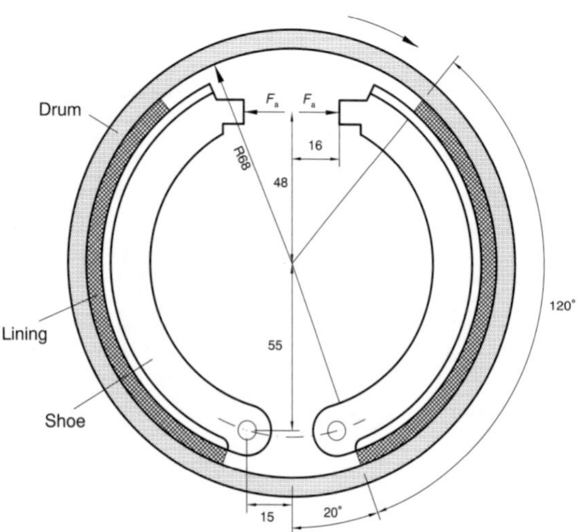

Fig. 7.17 Double long-shoe internal drum brake.

As $\theta_2 > 90°$, the maximum value of $\sin\theta$ is $\sin 90 = 1 = (\sin\theta)_{max}$.

For this brake with the direction of rotation as shown, the right hand shoe is self-energising. For the right hand shoe:

$$M_n = \frac{0.025 \times 0.068 \times 0.05701 \times 1.2 \times 10^6}{1} \times \left[\frac{1}{2}\left((124.7 - 4.745) \times \frac{2\pi}{360}\right) - \frac{1}{4}(\sin 249.4 - \sin 9.49)\right]$$

$$= 153.8\,\text{N}\cdot\text{m}$$

$$M_f = \frac{0.32 \times 0.025 \times 0.068 \times 1.2 \times 10^6}{1} \times \left[0.068(\cos 4.745 - \cos 124.7) + \frac{0.05701}{4}(\cos 249.4 - \cos 9.49)\right]$$

$$= 57.1\,\text{m}$$

$$a = 0.055 + 0.048 = 0.103\,\text{m}$$

The actuating force is

$$F_a = \frac{M_n - M_f}{a} = \frac{153.8 - 57.1}{0.103} = 938.9\,\text{N}$$

The torque applied by the right hand shoe is given by

$$T_{\text{right shoe}} = \frac{\mu w r^2 p_{max}}{(\sin\theta)_{max}}(\cos\theta_1 - \cos\theta_2)$$

$$= \frac{0.32 \times 0.025 \times 0.068^2 \times 1.2 \times 10^6}{1} \times (\cos 4.745 - \cos 124.7)$$

$$= 69.54\,\text{N}\cdot\text{m}$$

The torque applied by the left hand shoe cannot be determined until the maximum operating pressure p'_{max} for the left hand shoe has been calculated. As the left hand shoe is self-de-energising, the normal and frictional moments can be determined using equations 7.39 and 7.40:

$$M'_n = \frac{M_n p'_{max}}{p_{max}} = \frac{153.8 p'_{max}}{1.2 \times 10^6}$$

$$M'_f = \frac{M_f p'_{max}}{p_{max}} = \frac{57.1 p'_{max}}{1.2 \times 10^6}$$

The left hand shoe is self-de-energising, so

$$F_a = \frac{M_n + M_f}{a}$$

$F_a = 938.9$ N, as calculated earlier:

$$938.9 = \frac{153.8p'_{max} + 57.1p'_{max}}{1.2 \times 10^6 \times 0.103}$$

$$p'_{max} = 0.5502 \times 10^6 \text{ N/m}^2$$

The torque applied by the left hand shoe is given by

$$T_{\text{left shoe}} = \frac{\mu w r^2 p'_{max}}{(\sin\theta)_{max}}(\cos\theta_1 - \cos\theta_2)$$

$$= \frac{0.32 \times 0.025 \times 0.068^2 \times 0.5502 \times 10^6}{1} \times (\cos 4.745 - \cos 124.7)$$

$$= 31.89 \text{ N}\cdot\text{m}$$

The total torque applied by both shoes is

$$T_{\text{total}} = T_{\text{right shoe}} + T_{\text{left shoe}}$$

$$= 69.54 + 31.89 = 101.4 \text{ N}\cdot\text{m}$$

From this example, the advantage in torque capacity of using self-energising brakes is apparent. Both the left hand and the right hand shoes could be made self-energising by inverting the left hand shoe, having the pivot at the top. This would be advantageous if rotation occurred in just one direction. If, however, drum rotation is possible in either direction, it may be more suitable to have one brake self-energising for forward motion and one self-energising for reverse motion.

Band brakes One of the simplest types of braking device is the band brake. This consists of a flexible metal band lined with a frictional material wrapped partly around a drum. The brake is actuated by pulling the band against the drum, as illustrated in Fig. 7.18. For the clockwise rotation shown in Fig. 7.18, the friction forces increase F_1 relative to F_2. The relationship between the tight and slack sides of the band is given by

$$\frac{F_1}{F_2} = e^{\mu\theta} \tag{7.41}$$

where

F_1 = tension in the tight side of the band (N)
F_2 = tension in the slack side of the band (N)
μ = coefficient of friction
θ = angle of wrap (rad).

The point of maximum contact pressure for the friction material occurs at the tight end and is given by

$$p_{max} = \frac{F_1}{rw} \tag{7.42}$$

where w is the width of the band (m).

The torque braking capacity is given by

$$T = (F_1 - F_2)r \tag{7.43}$$

The relationship for the simple band brake shown in Fig. 7.18, between the applied lever force F_a and F_2 can be found by taking moments about the pivot point:

$$F_a c - F_2 a = 0$$

$$F_a = F_2 \frac{a}{c} \tag{7.44}$$

The brake configuration shown in Fig. 7.18 is self-energising for clockwise rotation. The level of self-energisation can be enhanced by using the differential band brake configuration shown in Fig. 7.19. Summation of the moments about the pivot gives

$$F_a c - F_2 a + F_1 b = 0$$

So the relationship between the applied load F_a and the band brake tension is given by

$$F_a = \frac{F_2 a - F_1 b}{c} \tag{7.45}$$

Note that the value of b must be less than a so that applying the lever tightens F_2 more than it loosens F_1. Substituting for F_1 in equation 7.45 gives

$$F_a = \frac{F_2(a - b\,e^{\mu\theta})}{c} \tag{7.46}$$

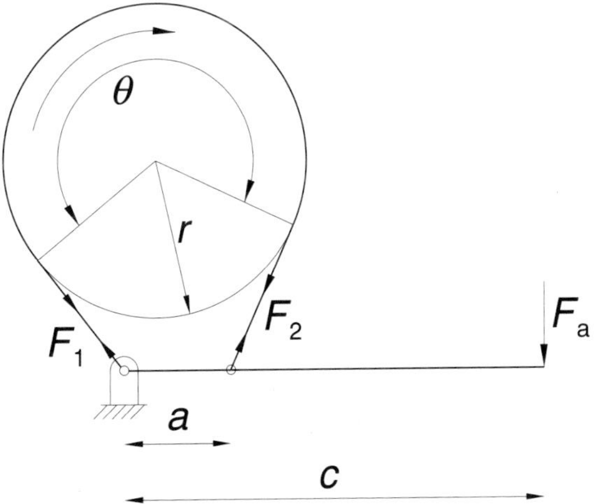

Fig. 7.18 Band brake.

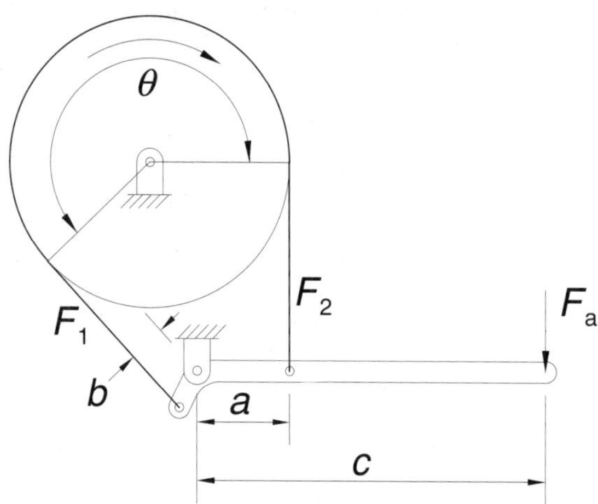

Fig. 7.19 Self-energising band brake.

The brake can be made self-locking if $a < b\,e^{\mu\theta}$ and the slightest touch on the lever would cause the brake to grab or lock abruptly. This principle can be used to permit rotation in one direction only as in hoist and conveyor applications.

Example Design a band brake to exert a braking torque of $85\,\text{N}\cdot\text{m}$. Assume the coefficient of friction for the lining material is 0.25 and the maximum permissible pressure is 0.345 MPa.

Solution Propose a trial geometry, say $r = 150\,\text{mm}$, $\theta = 225°$ and $w = 50\,\text{mm}$:

$$F_1 = p_{max} r w = 0.345 \times 10^6 \times 0.15 \times 0.05 = 2587\,\text{N}$$

$$F_2 = \frac{F_1}{e^{\mu\theta}} = \frac{2587.5}{e^{0.25(225\times 2\pi/360)}} = 969\,\text{N}$$

$$T = (F_1 - F_2)r = (2587.5 - 969)0.15 = 242.7\,\text{N}\cdot\text{m}$$

This torque is much greater than the $80\,\text{N}\cdot\text{m}$ desired, so try a different combination of r, θ and w until a satisfactory design is achieved.

Try $r = 0.1\,\text{m}$, $\theta = 225°$ and $w = 50\,\text{mm}$:

$$F_1 = p_{max} r w = 0.345 \times 10^6 \times 0.1 \times 0.05 = 1725\,\text{N}$$

$$F_2 = \frac{F_1}{e^{\mu\theta}} = \frac{1725}{e^{0.25(225\times 2\pi/360)}} = 646.3\,\text{N}$$

$$T = (F_1 - F_2)r = (1725 - 646.3)0.1 = 107.9\,\text{N}\cdot\text{m}$$

Try $r = 0.09\,\text{m}$, $\theta = 225°$ and $w = 50\,\text{mm}$:

$$F_1 = p_{max} r w$$

$$= 0.345 \times 10^6 \times 0.09 \times 0.05 = 1552.5\,\text{N}$$

$$F_2 = \frac{F_1}{e^{\mu\theta}} = \frac{1552.5}{e^{0.25(225\times 2\pi/360)}} = 581.7\,\text{N}$$

$$T = (F_1 - F_2)r = (1552.5 - 581.7)0.09$$

$$= 87.4\,\text{N}\cdot\text{m}$$

The actuating force is given by $F_a = F_2 a/c$. If $a = 0.08\,\text{m}$ and $c = 0.15\,\text{m}$, then

$$F_a = 581.7 \times \frac{0.08}{0.15} = 310.2\,\text{N}$$

Learning objectives checklist

- Can you determine the primary dimensions for a single disc clutch? ☐
- Can you determine the principal dimensions and number of discs for a multiple disc clutch for a given speed and power requirement? ☐
- Can you determine the torque capacity for short or long, internal or external brakes? ☐
- Can you determine the configuration of a brake to be self-energising? ☐

References and sources of information

Books and papers

Baker, A.K. 1992: *Industrial brake and clutch design*. Pentech Press.

Fazekas, G.A. 1972: On circular spot brakes. *Transactions of the ASME, Journal of Engineering for Industry*, 859–63.

Heisler, H. 1985: *Vehicle and engine technology*. Arnold.

Hindhede, U., Zimmerman, J.R., Hopkins, R.B., Erisman, R.J., Hull, W.C. and Lang, J.D. 1983: *Machine design fundamentals. A practical approach*. Wiley.

Juvinall, R.C. and Marshek, K.M. 1991: *Fundamentals of machine component design*. Wiley.

Neale, M.J. 1994: *Drives and seals: a tribology handbook*. Butterworth Heinemann.

Proctor, J. 1961: Selecting clutches for mechanical drives. *Product Engineering*, 43–58.

Vedamuttu, P. 1994: Clutches. In Hurst, K. (ed.), *Rotary power transmission design*. McGraw Hill.

Standards

BS AU 180: Part 4: 1982. Brake linings. Method for determining effects of heat on dimensions and form of disc brake pads. BSI.

BS 4639: 1987. Specification for brakes and braking systems for towed agricultural vehicles. BSI.

Web sites

http://www.warnernet.com/product.html

Clutches and brakes worksheet

1. Calculate the torque a clutch must transmit to accelerate a pulley with a moment of inertia

of $0.25\,\text{kg}\cdot\text{m}^2$ to: (a) 500 rpm in 2.5 s; (b) 1000 rpm in 2 s.

2. Calculate the energy that must be absorbed in stopping a 100 tonne airbus travelling at 250 km/h in an aborted take off stopping in 40 s.
3. A disc clutch has a single pair of mating surfaces of 300 mm outside diameter and 200 mm inner diameter. If the coefficient of friction is 0.3 and the actuating force is 4000 N, determine the torque capacity assuming: (a) uniform wear; (b) uniform pressure.
4. A multiple disc clutch running in oil is required for a touring motorcycle. The power demand is 75 kW at 9000 rpm. Space limitations restrict the maximum outer diameter of the clutch plates to 200 mm. Select an appropriate lining material and determine the radii for the discs, the number of discs required and the clamping force.
5. A multiple disc clutch is required for a high-performance motorcycle. The power demand is 100 kW at 12 000 rpm. Select an appropriate lining material and determine the radii for the discs, the number of discs required and the clamping force.
6. A multiple disc clutch is required for a small motorcycle. The power demand is 9 kW at 8500 rpm. Select an appropriate lining material and determine the radii for the discs, the number of discs required and the clamping force.
7. A caliper brake is required for the front wheels of a passenger car with a braking capacity of 320 N·m for each brake. Preliminary design estimates have set the brake geometry as $r_i = 100$ mm, $r_o = 140$ mm and $\theta = 40°$. Pads with a coefficient of friction of 0.35 have been selected. Each pad is actuated by means of a hydraulic cylinder of nominal diameter 25.4 mm. Determine the required actuating force, the average and the maximum contact pressures, and the required hydraulic pressure for brake actuation.
8. Determine the torque capacity for the single short-shoe brake shown in Fig. 7.20. The coefficient of friction is 0.35.
9. The maximum actuating force on the double short-shoe external drum brake illustrated in Fig. 7.21 is 1 kN. If the coefficient of friction for the shoe lining is 0.3, determine the torque capacity of the brake for clockwise rotation.
10. A double short-shoe external brake is illustrated in Fig. 7.22. The required actuating force to limit the drum rotation to 100 rpm is 2.4 kN. The coefficient of friction for the brake lining is 0.35. Determine the braking torque and the rate of heat generation.
11. The single long-shoe external brake illustrated in Fig. 7.23 operates on a drum of diameter 300 mm. The coefficient of friction for the brake lining is 0.3 and the face width of the shoes is 40 mm. If the actuating force is 500 N, determine the maximum shoe pressure and the braking torque.
12. A double long-shoe external drum brake is illustrated in Fig. 7.24. The face width of the

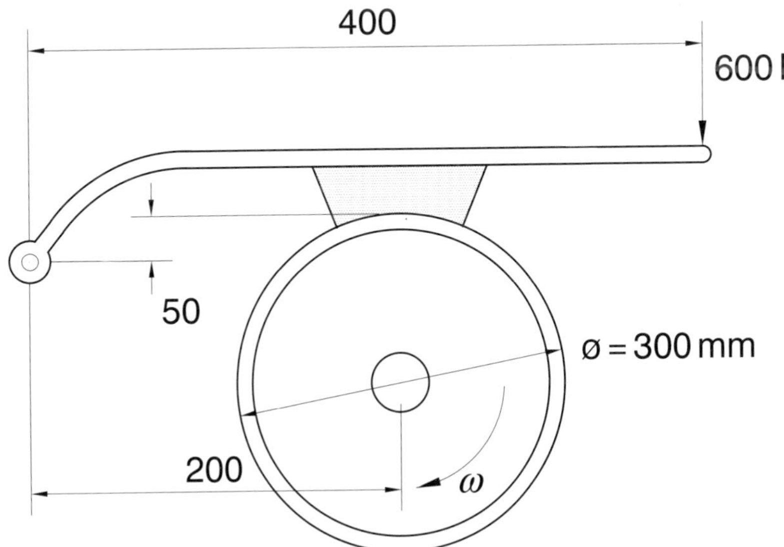

Fig. 7.20 Single short-shoe brake.

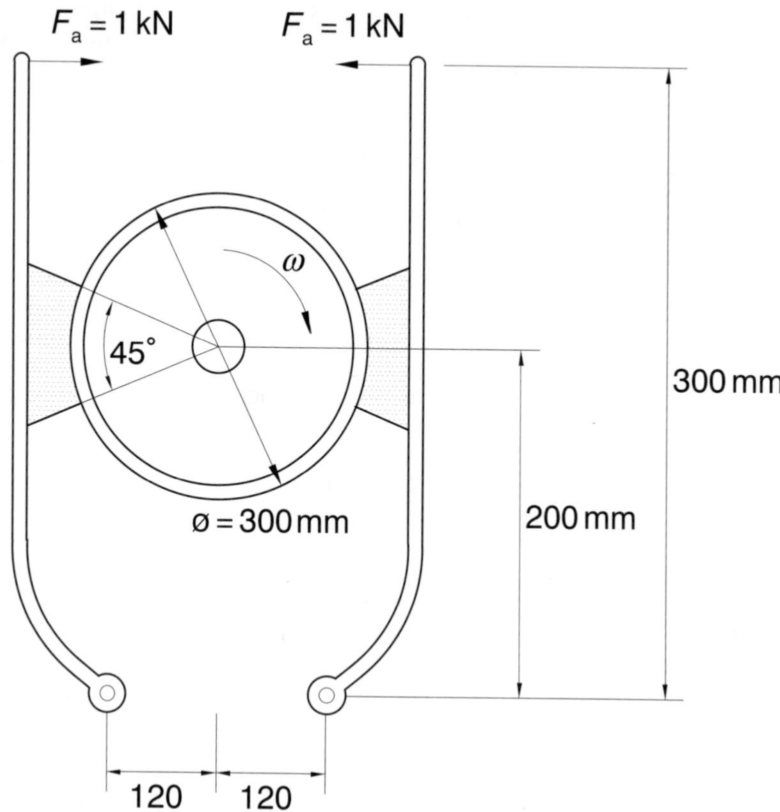

Fig. 7.21 Short-shoe external drum brake.

shoes is 50 mm and the maximum permissible lining pressure is 1 MPa. If the coefficient of friction is 0.32, determine the limiting actuating force and the torque capacity.

13. The double long-shoe internal drum brake illustrated in Fig. 7.25 has a shoe outer diameter of 280 mm and shoe width of 40 mm. Each shoe is actuated by a force of 2 kN. The coefficient of friction for the brake lining is 0.25. Find the magnitude of the maximum pressure and the braking torque.

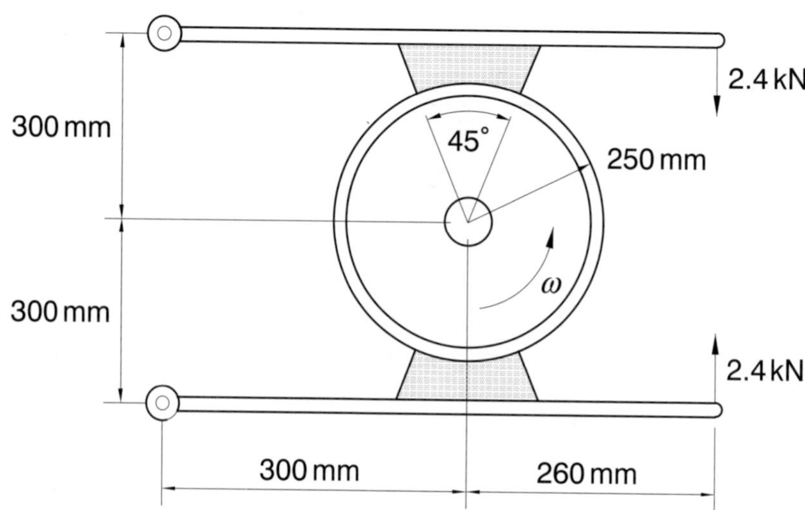

Fig. 7.22 Double short-shoe external brake.

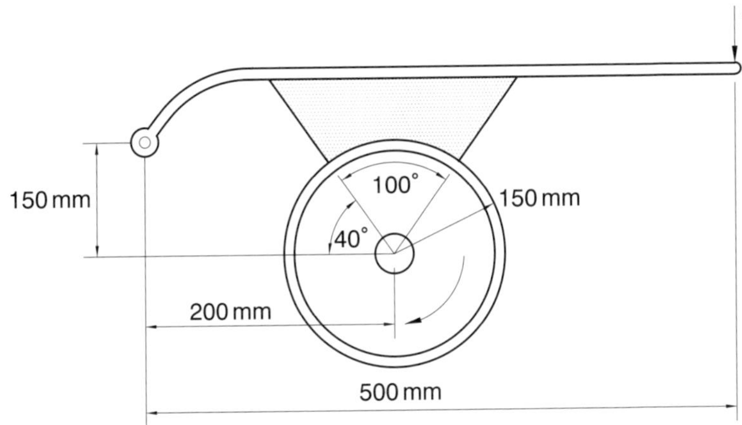

Fig. 7.23 Single long-shoe external brake.

14. A double long-shoe internal brake as illustrated in Fig. 7.26 has a diameter of 200 and a shoe face width of 30 mm. The coefficient of friction for the brake lining is 0.4 and the maximum permissible stress is 1.2 MPa. Each brake shoe is actuated by equal forces. Determine what the value is for the actuating force on each shoe and determine the torque capacity for the brake.
15. A simple band brake is shown in Fig. 7.27. The maximum permissible pressure for the brake lining is 0.6 MPa. The brake band is 100 mm wide and has a coefficient of friction of 0.3. The angle of contact between the band and the drum is 270°. Determine the tensions in the brake band and the torque capacity if the drum diameter is 0.36 m.

Answers

Worked solutions to the above problems are given in Appendix B.

1. 5.2 N · m, 13.1 N · m.
2. 241.1 MJ, 6.028 MW.
3. (a) 300 N · m; (b) 304 N · m.
4. No unique solution.

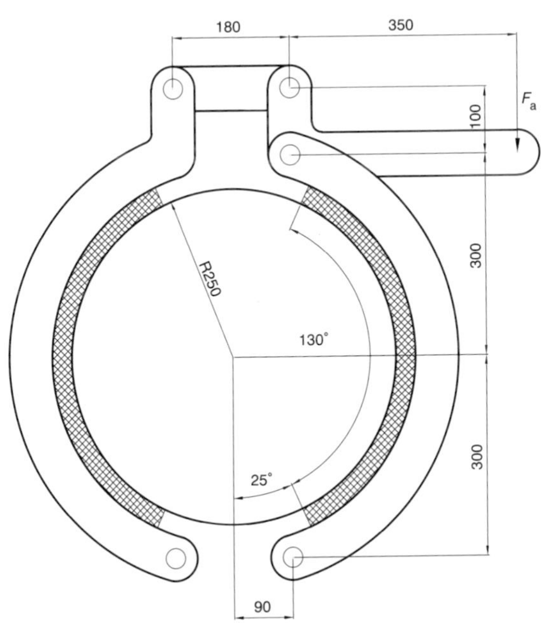

Fig. 7.24 Double long-shoe external drum brake.

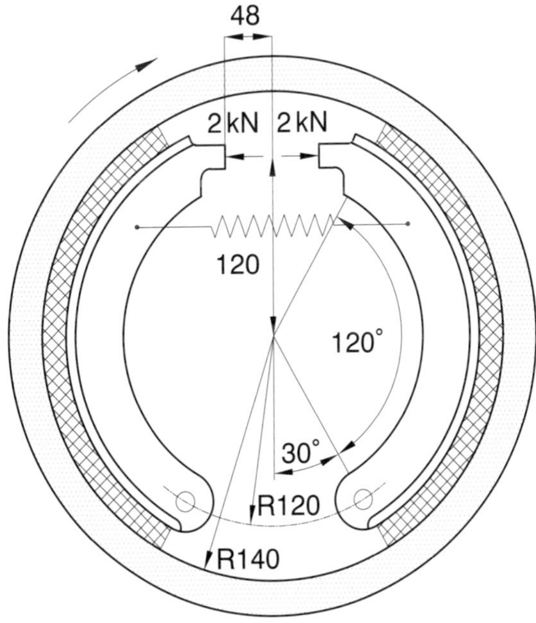

Fig. 7.25 Double long-shoe internal drum brake.

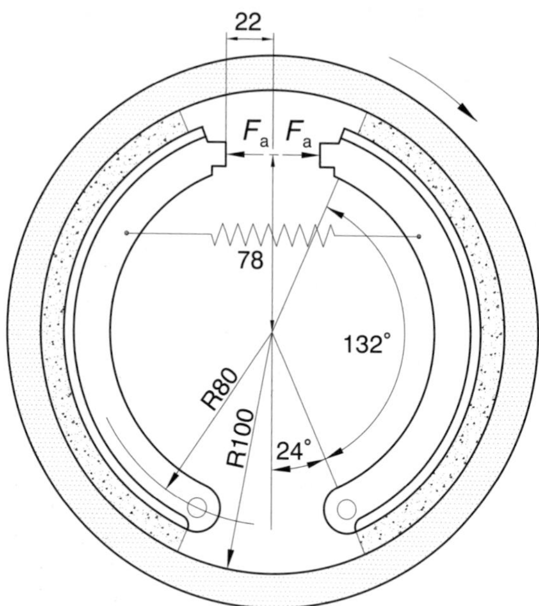

Fig. 7.26 Double long-shoe internal brake.

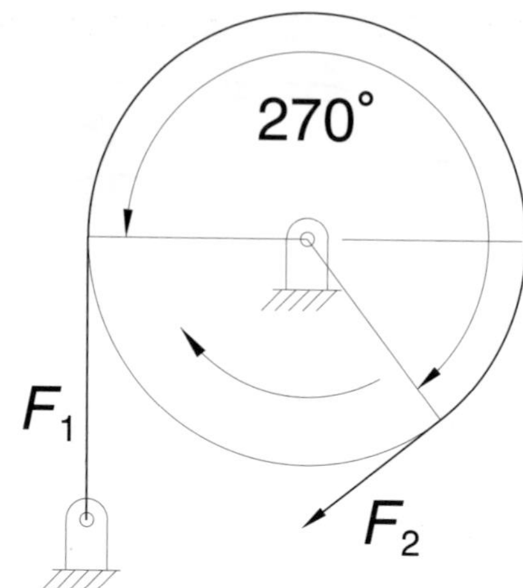

Fig. 7.27 Band brake.

5. No unique solution.
6. No unique solution.
7. 3810 N, 1.137×10^6 N/m^2, 1.364×10^6 N/m^2, 75 bar.
8. 69.04 N · m.
9. 135.4 N · m.
10. 393.4 N · m, 4.1 kW.
11. 0.179 MN/m^2, 59.2 N · m.
12. 1.7 kN, 3.2 kN · m.
13. 0.7244 MN/m^2, 334.9 N · m.
14. 1289 N, 317 N · m.
15. 1.47 kN · m.

8 Engineering tolerancing

The aims of this chapter are to introduce the concepts of component and process variability and the allocation of component tolerances. The ability to select suitable tolerances for the assembly of components should be developed, along with an understanding of geometric tolerancing and the capability to determine the quantitative variability of a process or assembly.

8.1 Introduction

A solid is defined by its surface boundaries. Designers typically specify a component's nominal dimensions such that it fulfils its requirements. In reality, components cannot be made repeatedly to nominal dimensions, due to surface irregularities and the intrinsic surface roughness. Some variability in dimensions must be allowed to ensure manufacture is possible. However, the variability permitted must not be so great that the performance of the assembled parts is impaired. The allowed variability on the individual component dimensions is called the tolerance.

The term tolerance applies not only to the acceptable range of component dimensions produced by manufacturing techniques, but also to the output of machines or processes. For example, the power produced by a given type of internal combustion engine varies from one engine to another. In practice, the variability is usually found to be modelled by a frequency distribution curve, for example the normal distribution (also called the Gaussian distribution). The modelling and analysis of this kind of variable output is outlined in Section 8.3. One of the tasks of the designer is to specify a dimension on a component and the allowable variability on this value that will give acceptable performance. A general introduction to the specification of component tolerances is given in Section 8.2.

8.2 Component tolerances

Control of dimensions is necessary in order to ensure assembly and interchangeability of components. Tolerances are specified on critical dimensions that affect clearances and interference fits. One method of specifying tolerances is to state the nominal dimension followed by the permissible variation, so a dimension could be stated as 40.000 ± 0.003 mm. This means that the dimension should be machined so that it is between 39.997 and 40.003 mm. Where the variation can vary either side of the nominal dimension, the tolerance is called a bilateral tolerance. For a unilateral tolerance, one tolerance is zero, e.g. $40.000^{+0.006}_{\ 0.000}$.

Most organisations have general tolerances that apply to dimensions when an explicit dimension is not specified on a drawing. For machined dimensions a general tolerance may be ± 0.5 mm. So a dimension specified as 15.0 mm may range between 14.5 mm and 15.5 mm. Other general tolerances can be applied to features such as angles, drilled and punched holes, castings, forgings, weld beads and fillets.

When specifying a tolerance for a component, reference can be made to previous drawings or general engineering practice. Tolerances are typically specified in bands as defined in British or ISO (International Organisation for Standardisation) standards. Table 8.1 gives a guide for the general applications of tolerances. For a given tolerance, e.g. H7/s6, a set of numerical values is available from a corresponding chart for the size of component under consideration. The section following gives specific examples of this for a shaft or cylindrical spigot fitting into a hole.

8.2.1 Standard fits for holes and shafts

A standard engineering task is to determine tolerances for a cylindrical component, e.g. a shaft, fitting or rotating inside a corresponding cylindrical component or hole. The tightness of fit will depend on the application. For example, a gear located on to a shaft would require a 'tight' interference fit, where the diameter of the shaft is actually slightly greater than the inside diameter of the gear hub in order to be able to transmit the desired torque. Alternatively, the diameter of a journal bearing must be greater than the diameter of the shaft to allow rotation. Given that it is not economically possible to manufacture components to exact dimensions, some variability in sizes of both the shaft and hole dimension must be specified. However, the range of variability should not be so large that the operation of the assembly is

Table 8.1 Example of tolerance bands and typical applications

Class	Description	Characteristic	ISO code	Assembly	Application
Clearance	Loose running fit	For wide commercial tolerances	H11/c11	Noticeable clearance	IC engine exhaust valve in guide
	Free running fit	Good for large temperature variations, high running speeds or heavy journal pressures	H9/d9	Noticeable clearance	Multiple bearing shafts, hydraulic piston in cylinder, removable levers, bearings for rollers
	Close running fit	For running on accurate machines and accurate location at moderate speeds and journal pressures	H8/f7	Clearance	Machine tool main bearings, crankshaft and connecting rod bearings, shaft sleeves, clutch sleeves, guide blocks
	Sliding fit	When parts are not intended to run freely, but must move and turn and locate accurately	H7/g6	Push fit without noticeable clearance	Push-on gear wheels and clutches, connecting rod bearings, indicator pistons
	Location clearance fit	Provides snug fit for location of stationary parts, but can be freely assembled	H7/h6	Hand pressure with lubrication	Gears, tailstock sleeves, adjusting rings, loose bushes for piston bolts and pipelines
Transition	Location transition fit	For accurate location (compromise between clearance and interference fit)	H7/k6	Easily tapped with hammer	Pulleys, clutches, gears, flywheels, fixed handwheels and permanent levers
	Location transition fit	For more accurate location	H7/n6	Needs pressure	Motor shaft armatures, toothed collars on wheels
Interference	Locational interference fit	For parts requiring rigidity and alignment with accuracy of location	H7/p6	Needs pressure	Split journal bearings
	Medium drive fit	For ordinary steel parts or shrink fits on light sections	H7/s6	Needs pressure or temperature difference	Clutch hubs, bearing bushes in blocks, wheels, connecting rods. Bronze collars on grey cast iron hubs

impaired. Rather than having an infinite variety of tolerance dimensions that could be specified, national and international standards have been produced defining bands of tolerances, examples of which are listed in Table 8.1, e.g. H11/c11. To turn this information into actual dimensions corresponding tables exist, defining the tolerance levels for the size of dimension under consideration (*see* Tables 8.2 and 8.3). In order to use this information the following list and Fig. 8.1 give definitions used in conventional tolerancing. Usually the hole-based system is used (data sheet BS 4500A, Table 8.2), as this results in a reduction in the variety of drill, reamer, broach and gauge tooling required within a company.

- **Size**: a number expressing in a particular unit the numerical value of a dimension.
- **Actual size**: the size of a part as obtained by measurement.
- **Limits of size**: the maximum and minimum sizes permitted for a feature.
- **Maximum limit of size**: the greater of the two limits of size.
- **Minimum limit of size**: the smaller of the two limits of size.
- **Basic size**: the size by reference to which the limits of size are fixed.
- **Deviation**: the algebraic difference between a size and the corresponding basic size.
- **Actual deviation**: the algebraic difference between the actual size and the corresponding basic size.
- **Upper deviation**: the algebraic difference between the maximum limit of size and the corresponding basic size.

Table 8.2 Selected ISO fits. Holes basis

Nominal sizes		Clearance fits												Transition fits				Interference fits			
		Tolerance (0.001 mm)		Tolerance (0.001 mm)		Tolerance (0.001 mm)		Tolerance (0.001 mm)		Tolerance (0.001 mm)		Tolerance (0.001 mm)		Tolerance (0.001 mm)		Tolerance (0.001 mm)		Tolerance (0.001 mm)		Tolerance (0.001 mm)	
Over (mm)	To (mm)	H11	c11	H9	d10	H9	e9	H8	f7	H7	g6	H7	h6	H7	k6	H7	n6	H7	p6	H7	s6
–	3	+60 0	−60 −120	+25 0	−20 −60	+25 0	−14 −39	+14 0	−6 −16	+10 0	−2 −8	+10 0	−6 0	+10 0	+6 0	+10 0	+10 +4	+10 0	+12 +6	+10 0	+20 +14
3	6	+75 0	−70 −145	+30 0	−30 −78	+30 0	−20 −50	+18 0	−10 −22	+12 0	−4 −12	+12 0	−8 0	+12 0	+9 +1	+12 0	+16 +8	+12 0	+20 +12	+12 0	+27 +19
6	10	+90 0	−80 −170	+36 0	−40 −98	+36 0	−25 −61	+22 0	−13 −28	+15 0	−5 −14	+15 0	−9 0	+15 0	+10 +1	+15 0	+19 +10	+15 0	+24 +15	+15 0	+32 +23
10	18	+110 0	−95 −205	+43 0	−50 −120	+43 0	−32 −75	+27 0	−16 −34	+18 0	−6 −17	+18 0	−11 0	+18 0	+12 +1	+18 0	+23 +12	+18 0	+29 +18	+18 0	+39 +28
18	30	+130 0	−110 −240	+52 0	−65 −149	+52 0	−40 −92	+33 0	−20 −41	+21 0	−7 −20	+21 0	−13 0	+21 0	+15 +2	+21 0	+28 +15	+21 0	+35 +22	+21 0	+48 +35
30	40	+160 0	−120 −280	+62 0	−80 −180	+62 0	−50 −112	+39 0	−25 −50	+25 0	−9 −25	+25 0	−16 0	+25 0	+18 +2	+25 0	+33 +17	+25 0	+42 +26	+25 0	+59 +43
40	50	+160 0	−130 −290																		
50	65	+190 0	−140 −330	+74 0	−100 −220	+74 0	−60 −134	+46 0	−30 −60	+30 0	−10 −29	+30 0	−19 0	+30 0	+21 +2	+30 0	+39 +20	+30 0	+51 +32	+30 0	+72 +53
65	80	+190 0	−150 −340																	+30 0	+78 +59
80	100	+220 0	−170 −390	+87 0	−120 −260	+87 0	−72 −159	+54 0	−36 −71	+35 0	−12 −34	+35 0	−22 0	+35 0	+25 +3	+35 0	+45 +23	+35 0	+59 +37	+35 0	+93 +71
100	120	+220 0	−180 −400																	+35 0	+101 +79
120	140	+250 0	−200 −450																	+40 0	+117 +92
140	160	+250 0	−210 −460	+100 0	−145 −305	+100 0	−84 −185	+63 0	−43 −83	+40 0	−14 −39	+40 0	−25 0	+40 0	+28 +3	+40 0	+52 +27	+40 0	+68 +43	+40 0	+125 +100
160	180	+250 0	−230 −480																	+40 0	+133 +108
180	200	+290 0	−240 −530																	+46 0	+151 +122
200	225	+290 0	−260 −550	+115 0	−170 −355	+115 0	−100 −215	+72 0	−50 −96	+46 0	−15 −44	+46 0	−29 0	+46 0	+33 +4	+46 0	+60 +31	+46 0	+79 +50	+46 0	+159 +130
225	250	+290 0	−280 −570																	+46 0	+169 +140
250	280	+320 0	−300 −620	+130 0	−190 −400	+130 0	−110 −240	+81 0	−56 −108	+52 0	−17 −49	+52 0	−32 0	+52 0	+36 +4	+52 0	+66 +34	+52 0	+88 +56	+52 0	+190 +158
280	315	+320 0	−330 −650																	+52 0	+202 +170
315	355	+360 0	−360 −720	+140 0	−210 −440	+140 0	−125 −265	+89 0	−62 −119	+57 0	−18 −54	+57 0	−36 0	+57 0	+40 +4	+57 0	+73 +37	+57 0	+98 +62	+57 0	+226 +190
355	400	+360 0	−400 −760																	+57 0	+244 +208
400	450	+400 0	−440 −840	+155 0	−230 −480	+155 0	−135 −290	+97 0	−68 −131	+63 0	−20 −60	+63 0	−40 0	+63 0	+45 +5	+63 0	+80 +40	+63 0	+108 +68	+63 0	+272 +232
450	500	+400 0	−480 −880																	+63 0	+292 +252

Table 8.3 Selected ISO fits. Shafts basis

Nominal sizes		Clearance fits												Transition fits				Interference fits			
		Tolerance (0.001 mm)		Tolerance (0.001 mm)		Tolerance (0.001 mm)		Tolerance (0.001 mm)		Tolerance (0.001 mm)		Tolerance (0.001 mm)		Tolerance (0.001 mm)		Tolerance (0.001 mm)		Tolerance (0.001 mm)		Tolerance (0.001 mm)	
Over (mm)	To (mm)	h11	C11	h9	D10	h9	E9	h7	F8	h6	G7	h6	H7	h6	K7	h6	N7	h6	P7	h6	S7
–	3	0 −60	+120 +60	0 −25	+60 +20	0 −25	+39 +14	0 −10	+20 +6	0 −6	+12 +2	0 −6	+10 0	0 −6	0 −10	0 −6	−4 −14	0 −6	−6 −16	0 −6	−14 −24
3	6	0 −75	+145 +70	0 −30	+78 +30	0 −30	+50 +20	0 −12	+28 +10	0 −8	+16 +4	0 −8	+12 0	0 −8	+3 −9	0 −8	−4 −16	0 −8	−8 −20	0 −8	−15 −27
6	10	0 −90	+170 +80	0 −36	+98 +40	0 −36	+61 +25	0 −15	+35 +13	0 −9	+20 +5	0 −9	+15 0	0 −9	+5 −10	0 −9	−4 −19	0 −9	−9 −24	0 −9	−17 −32
10	18	0 −110	+205 +95	0 −43	+120 +50	0 −43	+75 +32	0 −18	+43 +16	0 −11	+24 +6	0 −11	+18 0	0 −11	+6 −12	0 −11	−5 −23	0 −11	−11 −29	0 −11	−21 −39
18	30	0 −130	+240 +110	0 −52	+149 +65	0 −52	+92 +40	0 −21	+53 +20	0 −13	+28 +7	0 −13	+21 0	0 −13	+6 −15	0 −13	−7 −28	0 −13	−14 −35	0 −13	−27 −48
30	40	0 −160	+280 +120	0 −62	+180 +80	0 −62	+112 +50	0 −25	+64 +25	0 −16	+34 +9	0 −16	+25 0	0 −16	+7 −18	0 −16	−8 −33	0 −16	−17 −42	0 −16	−34 −59
40	50	0 −160	+290 +130																		
50	65	0 −190	+330 +140	0 −74	+220 +100	0 −74	+134 +60	0 −30	+76 +30	0 −19	+40 +10	0 −19	+30 0	0 −19	+9 −21	0 −19	−9 −39	0 −19	−21 −51	0 −19	−42 −72
65	80	0 −190	+340 +150																	0 −19	−48 −78
80	100	0 −220	+390 +170	0 −87	+260 +120	0 −87	+159 +72	0 −35	+90 +36	0 −22	+47 +12	0 −22	+35 0	0 −22	+10 −25	0 −22	−10 −45	0 −22	−24 −59	0 −22	−58 −93
100	120	0 −220	+400 +180																	0 −22	−66 −101
120	140	0 −250	+450 +200																	0 −25	−77 −117
140	160	0 −250	+460 +210	0 −100	+305 +145	0 −100	+185 +85	0 −40	+106 +43	0 −25	+54 +14	0 −25	+40 0	0 −25	+12 −28	0 −25	−12 −52	0 −25	−28 −68	0 −25	−85 −125
160	180	0 −250	+480 +230																	0 −25	−93 −133
180	200	0 −290	+530 +240																	0 −29	−105 −151
200	225	0 −290	+550 +260	0 −115	+355 +170	0 −115	+215 +100	0 −46	+122 +50	0 −29	+61 +15	0 −29	+46 0	0 −29	+13 −33	0 −29	−14 −60	0 −29	−36 −79	0 −29	−113 −159
225	250	0 −290	+570 +280																	0 −29	−123 −169
250	280	0 −320	+620 +300	0 −130	+400 +190	0 −130	+240 +110	0 −52	+137 +56	0 −32	+62 +17	0 −32	+52 0	0 −32	+16 −36	0 −32	−14 −66	0 −32	−33 −88	0 −32	−138 −190
280	315	0 −320	+650 +330																	0 −32	−150 −202
315	355	0 −360	+720 +360	0 −140	−440 +210	0 −140	+265 +125	0 −57	+151 +62	0 −36	+75 +18	0 −36	+57 0	0 −36	+17 −40	0 −36	−16 −73	0 −36	−41 −98	0 −36	−169 −226
355	400	0 −360	+760 +400																	0 −36	−187 −244
400	450	0 −400	+840 +440	0 −155	−480 +230	0 −155	+290 +135	0 −63	+165 +68	0 −40	+83 +20	0 −40	+63 0	0 −40	+18 −45	0 −40	−17 −80	0 −40	−45 −108	0 −40	−209 −272
450	500	0 −400	+880 +480																	0 −40	−229 −292

Reproduced from BS 4500, data sheet 4500B.

- **Lower deviation**: the algebraic difference between the minimum limit of size and the corresponding basic size.
- **Tolerance**: the difference between the maximum limit of size and the minimum limit of size.
- **Shaft**: the term used by convention to designate all external features of a part (including parts that are not cylindrical).
- **Hole**: the term used by convention to designate all internal features of a part.

BS 4500 data sheets 4500A and 4500B list selected ISO clearance, transitional and interference fits for internal and external components. When using the Standard, upper case letters refer to the hole and lower case letters to the shaft.

Example Find the shaft and hole dimensions for a loose running fit with a 35 mm diameter basic size.

Solution From Table 8.1 the ISO symbol is 35H11/c11 (the 35 designating a 35 mm nominal tolerance). From data sheet BS 4500A (Table 8.2) the hole tolerance, for a 35 mm hole, is $^{+0.16}_{\ 0.00}$ mm and the shaft tolerance, for a 35 mm shaft, is $^{-0.12}_{-0.28}$ mm.

So the hole should be given dimensions $35.00^{+0.16}_{\ 0.00}$ mm and the shaft $35.00^{-0.12}_{-0.28}$ mm.

Example Find the shaft and hole dimensions for a medium drive fit using a basic hole size of 60 mm.

Solution From Table 8.1 the ISO symbol is 60H7/s6. From data sheet BS 4500A (Table 8.2) the hole tolerance is $^{+0.030}_{\ 0.000}$ mm and the shaft tolerance is $^{+0.072}_{+0.053}$ mm.

So the hole should be given dimensions $60.00^{+0.030}_{\ 0.000}$ mm and the shaft $60.00^{+0.072}_{+0.053}$ mm.

8.2.2 Interference fits

Interference fits are those for which, prior to assembly, the inside component is larger than the outside component. There is some deformation of the components after assembly and a pressure exists at the mating surfaces. This pressure is given by equation 8.1 if both components are of the same material, or by equation 8.2 if the two components are of differing materials (*see* Fig. 8.2 for definition of radii):

$$p = \frac{E\delta}{2b}\frac{(c^2 - b^2)(b^2 - a^2)}{2b^2(c^2 - a^2)} \qquad (8.1)$$

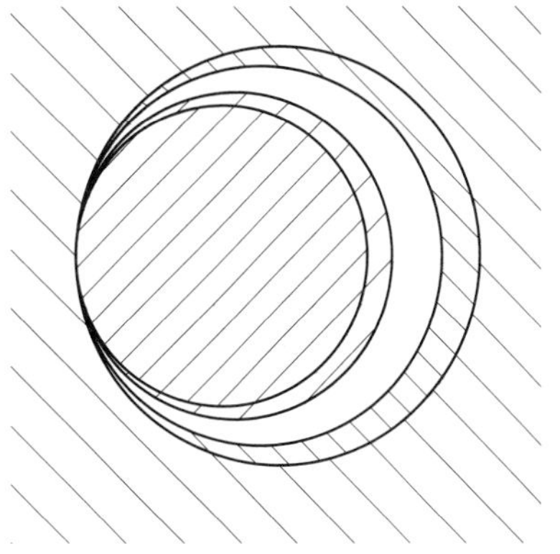

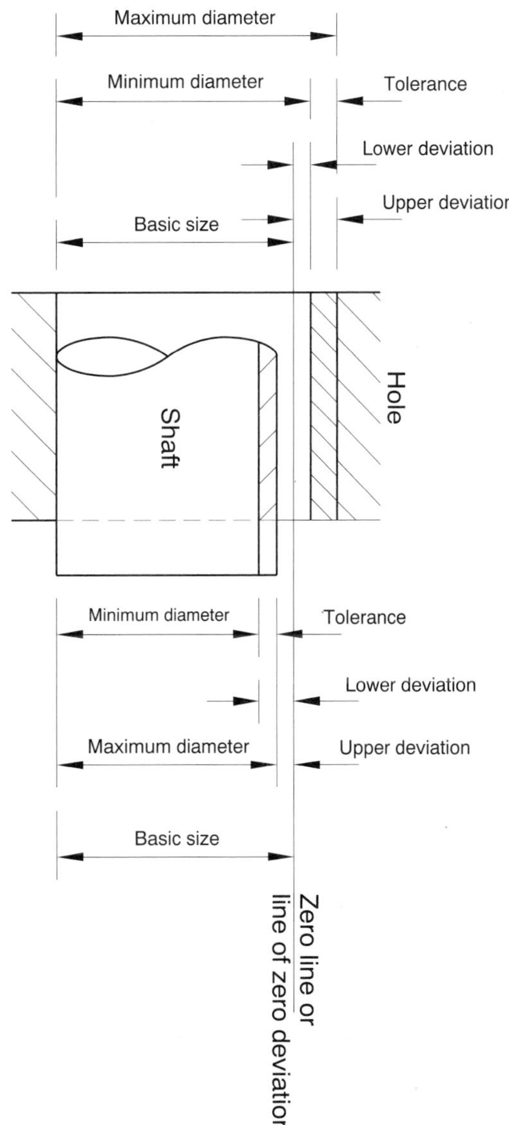

Fig. 8.1 Definitions of terms used in conventional tolerancing (after BS 4500).

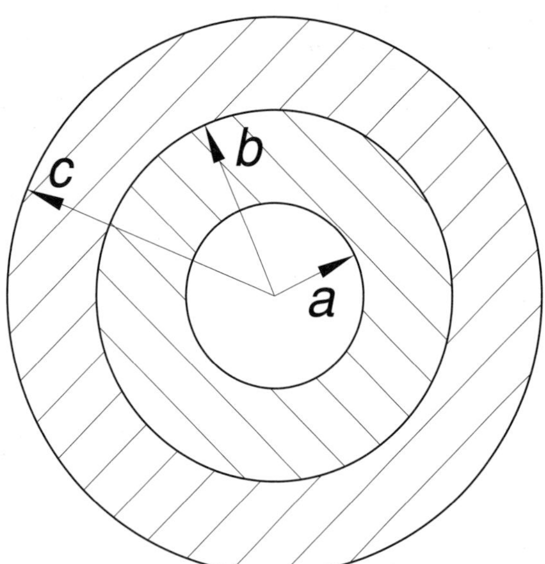

Fig. 8.2 Terminology for cylindrical interference fit.

$$p = \frac{\delta}{2b\left[\frac{1}{E_o}\left(\frac{c^2+b^2}{c^2-b^2}+\mu_o\right)+\frac{1}{E_i}\left(\frac{b^2+a^2}{b^2-a^2}-\mu_i\right)\right]} \quad (8.2)$$

where

p = pressure at the mating surface (N/m^2)
δ = total diameter interference (m)
a = internal radius of inner cylinder (m)
b = outer radius of inner cylinder (m)
c = outer radius of outer cylinder (m)
E = Young's modulus (N/m^2)
μ = Poisson's ratio.

The subscripts i and o refer to the inner and outer components respectively.

The tensile stress in the outer component, σ_o, can be calculated from

$$\sigma_o = p\left(\frac{c^2+b^2}{c^2-b^2}\right) \quad (8.3)$$

and the compressive stress in the inner component, σ_i, from

$$\sigma_i = -p\left(\frac{b^2+a^2}{b^2-a^2}\right) \quad (8.4)$$

The increase in diameter of the outer component, δ_o, due to the tensile stress, can be calculated from

$$\delta_o = \frac{2bp}{E_o}\left(\frac{c^2+b^2}{c^2-b^2}+\mu_o\right) \quad (8.5)$$

and the decrease in diameter of the inner component due to the compressive stress, δ_i, from

$$\delta_i = -\frac{2bp}{E_i}\left(\frac{b^2+a^2}{b^2-a^2}-\mu_i\right) \quad (8.6)$$

The torque which can be transmitted between cylindrical components by an interference fit can be estimated by

$$T = 2fp\pi b^2 L \quad (8.7)$$

where

T = torque (N·m)
f = coefficient of friction
p = interference pressure (N/m^2)
b = interface radius (m)
L = length of interference fit (m).

Example What interference fit is required in order to transmit 18 kN·m torque between a 150 mm diameter steel shaft and a 300 mm outside diameter cast iron hub which is 250 mm long? (Take Young's modulus for the steel and cast iron to be 200 GPa and 100 GPa respectively; assume the coefficient of friction is 0.12 and that Poisson's ratio is 0.3 for both materials.)

Solution The torque is given by $T = 2fp\pi b^2 L$ and this equation can be rearranged to evaluate the pressure required to transmit the torque:

$$p = \frac{T}{2f\pi b^2 L} = \frac{18\times10^3}{2\times0.12\pi(0.075)^2\times0.25}$$
$$= 16.97\times10^6\ \text{N/m}^2$$

The interference required to generate this pressure is given by

$$\delta = 2bp\left[\frac{1}{E_o}\left(\frac{c^2+b^2}{c^2-b^2}+\mu_o\right)+\frac{1}{E_i}(1-\mu_i)\right]$$
$$= 2\times0.075\times16.97\times10^6$$
$$\times\left[\frac{1}{100\times10^9}\left(\frac{0.15^2+0.075^2}{0.15^2-0.075^2}+0.3\right)+\frac{1}{200\times10^9}(1-0.3)\right]$$
$$= 5.89\times10^{-5}\ \text{m}$$

So the required interference fit is 0.059 mm. This could be specified using the H7/s6 tolerance band, with the hole diameter dimension as $150.000^{+0.04}_{+0}$ mm and the shaft diameter as $150.000^{+0.125}_{+0.100}$ mm.

8.2.3 Machine capability

A given manufacturing technique can only produce a component within a range of accuracy around the specified nominal dimension. For instance, centre lathes are capable of machining 10 mm diameter components to tolerances of 0.01 mm, as illustrated in Fig. 8.3. In addition, different manufacturing techniques produce differing levels of surface roughness, as shown in Table 8.4. Given that drilling cannot be normally used to give a surface roughness of better than 0.8 µm, tolerances for drilled components should not be specified smaller than 0.0008 mm. When defining a tolerance the designer should have in mind the 'total design' of the product. This encompasses efficient operation of the product, the ability of the company to manufacture the component and total costs (design, manufacture, servicing and performance etc.). Different manufacturing techniques cost different amounts, as illustrated in Fig. 8.4, and this should be considered.

8.2.4 Geometric tolerancing

The location for a drilled hole has been specified in Fig. 8.5 by means of a dimension with bilateral tolerances. The dimensions given imply a square rectangular tolerance zone as identified in Fig. 8.6(a), anywhere within which it would be permissible to set the hole centre. Examination of the tolerance zone shows that the maximum deviation in location of the hole centre actually occurs along the diagonals, as shown in Fig. 8.6(b). Thus, although the drawing has superficially specified the maximum tolerance deviation as ±0.1 mm or 0.2 mm, the maximum actual deviation within the specification of the drawing is 0.28 mm. If the deviation of 0.28 mm provides correct functioning of the device in the diagonal directions, it is likely to be acceptable in all directions. If this is the case then a circular tolerance could be specified with a diameter of 0.28 mm, as shown in Fig. 8.7. The area of this circular zone is 57 per cent larger than the square tolerance zone and could have

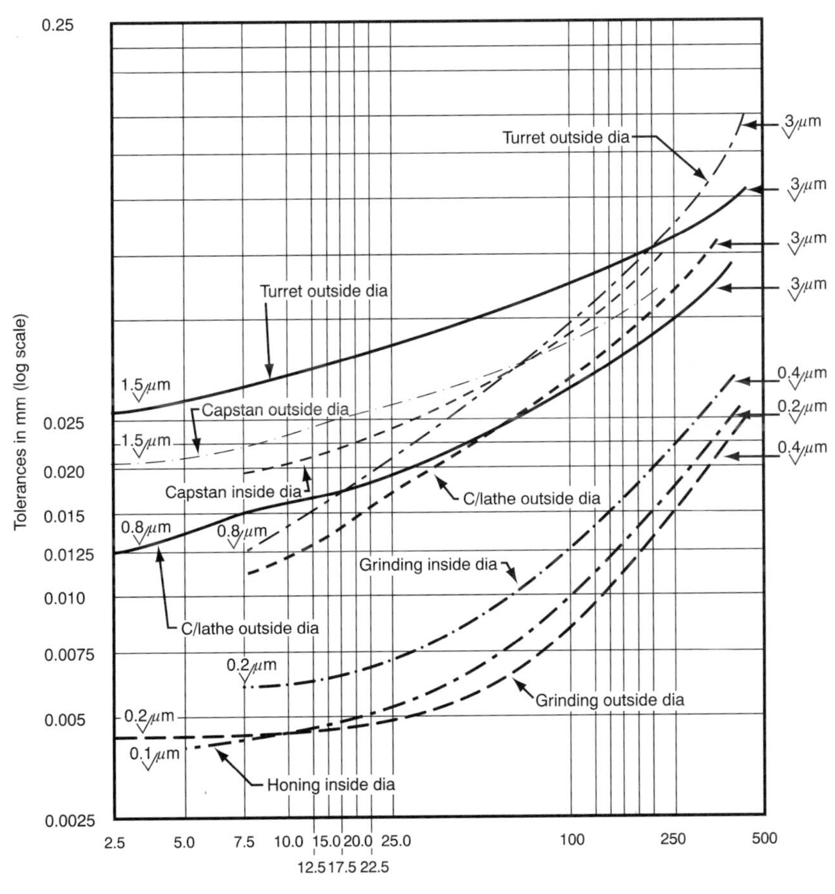

Fig. 8.3 Accuracy to be expected from machining operators using various machines (BSI, 1991).

Table 8.4 Surface roughness values produced by common production processes

Process	Roughness value, R_a (μm): 50–25	25–12.5	12.5–6.3	6.3–3.2	3.2–1.6	1.6–0.8	0.8–0.4	0.4–0.2	0.2–0.1	0.1–0.05	0.05–0.025	0.025–0.0125
Flame cutting	○	●	○									
Sawing	○	●	●	●	○							
Planing, shaping	○	●	●	●	●	●	○					
Drilling			○	●	●	○						
Electric discharge machining			○	○	●	○						
Milling		○	○	●	●	●	○	○				
Broaching				○	●	●	○					
Reaming				○	●	●	○					
Boring, turning		○	○	●	●	●	●	○	○	○		
Roller burnishing							○	●	○			
Grinding				○	○	●	●	●	●	○	○	
Honing						○	●	●	●	○	○	
Polishing							○	●	●	○	○	○
Lapping							○	●	●	●	○	○
Superfinishing							○	○	●	●	○	○
Sand casting	○	●	○									
Hot rolling	○	●	○									
Forging		○	●	●	○							
Permanent mould casting				○	●	○						
Investment casting				○	●	○	○					
Extruding			○	○	●	●	○					
Cold rolling, drawing				○	●	●	○	○				
Die casting					○	●	○					

After BSI (1991).
●: Average application.
○: Less frequent application.

substantial implications for cost reduction in the manufacturing process.

The specification of a circular tolerance zone can be achieved by specifying the exact desired location of the hole and a permissible circular tolerance zone centred at the exact location, as illustrated in Fig. 8.8. Here the exact location of the hole is indicated by the boxed dimensions and the tolerance zone is indicated by the 'crossed circle' symbol within the rectangular box. The diameter of the tolerance zone is specified to the right of the symbol.

The symbol used for the positional tolerance is one of a series used in engineering drawing, as illustrated in Table 8.5, and their use is referred to as geometric tolerancing. Geometric tolerancing allows the engineer to specify control over feature locations and deviation of form such as flatness, parallelism and concentricity. The various symbols used and brief descriptions are given in Table 8.5. Figure 8.9 illustrates their use for defining the locating radius and shoulder for a centrifugal compressor impeller. In order to avoid significant out-of-balance forces and achieve the correct clearance between the compressor blade tips and the stationary shroud, it is critical that the impeller is located concentrically on the shaft and the back face is perpendicular to the shaft. Control over the shaft radial dimensional variation has been specified by means of the runout symbol, which defines the limits of variability for the cylindrical surface and the squareness symbol for the shoulder.

The *Manual of British Standards in Engineering Drawing and Design* (BSI, 1991) defines the standard use of engineering drawing methods and symbols, and the text by Spotts (1983) provides an excellent introduction to tolerancing.

8.3 Statistical tolerancing

Many processes produce outputs, such as component dimensions or power, which follow a frequency distribution such as the 'normal' or Gaussian distribution illustrated in Fig. 8.10. The average value of the output, say a component dimension, is called the mean. The spread of the output values around the mean represents the variability of the process and is represented by a

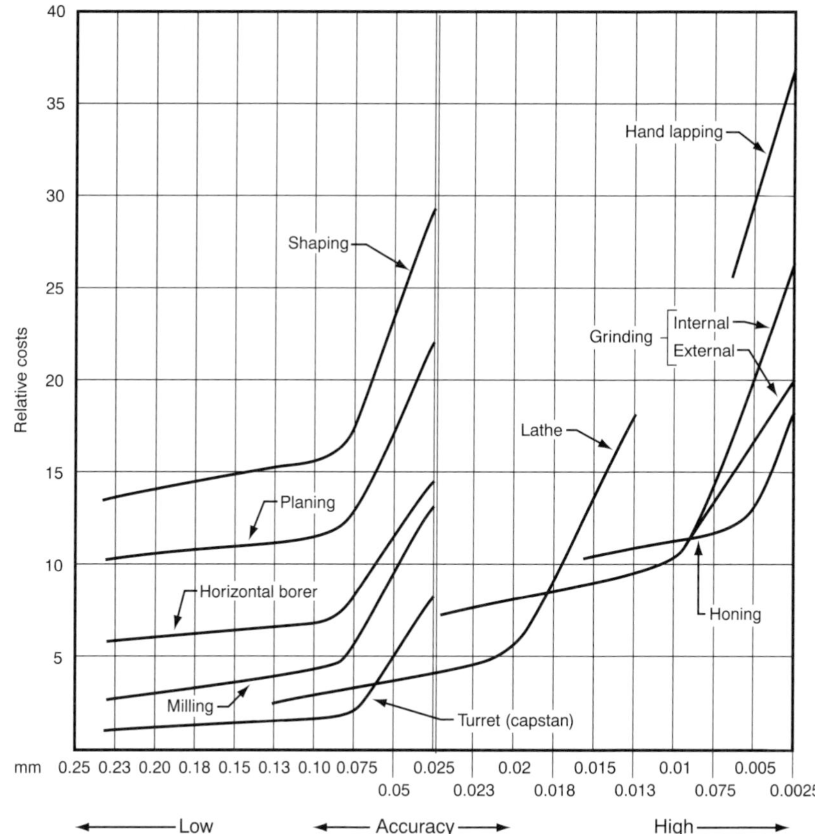

Fig. 8.4 Relative cost of various machine and hand processes for achieving set tolerances (BSI, 1991).

quantity called the standard deviation. The normal distribution is defined such that 68.26 per cent of all the output is within one standard deviation of the mean, 95.44 per cent of all the output is within two standard deviations of the mean and 99.73 per cent of all the output is within three standard deviations of the mean.

The normal distribution is defined by

$$f(x) = \frac{1}{\sigma\sqrt{2\pi}} \exp\left[\frac{1}{2}\left(\frac{x-\mu}{\sigma}\right)^2\right] \qquad -\infty < x < \infty \quad (8.8)$$

where

$f(x)$ is the probability density function of the continuous random variable x
μ is the mean
σ is the standard deviation.

Equation 8.8 can be solved numerically or by use of a table of solutions (as given in Table 8.6). As an illustration of use of the normal table, $z = 3$ gives $F(z) = 0.998650$. So $2 \times (1 - 0.998650) = 0.0027 = 0.27$ per cent of all items can be expected to fall outside of the tolerance limits. In other words, 99.73 per cent of all items produced will have dimensions within the tolerance limits.

In practice, 80 per cent of all processes, e.g. many machined dimensions and engine power outputs, are well modelled by the normal distribution (BSI (1983)). Given a set of data, a statistical test can be undertaken to determine whether the data is

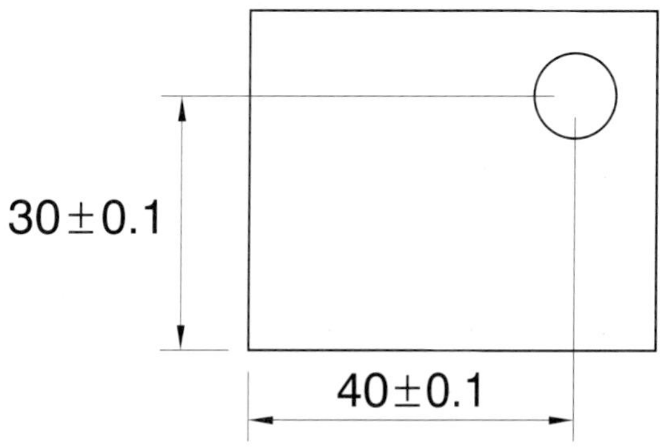

Fig. 8.5 Drilled hole location.

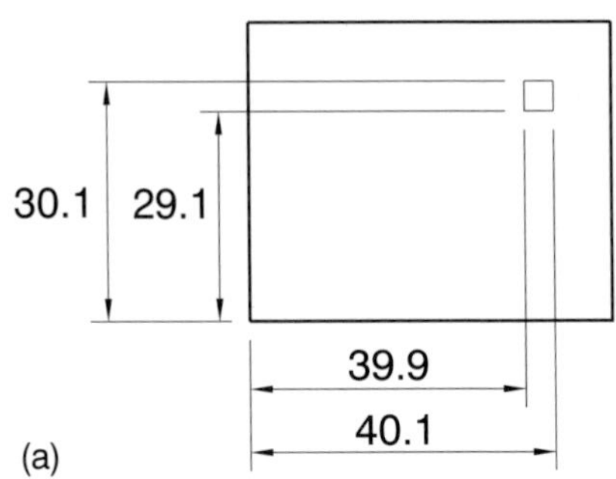

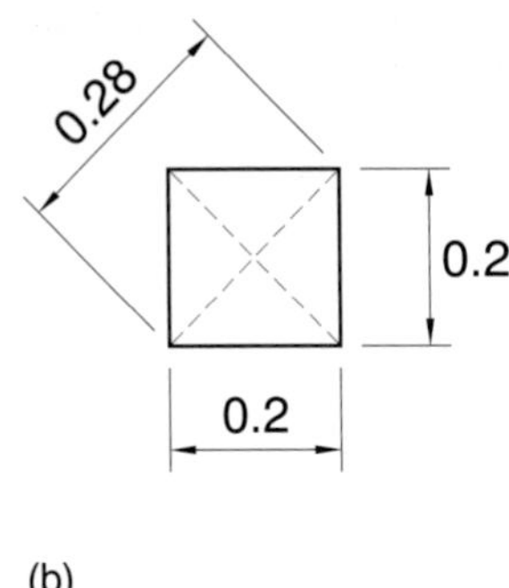

Fig. 8.6 Implied tolerance zone.

normally distributed. If measurements of a dimension for a particular component are found to be normally distributed with mean μ and standard deviation σ, then the tolerance limits are often taken as $\mu \pm 3\sigma$. These tolerance limits are called 'natural tolerance limits'. So if the mean and standard deviation of a component are 8.00 and 0.01 respectively, the dimension and bilateral tolerances would be defined as $8.00^{+0.03}_{-0.03}$ (i.e. the natural tolerance $= 6\sigma = 6 \times 0.01 = 0.06$, so the bilateral tolerance is $0.06/2 = 0.03$).

8.3.1 Sure-fit or extreme variability

The sure-fit or extreme variability of a function of several uncorrelated random variables could be found by substituting values for the variables to find the maximum and the minimum value of the function, e.g. for the function

$$y = \frac{x_1 + 2x_2}{x_3^3}$$

the maximum value of y_{max} is given by

$$y_{max} = \frac{x_{1,max} + 2x_{2,max}}{x_{3,min}^3}$$

Similarly, for the minimum value,

$$y_{min} = \frac{x_{1,min} + 2x_{2,min}}{x_{3,max}^3}$$

The sure-fit or extreme variability can then be determined as

$$\Delta y = y_{max} - y_{min}$$

If $x_1 = 12 \pm 0.1$, $x_2 = 15 \pm 0.2$, $x_3 = 2 \pm 0.15$, then $x_{1,max} = 12.1$, $x_{1,min} = 11.9$, $x_{2,max} = 15.2$, $x_{2,min} = 14.8$, $x_{3,max} = 2.15$, $x_{3,min} = 1.85$. $\bar{y} = 5.25$:

$$y_{max} = \frac{x_{1,max} + 2x_{2,max}}{x_{3,min}^3} = \frac{12.1 + (2 \times 15.2)}{1.85^3} = 6.712$$

$$y_{min} = \frac{x_{1,min} + 2x_{2,min}}{x_{3,max}^3} = \frac{11.9 + (2 \times 14.8)}{2.15^3} = 4.176$$

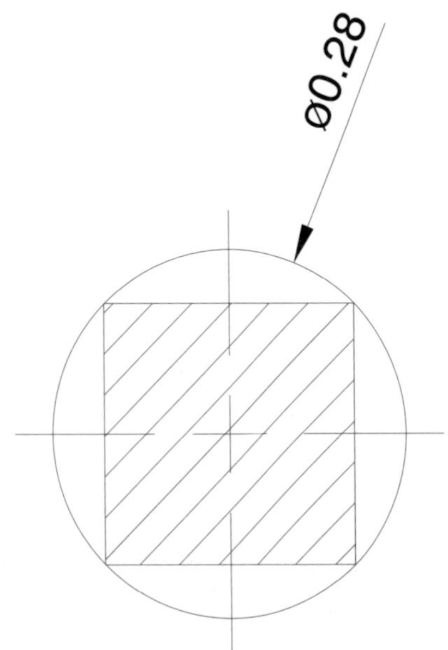

Fig. 8.7 Circular tolerance zone.

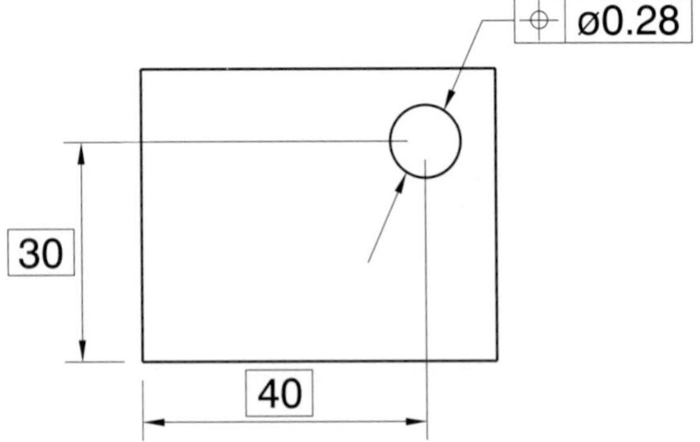

Fig. 8.8 Circular tolerance zone specification.

Table 8.5 Geometric tolerancing symbols

Symbol	Definition	Description
⏤	Straightness	Defines the straightness tolerance for an edge, axis of revolution or line on a surface. The tolerance zone is either the area between two parallel straight lines in the plane containing the toleranced edge or surface line, or a cylindrical region for an axis of revolution
⏥	Flatness	Condition of a surface having all elements in one plane
○	Roundness	Condition on a surface of revolution (cylinder or cone or sphere) where all the points of the surface are intersected by any plane: (i) perpendicular to a common axis (cylinder/cone); (ii) passing through a common centre (sphere), equidistant from the centre
⌭	Cylindricity	Tolerance zone is the annular space (radial separation) between two coaxial cylinders
⌒	Profile of a line	Condition permitting a uniform amount of profile variation, either unilaterally or bilaterally, along a line element of a feature
⌓	Profile of a surface	Defines the tolerance zone of variability for a surface relative to the perfect or exact form
∥	Parallelism	Condition of a surface, line or axis which is equidistant at all points from a datum plane or axis
⊥	Squareness	Defines the angular tolerance for perpendicularity (90°)
∠	Angularity	Defines the allowable variability for angular location. The angular tolerance zone is defined by either two parallel lines or surfaces
⌖	Position	Defines a zone within which the axis or centre plane of a feature is permitted to vary from the true, theoretically correct, position
◎	Concentricity	Limits the deviation of the centre or axis of a feature from its true position
⌯	Symmetry	Typically defines the area between two parallel lines or the space between two parallel surfaces which are symmetrically located about a datum feature
↗	Runout	Deviation in position of a surface of revolution as a part is revolved about a datum axis

$\Delta y = y_{max} - y_{min} = 6.712 - 4.176 = 2.536$, so the variable $y = \bar{y} \pm (\Delta y/2) = 5.25 \pm 1.268$.

Alternatively, the expression given in equation 8.13 can be used to approximate this extreme variability:

$$\Delta y \approx \left|\frac{\partial y}{\partial x_1}\right|\Delta x_1 + \left|\frac{\partial y}{\partial x_2}\right|\Delta x_2 + \left|\frac{\partial y}{\partial x_3}\right|\Delta x_3$$

where

$$\Delta x_1 = x_{1,max} - x_{1,min}$$

$$\Delta x_2 = x_{2,max} - x_{2,min}$$

$$\Delta x_3 = x_{3,max} - x_{3,min}$$

For the numerical example:

$$\frac{\partial y}{\partial x_1} = \frac{1}{x_3^3} = \frac{1}{2^3} = 0.125$$

$$\frac{\partial y}{\partial x_2} = \frac{2}{x_3^3} = \frac{2}{2^3} = 0.25$$

$$\frac{\partial y}{\partial x_3} = -\frac{3(x_1 + 2x_2)}{x_3^4} = 7.875$$

$$\Delta x_1 = 0.2,\ \Delta x_2 = 0.4,\ \Delta x_3 = 0.3$$

$$\Delta y_{approx} = 0.125 \times 0.2 + 0.25 \times 0.4 + 7.875 \times 0.3$$

$$= 2.4875$$

So $y = 5.25 \pm 1.244$ (a value within a few per cent of the value calculated in the above exact method).

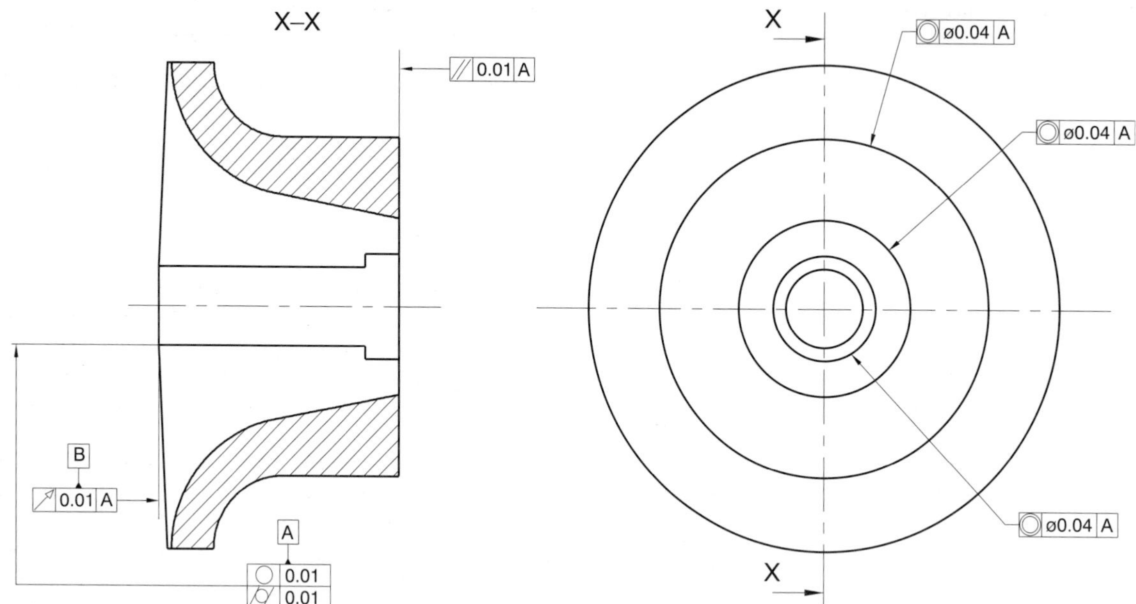

Fig. 8.9 Example of the use of geometric tolerancing for specifying control over centrifugal compressor shaft dimensions.

8.3.2 Linear functions or tolerance chains

Often we require to know the overall tolerance in a dimension chain, or knowledge of which dimensions can be slackened without a deleterious effect on performance, or which tolerances need to be tightened to improve the overall tolerance.

Taking a simple linear chain of components of overall dimension z:

$$z = x_1 + x_2 + x_3 + \cdots + x_n \tag{8.9}$$

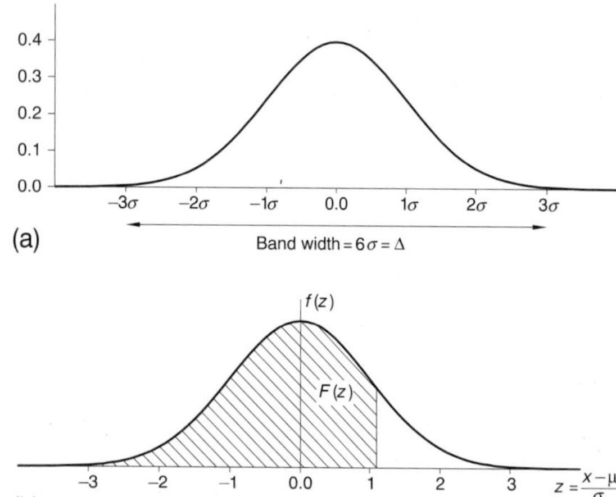

Fig. 8.10 The normal distribution: (a) band width; (b) $f(z)$ and $F(z)$.

The expected value of z is given by

$$E(z) = \mu_1 + \mu_2 + \cdots + \mu_n = \mu_z \tag{8.10}$$

and the variance by

$$\text{variance}(z) = \sigma_1^2 + \sigma_2^2 + \sigma_3^2 + \cdots + \sigma_n^2 = \sigma_z^2 \tag{8.11}$$

If the response variable is a linear function of several measured variables, each of which is normally distributed, then the response variable will also be normally distributed.

Example A product is made by aligning four components (*see* Fig. 8.11) whose lengths are x_1, x_2, x_3 and x_4. The overall length is denoted by z. The tolerance limits of the lengths are known to be 8.000 ± 0.030 mm, 10.000 ± 0.040, 15.000 ± 0.050 and 7.000 ± 0.030 respectively. If the lengths of the four components are independently normally distributed, estimate the tolerance limits for the length of the overall assembly.

Solution Assuming $\pm 3\sigma$ natural tolerance limits:

$$\sigma_1 = 0.03/3 = 0.01$$

$$\sigma_2 = 0.04/3 = 0.0133333$$

$$\sigma_3 = 0.05/3 = 0.0166667$$

$$\sigma_4 = 0.03/3 = 0.01$$

Table 8.6 The normal distribution

z	$f(z)$	$F(z)$
0.0	0.398942	0.5
0.1	0.396952	0.539827
0.2	0.391043	0.559260
0.3	0.381388	0.617911
0.4	0.368270	0.655422
0.5	0.352065	0.691465
0.6	0.333225	0.725747
0.7	0.312254	0.758036
0.8	0.289692	0.788145
0.9	0.266085	0.815940
1.0	0.241971	0.841345
1.1	0.217852	0.864334
1.2	0.194186	0.884930
1.3	0.171369	0.903195
1.4	0.149727	0.919243
1.5	0.129518	0.933193
1.6	0.110921	0.945201
1.7	0.094049	0.955435
1.8	0.078950	0.964069
1.9	0.065616	0.971284
2.0	0.053991	0.977250
2.1	0.043984	0.982136
2.2	0.035475	0.986097
2.3	0.028327	0.989276
2.4	0.022395	0.991803
2.5	0.017528	0.993791
2.6	0.013583	0.995339
2.7	0.010421	0.996533
2.8	0.007915	0.997495
2.9	0.005952	0.998134
3.0	0.004432	0.998650
3.5	0.000873	0.999768
4.0	0.000134	0.999968
4.5	0.000016	0.999996
5.0	0.0000015	0.9999997

$$\sigma_z^2 = \sigma_1^2 + \sigma_2^2 + \sigma_3^2 + \sigma_4^2$$

$$= 0.01^2 + 0.013333^2 + 0.0166667^2 + 0.01^2$$

$$= 6.555 \times 10^{-4}$$

$$\sigma_z = 0.0256$$

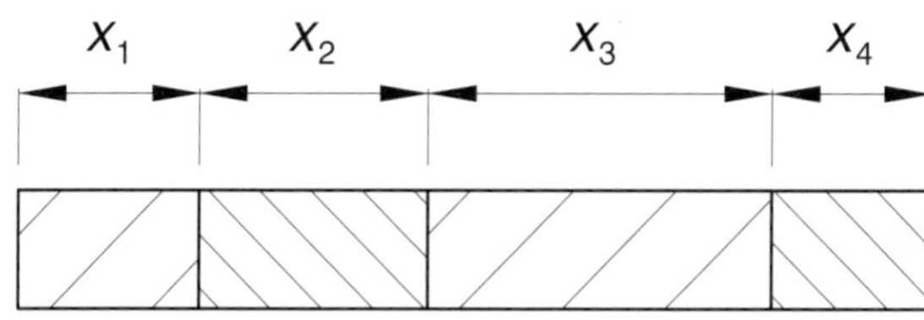

Fig. 8.11 Dimension chain.

Overall variability $= 6\sigma_z$
Tolerance limits $= \pm(6\sigma_z/2) = \pm 3\sigma_z = \pm 0.077$

The mean length of the overall assembly is given by

$$\mu_z = \mu_{x_1} + \mu_{x_2} + \mu_{x_3} + \mu_{x_4} = 8 + 10 + 15 + 7$$

$$= 40\,\text{mm}$$

$$z = 40.000 \pm 0.077\,\text{mm}$$

So 99.73 per cent of all items would have an overall dimension within these tolerance limits. (By comparison, the sure fit or extreme variability tolerance limits would be $z = 40.000 \pm 0.150$ mm, i.e. 100 per cent of all items would be within this tolerance limit.)

The basic normal model indicated a better, or more attractive (certainly for marketing purposes) tolerance band. If the sure-fit model was used and an assembly tolerance of ± 0.077 mm was desired the tolerances on the individual components would need to be 8.000 ± 0.019, 10.000 ± 0.019, 15.000 ± 0.019 and 7.000 ± 0.019 mm. This is a very fine or tight tolerance and would be costly to achieve. As the statistical model takes account of 99.73 per cent of the assemblies, use of the sure-fit approach only gives 0.27 per cent of the assemblies with better quality and would require considerable extra effort and expense.

Example Suppose the statistical tolerance bandwidth for the overall length of the previous example (± 0.077 mm) is too large for the particular application. An overall statistical tolerance of ± 0.065 mm is required. How can this be achieved?

Solution This can be achieved in several different ways. One way would be to set the tolerance (and hence the standard deviation) equal, on all the components:

$$\sigma_1 = \sigma_2 = \sigma_3 = \sigma_4 = \sigma$$

Set

$$\sigma_z^2 = (0.065/3)^2 = 4\sigma^2$$

Hence

$$\sigma = 0.0108\,\text{mm}$$

Hence the tolerance limits for each individual component would be $\pm 3\sigma = \pm 0.033$ mm.

Alternatively, the tolerance limits on say components 1 and 4 could be kept the same and the tolerance limits on components 2 and 3 could be tightened.

Suppose we decide that the tolerance limits on items 2 and 3 should be equal ($\sigma_2 = \sigma_3$):

$$\sigma_z^2 = \sigma_1^2 + \sigma_4^2 + 2\sigma_2^2$$

$$2\sigma_2^2 = 2.69444 \times 10^{-4}$$

$$\sigma_2 = 0.0116\,\text{mm}$$

So the tolerance limits on components 2 and 3 would need to be ± 0.035 mm in order for the overall tolerance of the assembly to be ± 0.065 mm.

Example Which of the two following methods of manufacturing a product 150 mm in length would give the best overall tolerance:

1. an assembly of ten components, each of length 15.000 mm with tolerance limits of ± 0.050 mm;
2. a single component of 150 mm, also with tolerance limits of ± 0.050 mm.

Solution

1. $\sigma_1 = 0.05/3$

   $\sigma_z^2 = 10\sigma_1^2$

   $\sigma_z = 0.0527$

   Tolerance limits for overall length $\pm 3\sigma = \pm 0.158$ mm:

   $\mu_z = 10\mu_1 = 150.000$ mm

   $z = 150.000 \pm 0.158$ mm

2. $z = 150.000 \pm 0.050$ mm. As this method gives smaller tolerance limits, this is the most suitable way of achieving the tighter tolerance.

8.3.3 Several independent, uncorrelated random variables

Statistical analysis of tolerances can be applied to calculate the variation in physical phenomena such as stress, fluid flow rates, compression ratio etc., as well as component dimension chains. BS 5760 states that the normal distribution model can be considered to be an appropriate distribution for modelling 80 per cent or more cases of distributional analysis in non-life circumstances. As such this analysis has wide-ranging capability.

The theoretical preliminaries that enable analysis of functions of several independent, uncorrelated random variables are given below (after Furman, 1981).

Let y be a function of several independent, uncorrelated random variables:

$$y = f(x_1, x_2, \ldots, x_j, \ldots, x_n) \tag{8.12}$$

The sure-fit or extreme variability of the variable y is given by

$$\delta y \approx \sum_{j=1}^{n} \frac{\partial f}{\partial x_j} \delta x_j \tag{8.13}$$

The statistical mean and standard deviation of a function of several independent, uncorrelated random variables are given by

$$\mu_y \approx f(\mu_1, \mu_2, \ldots, \mu_n) + \frac{1}{2}\sum_{j=1}^{n} \frac{\partial^2 f}{\partial x_j^2} \sigma_{x_j}^2 \tag{8.14}$$

$$\sigma_y^2 \cong \sum_{j=1}^{n} \left(\frac{\partial f}{\partial x_j}\right)^2 \sigma_{x_j}^2 + \frac{1}{2}\sum_{j=1}^{n} \left(\frac{\partial^2 f}{\partial x_j^2}\right)^2 \sigma_{x_j}^4$$

$$\approx \sum_{j=1}^{n} \left(\frac{\partial f}{\partial x_j}\right)^2 \sigma_{x_j}^2 \tag{8.15}$$

These equations may appear complex; however, their application is procedural and various examples are given to show their use.

Example The stiffness of a helical compression spring (*see* Fig. 8.12) can be calculated from

$$k = \frac{d^4 G}{8D^3 n}$$

where

d = wire diameter
G = modulus of rigidity
D = coil diameter

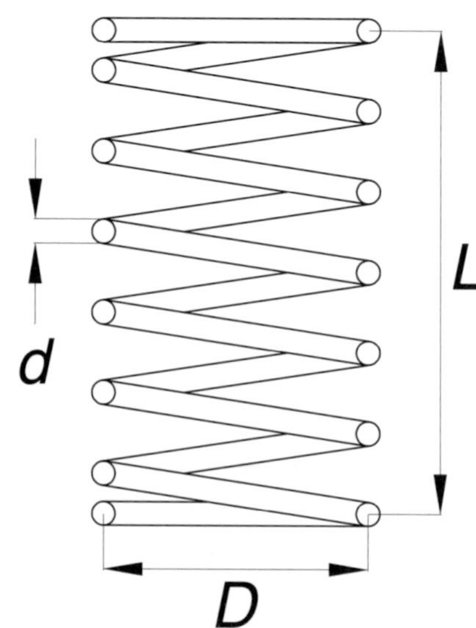

Fig. 8.12 Helical compression spring.

n = number of coils
k = spring stiffness.

The mean and standard deviation for each variable are known:

$\mu_d = 2.34\,\text{mm},\ \mu_D = 16.71\,\text{mm},$

$\mu_G = 79.29 \times 10^3\,\text{N/mm}^2,\ \mu_n = 14\,\text{coils}$

$\sigma_d = 0.010\,\text{mm},\ \sigma_D = 0.097\,\text{mm},$

$\sigma_G = 1.585 \times 10^3\,\text{N/mm}^2,\ \sigma_n = 0.0833\,\text{coils}$

Calculate the sure-fit extreme tolerance limits and the statistical basic normal tolerance limits for the spring stiffness. Assume $\pm 3\sigma$ natural tolerance limits: i.e. 99.73% of measurements lie within plus or minus three standard deviations.

Solution Substituting values into the equation for k to give the average value of the spring constant:

$$\bar{k} = \frac{2.34^4 \times 79.29 \times 10^3}{8 \times 16.71^3 \times 14} = 4.549\,\text{N/mm}$$

$$\sigma_d = 0.01\,\text{mm},\ \Delta d = 6 \times \sigma_d = 0.06\,\text{mm}$$

$$\sigma_G = 1.585 \times 10^3,\ \Delta G = 6 \times \sigma_G = 9510\,\text{N/mm}^2$$

$$\sigma_D = 0.097,\ \Delta D = 6 \times \sigma_D = 0.582\,\text{mm}$$

$$\sigma_n = 0.0833\ \text{coils},\ \Delta n = 6 \times \sigma_n = 0.5$$

$$\frac{\partial k}{\partial d} = \frac{4d^3G}{8D^3n} = \frac{2.34^3 \times 79.29 \times 10^3}{2 \times 16.71^3 \times 14} = 7.776$$

$$\frac{\partial k}{\partial G} = \frac{d^4}{8D^3n} = 5.7374 \times 10^{-5}$$

$$\frac{\partial k}{\partial D} = -\frac{3d^4G}{8D^4n} = -8.1673 \times 10^{-1}$$

$$\frac{\partial k}{\partial n} = -\frac{d^4G}{8D^3n^2} = -3.2494 \times 10^{-1}$$

$$\Delta k_{\text{sure-fit}} = \left|\frac{\partial k}{\partial d}\right|\Delta d + \left|\frac{\partial k}{\partial G}\right|\Delta G + \left|\frac{\partial k}{\partial D}\right|\Delta D + \left|\frac{\partial k}{\partial n}\right|\Delta n$$

$$= 7.776 \times 0.06 + 5.7374 \times 10^{-5} \times 9510 + 0.81673 \times 0.582 + 0.32494 \times 0.5$$

$$= 0.46656 + 0.5456 + 0.4753 + 0.16247$$

$$= 1.65\,\text{N/mm}$$

This figure ($\Delta k_{\text{sure-fit}}$) gives the overall worst case variability of the spring stiffness (i.e. the total tolerance). The bilateral tolerance can be obtained by dividing Δk by two (i.e. $1.65/2 = 0.825$).

So the spring stiffness can be stated as

$$k = 4.549 \pm 0.825\,\text{N/mm}$$

Calculation of the basic normal variability:

$$\Delta k^2_{\text{basic normal}} = \left|\frac{\partial k}{\partial d}\right|^2 \Delta d^2 + \left|\frac{\partial k}{\partial G}\right|^2 \Delta G^2 + \left|\frac{\partial k}{\partial D}\right|^2 \Delta D^2 + \left|\frac{\partial k}{\partial n}\right|^2 \Delta n^2$$

$$= 0.21768 + 0.297708 + 0.22594 + 0.026397 = 0.7677$$

$$\Delta k_{\text{basic normal}} = 0.8762\,\text{N/mm}$$

The bilateral tolerance can be calculated by dividing this value in two, i.e. the spring stiffness $k = 4.549 \pm 0.4381$.

The basic normal variation model implies a narrower band width of variation $k = 4.549 \pm 0.438\,\text{N/mm}$ versus $k = 4.549 \pm 0.825\,\text{N/mm}$ for sure-fit. This is more attractive from a sales perspective and accounts for 99.73 per cent (i.e. nearly all) of all items.

If we wanted to decrease the variation in k we would need to tighten the individual tolerances on d, G, D and n, with particular attention to d and D as these are controllable and influential on the overall variability.

The standard deviation of the spring stiffness can be calculated, if required, by $\sigma_k = \Delta k/6 = 0.146$.

Example In the design work for a model single cylinder reciprocating engine, the work done during the expansion stroke is to be represented as

$$W = \frac{P\pi D^2 C(1 - Q^{n-1})}{4(n-1)}$$

where

$$Q = \frac{C}{C + 2R}$$

and

$$C = M - R - L - H$$

The quantities are defined as:

W = work done during the expansion stroke
P = cylinder pressure at start of stroke
D = piston diameter
C = length of the clearance volume

Q = ratio of clearance to clearance-plus-swept volumes
n = expansion stroke polytropic exponent
R = crank radius
M = distance from crank centre to underside of cylinder head
L = connecting rod length
H = distance from gudgeon pin to piston crown.

It is recognised that the independent quantities are subject to variability and it is desired to determine the consequent extreme and probable variability in the work done.

If

$$P = 20 \times 10^5 \pm 0.5 \times 10^5\ \mathrm{N/m^2}$$

$$n = 1.3 \pm 0.05$$

$$D = 4 \times 10^{-2} \pm 25 \times 10^{-6}\ \mathrm{m}$$

$$R = 2 \times 10^{-2} \pm 50 \times 10^{-6}\ \mathrm{m}$$

$$M = 10.5 \times 10^{-2} \pm 50 \times 10^{-6}\ \mathrm{m}$$

$$H = 2 \times 10^{-2} \pm 25 \times 10^{-6}\ \mathrm{m}$$

$$L = 6 \times 10^{-2} \pm 50 \times 10^{-6}\ \mathrm{m}$$

determine the extreme and probable (normal model) limits of the work (Ellis (1990)).

To help with your solution, certain partial derivatives have already been calculated:

$$\frac{\partial W}{\partial R} = -2599.46$$

$$\frac{\partial W}{\partial M} = 2888.39$$

$$\frac{\partial W}{\partial H} = -2888.39$$

$$\frac{\partial W}{\partial L} = -2888.39$$

Note: it may be useful to recall that if $y = a^x$, then

$$\frac{\mathrm{d}y}{\mathrm{d}x} = x \log_e(a)$$

Solution The sure-fit 'extreme' limit of the work done is given by

$$\Delta W_{\text{sure-fit}} = \left|\frac{\partial W}{\partial P}\right|\Delta P + \left|\frac{\partial W}{\partial D}\right|\Delta D + \left|\frac{\partial W}{\partial M}\right|\Delta M + \left|\frac{\partial W}{\partial R}\right|\Delta R + \left|\frac{\partial W}{\partial L}\right|\Delta L + \left|\frac{\partial W}{\partial H}\right|\Delta H + \left|\frac{\partial W}{\partial n}\right|\Delta n$$

The average values of C, Q, and W are: $\bar{C} = 0.005$, $\bar{Q} = 0.1111$, $\bar{W} = 20.22$:

$$\frac{\partial W}{\partial P} = \frac{\pi D^2 C(1 - Q^{n-1})}{4(n-1)} = \pi 0.04^2 \times 0.005 \frac{1 - 0.1111^{0.3}}{4 \times 0.3} = 1.011035 \times 10^{-5}$$

$$\Delta P = 2 \times 0.5 \times 10^5 = 1 \times 10^5$$

$$\left|\frac{\partial W}{\partial P}\right|\Delta P = 1.011035$$

$$\frac{\partial W}{\partial D} = 2P\pi DC\frac{1 - Q^{n-1}}{4(n-1)} = 2\pi 20 \times 10^5 \times 0.04 \times 0.005 \times \frac{1 - 0.1111^{0.3}}{4 \times 0.3} = 1011.035$$

$$\Delta D = 50 \times 10^{-6}$$

$$\left|\frac{\partial W}{\partial D}\right|\Delta D = 0.0505517$$

$$\frac{\partial W}{\partial M} = 2888.39 \quad \Delta M = 100 \times 10^{-6}$$

$$\left|\frac{\partial W}{\partial M}\right|\Delta M = 0.288839$$

$$\frac{\partial W}{\partial R} = -2599.46 \quad \Delta R = 100 \times 10^{-6}$$

$$\left|\frac{\partial W}{\partial R}\right|\Delta R = 0.259946$$

$$\frac{\partial W}{\partial L} = -2888.39 \quad \Delta L = 100 \times 10^{-6}$$

$$\left|\frac{\partial W}{\partial L}\right|\Delta L = 0.288839$$

$$\frac{\partial W}{\partial H} = -2888.39 \quad \Delta H = 50 \times 10^{-6}$$

$$\left|\frac{\partial W}{\partial H}\right|\Delta H = 0.1444195$$

Let

$$W = K_3\left(\frac{1 - Q^{n-1}}{n-1}\right) \quad K_3 = \frac{P\pi D^2 C}{4}$$

$$\left|\frac{\partial W}{\partial n}\right| = K_3\left(\frac{-(n-1)\frac{\partial}{\partial n}Q^{n-1} - (1-Q^{n-1})1}{(n-1)^2}\right)$$

$$= K_3\left(\frac{-(n-1)Q^{n-1}\ln Q - (1-Q^{n-1})}{(n-1)^2}\right)$$

which on substituting for K_3 gives

$$\frac{W(n-1)}{1-Q^{n-1}}\left(\frac{-(n-1)Q^{n-1}\ln Q - (1-Q^{n-1})}{(n-1)^2}\right)$$

so that

$$\frac{\partial W}{\partial n} = -\frac{W}{n-1}\left(\frac{(n-1)Q^{n-1}\ln Q}{1-Q^{n-1}} + 1\right)$$

$$= \frac{20.22}{0.3}\left(\frac{0.3 \times 0.1111^{0.3}\ln 0.1111}{1-0.1111^{0.3}} + 1\right)$$

$$= 19.7909$$

$$\Delta n = 2 \times 0.05 = 0.1$$

$$\left|\frac{\partial W}{\partial n}\right|\Delta n = 1.979$$

$$\Delta W_{\text{sure-fit}} = 1.011035 + 0.0505517 + 0.288839 + 0.259946 + 0.288839 + 0.1444195 + 1.979 = 4.022$$

$$W = \bar{W} \pm (\Delta W/2) = 20.22 \pm 2.011\ \text{J}$$

Statistical tolerance limits:

$$\Delta W^2_{\text{normal}} = \left(\frac{\partial W}{\partial P}\right)^2 \Delta P^2 + \left(\frac{\partial W}{\partial D}\right)^2 \Delta D^2 + \left(\frac{\partial W}{\partial M}\right)^2 \Delta M^2 + \left(\frac{\partial W}{\partial R}\right)^2 \Delta R^2 + \left(\frac{\partial W}{\partial L}\right)^2 \Delta L^2 + \left(\frac{\partial W}{\partial H}\right)^2 \Delta H^2 + \left(\frac{\partial W}{\partial n}\right)^2 \Delta n^2 = 5.196$$

$$\Delta W = 2.2796$$

So

$$W = \bar{W} \pm (\Delta W/2) = 20.22 \pm 1.1398\ \text{J}$$

If required:

$$\sigma_W = 1.1398/3 = 0.38$$

8.3.4 Modern statistical design techniques and quality assurance

The traditional approach to handling quality problems is based upon defining quality as conformance to specifications. For example, if the dimension and tolerance for a component was defined as 40.0 ± 0.1 mm, it would not matter whether the component dimension was 40.1, 40.5, 40.0 or 39.9 mm, the specification would have been satisfied. A Japanese engineer called Genichi Taguchi has defined a new approach for solving quality problems based on product uniformity around the target value. In the example above, the target value is 40.0 and the approach proposed by Taguchi is to endeavour to ensure as many components as possible are near to this value. Essentially, Taguchi methods are based on the premise that quality is the avoidance of financial loss to society after the article is shipped. Within this definition, quality is related to quantifiable monetary loss, not emotional gut reaction. Taguchi techniques of ensuring good quality require a three-step approach to product or process design consisting of system, parameter and tolerance design. These are defined below.

1. **System design**: first, all the possible systems that can perform the required functions must be considered, including new ones that have not yet been developed.
2. **Parameter design**: during this stage the appropriate system parameters should be specified to improve quality and reduce costs.
3. **Tolerance design**: decisions concerning the tolerance specifications for all the components and appropriate grades of materials must be made. The objective of tolerance design is to decide trade-offs between quality levels and cost in designing new systems.

As a qualitative illustration of this technique, a case study is outlined of the production of the same design of an automotive transmission by a multinational corporation but at two different sites, one in Japan and one in the USA. Both factories manufactured the transmission within a tolerance band specified by the designer, but in market surveys it was found that American customers preferred the Japanese-manufactured transmissions. Various factors were investigated, including the dimensional tolerance chain, which is shown for the two factories in Fig. 8.13. Too tight a tolerance chain results in high wear or binding of the transmission and too loose a tolerance chain results in high noise and component impact levels. The Japanese-manufactured transmission was found to be approximately

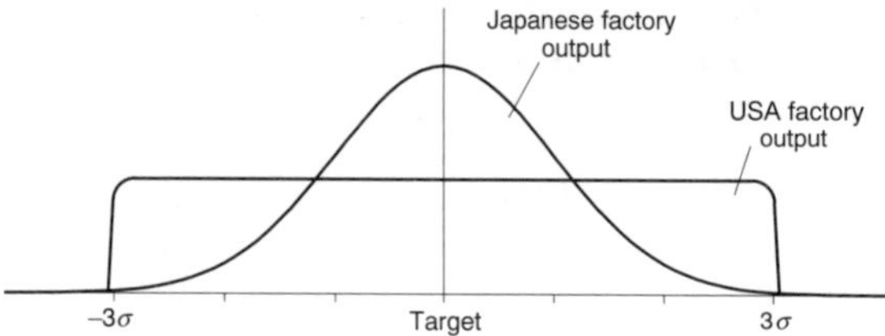

Fig. 8.13 Distribution of automotive transmission tolerance.

normally distributed about the target value, with 0.27 per cent of items falling outside the tolerance bandwidth. The output from the US factory was rectangularly distributed with all the items within the tolerance band. So the product from the US factory was always within the tolerance specifications, unlike the output from the Japanese factory which was occasionally outside the tolerance specification. However, the market surveys showed that customers had preferred the product that had originated from the Japanese factory. This initially surprising result, considering the on-target nature of the US product, can be identified as being due to proportionally more customers getting a higher quality, or nearer the target, product if supplied from the Japanese factory. The customers' perception of quality responded to the Japanese factory paying more attention to meeting the target, as opposed to the USA factory's approach of meeting the tolerances.

Application of Taguchi methods requires an understanding of statistical information along with budgetary information for each stage of the design, production and marketing process. Further information on these methods can be found in the texts by Taguchi (1993), Taguchi *et al.* (1989) and Phadke (1989).

Learning objectives checklist

Can you select a tolerance band for different practicable applications? ☐

Can you determine an interference fit torque capacity? ☐

Can you recognise a geometric tolerance symbol? ☐

Can you determine the sure-fit and statistical variability of a process? ☐

References and sources of information

Books

Chatfield, C. 1983: *Statistics for technology*. Chapman and Hall.

Ellis, J. 1990: *Mechanical design*. BEng final examination, University of Sussex.

Furman, T.T. 1981: *Approximate methods in engineering design*. Academic Press.

Haugen, E.B. 1968: *Probabilistic approaches to design*. Wiley.

Phadke, M.S. 1989: *Quality engineering using robust design*. Prentice Hall.

Spotts, M.F. 1983: *Dimensioning and tolerancing for quantity production*. Prentice Hall.

Taguchi, G. 1993: *Taguchi on robust technology development*. ASME Press.

Taguchi, G., Elsayed, E.A. and Hsiang, T. 1989: *Quality engineering in production systems*. McGraw Hill.

Standards

BSI 1983: *Quality assurance*. BSI Handbook 22.

BSI 1991: *Manual of British Standards in engineering drawing and design*, 2nd edition.

BS 1916: Part 1: 1953. Specification for limits and fits for engineering. Limits and tolerances. BSI.

BS 4500A: 1970. Specification for ISO limits and fits. Data sheet: selected ISO fits – holes basis. BSI.

BS 4500B: 1970. Specification for ISO limits and fits. Data sheet: selected ISO fits – shafts basis. BSI.

BS 5760: Part 2: Guide to the assessment of reliability. BSI.

Engineering tolerancing worksheet

1. A bronze bushing is to be installed in a steel shaft. The bushing has an inner diameter of 50 mm and a nominal outer diameter of 62.5 mm. The steel sleeve has a nominal inner diameter of 62.5 mm and an outer diameter of 87.5 mm.
   (a) Specify the limits of size for the outer diameter of the bushing and the inner diameter of the sleeve so as to provide a medium drive fit.
   (b) Determine the limits of interference that would result.
   (c) For the maximum resulting interference calculate the pressure generated between the bushing and the sleeve, the stresses in both components and the deformations.

   Take $E = 200$ GPa for steel and $E = 113$ GPa for bronze. Take $\mu = 0.27$ for both materials.
2. Determine the interference fit required to transmit 6 kW of power at 30 rpm between a 40 mm diameter steel shaft and a 100 mm outer diameter cast iron hub which is 80 mm long. Take $E = 207$ GPa for steel and $E = 100$ GPa for cast iron, the coefficient of friction as 0.12 and Poisson's ratio as 0.3 for both materials.

3. Describe the differences between clearance, transition and interference fits and outline examples of usage.
4. A device is assembled by combining three components in line whose lengths are 40 ± 0.05, 20 ± 0.05 and 35 ± 0.03 mm. If the lengths of the three components are independently normally distributed, calculate the statistical and sure-fit tolerance limits. If the tolerance limit for the overall length is deemed too wide and the tolerance limit for the third length cannot be varied, find the reduced statistical tolerance limits for length one (and length two if necessary) such that the overall tolerance limits are ± 0.07 mm.
5. Three measured variables x_1, x_2 and x_3 have small independent errors. The precisions are $\sigma_{x1} = 0.02$, $\sigma_{x2} = 0.03$, $\sigma_{x3} = 0.04$. The response variable is given by

$$z = \frac{x_1 + x_2}{4x_3}$$

Estimate the precision of z when the measurements of x_1, x_2 and x_3 are 2.49, 3.05, 2.05 respectively.
6. A floating bush seal is to be used to limit the escape of oil around a shaft. The leakage flow is assumed to be laminar and can be estimated by

$$Q = \frac{\pi \phi \Delta r^3 \Delta P}{12 \mu L}$$

where

ϕ is the diameter of the shaft
Δr is the radial clearance
ΔP is the pressure drop across the seal
μ is the dynamic viscosity
L is the axial length of the bush.

These independent quantities are subject to variability and a knowledge of the extreme and probable variability in the leakage flow rate is desired.

If

$$\phi = 0.15 \pm 5.0 \times 10^{-5}\ \text{m}$$

$$\Delta r = 1.0 \pm 0.06\ \text{mm}$$

$$\Delta P = 0.5 \times 10^5 \pm 0.025 \times 10^5\ \text{Pa}$$

$$\mu = 0.02 \pm 1.0 \times 10^{-4}\ \text{Pa} \cdot \text{s}$$

$$L = 0.01 \pm 0.1 \times 10^{-3}\ \text{m}$$

determine the extreme and probable (normal model) limits on the leakage flow.
7. Describe the differences between the sure-fit (extreme variability) and the basic normal model for accounting for variability in component assemblies or processes.
8. For a standard series 'A' Belleville spring, the spring rate or stiffness can be determined by the equation

$$k = \frac{4E}{(1 - \mu^2)} \times \frac{t^3}{K_1 D_e^2}$$

where E is the Young's modulus of the washer material, t is the washer thickness, μ is Poisson's ratio, K_1 is a dimensionless constant and D_e is the external diameter.

Determine the sure-fit and probable limits for the spring stiffness, stating any assumptions made if

$$t = 2.22 \pm 0.03\ \text{mm}$$

$$D_e = 40.00 \pm 0.08\ \text{mm}$$

$$\mu = 0.30 \pm 0.003$$

$$E = 207 \times 10^9 \pm 2 \times 10^9\ \text{N/m}^2$$

$$K_1 = 0.69$$

Comment on which tolerance could be altered to best reduce the overall variability.

Answers

Worked solutions to the above problems are given in Appendix B.

1. H7/s6 bush $62.50^{+0.03}_{+0}$ mm, shaft $62.50^{+0.072}_{+0.053}$ mm, $\delta = 0.072$ mm, $p = 21.06$ MPa, $\sigma_o = 64.9$ MPa, $\sigma_i = 95.94$ MPa, $\delta_o = 0.022$ mm, $\delta_i = 0.05$ mm.
2. 64 μm.
4. Sure-fit 95 ± 0.13, basic normal 95 ± 0.077, 40 ± 0.045, 20 ± 0.045.
5. $\sigma_z = 0.0139$.
6. $Q_{\text{sure-fit}} = 0.009817 \pm 2.409 \times 10^{-3}\ \text{m}^3/\text{s}$, $Q_{\text{basic normal}} = 0.009817 \pm 1.837 \times 10^{-3}\ \text{m}^3/\text{s}$.
8. $k_{\text{sure-fit}} = 9\,017\,000 \pm 506\,700$ N/m, $k_{\text{basic normal}} = 9\,017\,000 \pm 378\,100$ N/m.

9 Design

The aim of this chapter is to explore design in its broad context. Specifically the process of 'total design' is described, as well as design tools such as computer-aided design, optimisation and design management. A series of case studies are presented, along with a guide to the extensive literature on design.

9.1 Introduction

The term design conveys different meanings to different people. To some, design is the use of drawing tools, pen and paper or mouse and computer, to draw details of a component or device. To others, design is a sophisticated process resulting in the detailed design of an overall system such as a car or computer. The word design can be taken as either a verb or a noun and is used here in both forms.

The principal content of this text, Chapters 2–7, has been centred on detailed design, or selection, of mechanical components; devices which when used in conjunction can form a machine or other useful mechanical device. Detailed design is a part of the wider process of design. The role of the designer is to create a new device or product. This involves the application of many skills and processes from the identification of the market need, product specification, the generation of concept solutions, detailed design, manufacture and marketing. These processes can be considered as forming the core of the design activity and their collective consideration is called 'total design' – as discussed in Chapter 1 (SEED, 1985; Pugh, 1990). The process of total design is covered in Section 9.2. This design model and methodology has been adopted by many organisations, as it provides a framework allowing the design process to be competently undertaken and exploited.

Traditional design approaches are no longer effective as products have become more sophisticated, consumer driven and government influenced, and as an increasing number of concepts are protected by patenting, precluding their economic usage. Products based on 'blank sheet of paper' ideas are not that likely to be able to compete with today's global industries unless the originators have raised their comprehensive competence to such a level that they can undertake design from consideration of the system to detailed design, marketing and costing. A good example of just this is recounted by James Dyson (1997). Sustained success requires systematic thoroughness and meticulous attention to detail from the beginning to the end of the design process.

It is worth noting here the relevance of many of the subjects covered in modern engineering education. The qualities demanded of designers included detailed scientific knowledge, logical thought process, creative, conceptual and innovative abilities and communication skills. Traditional engineering science courses such as thermodynamics, heat transfer and fluid mechanics, statics and dynamics, satisfy the demands for detailed technical scientific knowledge and development of logical thought process. In addition, the cohort of courses in engineering management and technical communication satisfy requirements for organisation and communication. The remaining demands must be satisfied by studies in design and personal professional development. This requires a degree of self-cultivation and provision of the environment for creative, innovative and conceptual skills to develop. This is of course aided by a good general knowledge of what has been done, what is possible and what is scientifically feasible.

9.2 Total design

Any product, for example a car or a jet engine, requires coordinated input from several different specialisms, as illustrated in Fig. 9.1(a). Any one input considered in isolation can be regarded as partial design (*see* Fig. 9.1(b)). If it occurred on its own, no product would result. Industry is principally concerned with total design or the sum of the partial design inputs, as this results in identifiable and marketable products.

Total design is the systematic activity from the identification of the market need to the selling of the product or process to satisfy this need. Total design can be regarded as having a central core of activities consisting of market need, product design specification, conceptual design, detailed design, manufacture and marketing, and is illustrated again in Fig. 9.2 (*see* also Chapter 1).

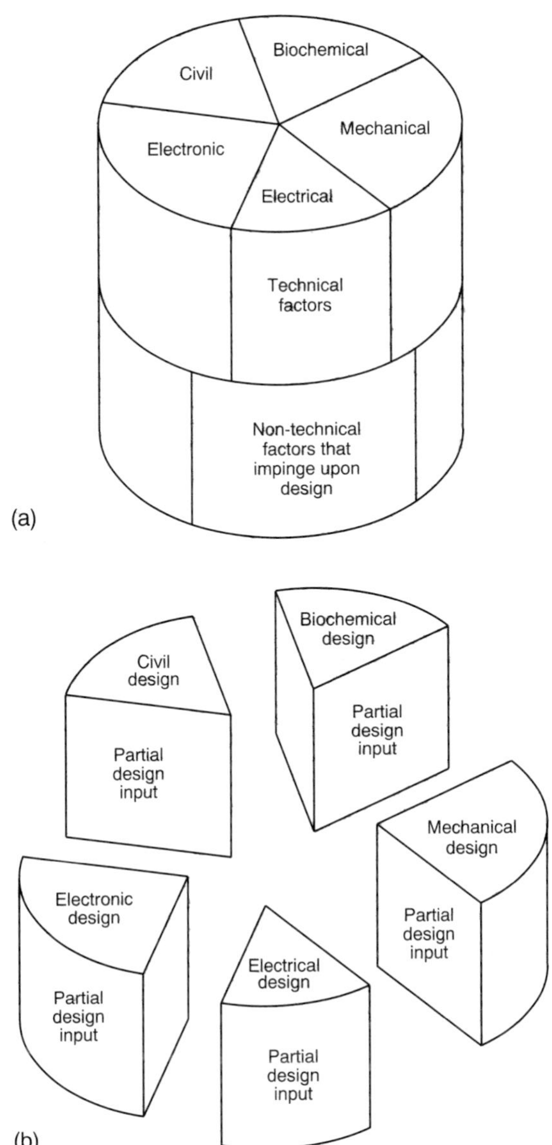

Fig. 9.1 (a) Total design inputs. (b) Partial design input (after Pugh, 1990).

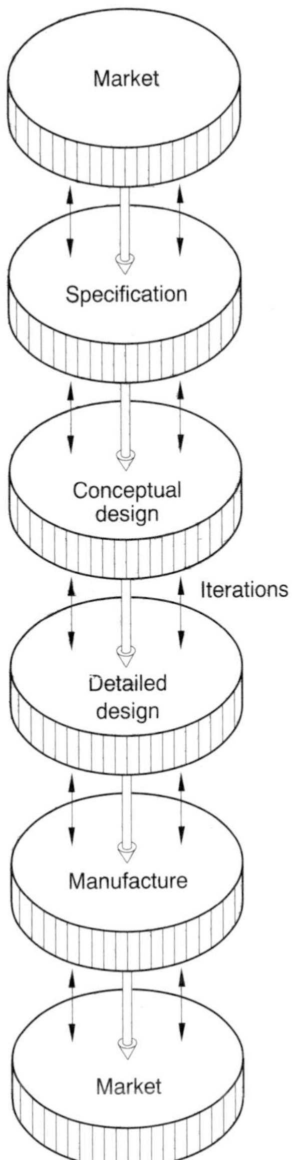

Fig. 9.2 Core design activities (after Pugh, 1990).

9.2.1 Market requirement

Design usually starts with a need or requirement which when satisfied will allow a product or process to be offered to an existing or new market. The market need may be to meet competition, a response to social changes, opportunities to reduce hazards, inconvenience or costs, or exploitation of a new market opportunity. This is the 'front end' of design. Front-end people such as market researchers and information scientists are fundamental to the design process. The starting point for any design should be the establishment of the market need situation. Common practice is to produce a document, called 'the brief', at this stage. The brief varies from a simple statement of the requirements to a comprehensive document that describes the users' true needs.

9.2.2 Product design specification

Having identified a general need or requirement and documented this in the form of a brief, the specification of the product to be developed

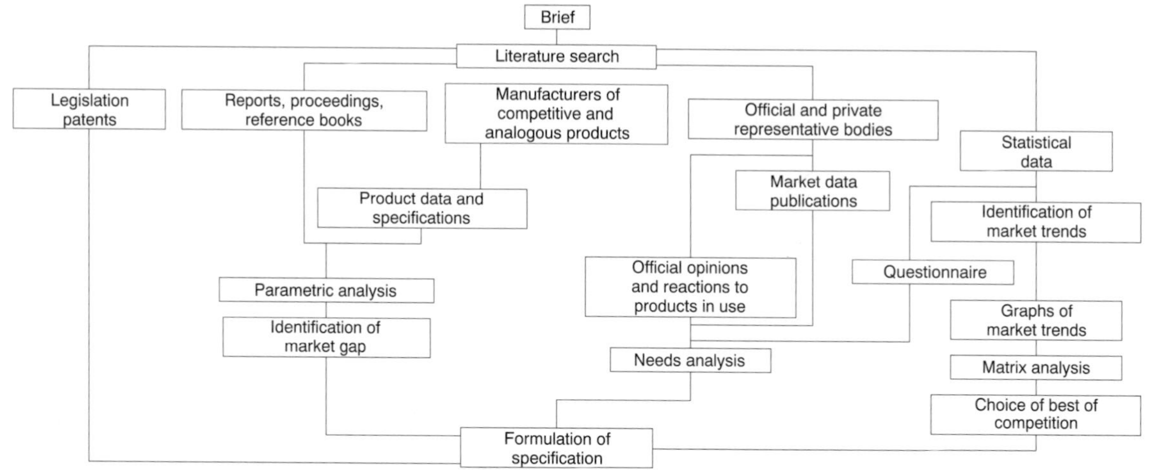

Fig. 9.3 Information required for the production of the product design specification (after Pugh, 1990).

should be generated. This is called the product design specification (PDS). The PDS acts as the controlling mechanism, mantle or envelope for the total design activity. Whatever you are concerned with, the PDS gives you the terms of reference. The starting point for design activity is market research, analysis of the competition, literature and patent searching. Once this has been undertaken the PDS can be developed. Figure 9.3 shows the areas of information and research required to produce the PDS and Fig. 9.4 shows the information content of a typical PDS. In a new design it is preferable to consider each aspect identified in Fig. 9.4; however, for the re-engineering of a current product, engineering expediency dictates that some aspects will be passed over. The PDS thus consists of a document covering a wide range of considerations. A possible format for the PDS is shown in Fig. 9.5. This shows just two of the 30-plus items listed in Fig. 9.4, but gives a sensible format for documentation of the PDS.

The PDS is dynamic. If during the design of a product there is a good reason for changing the basic PDS, this can be done. It should be noted, however, that the PDS may actually form part of contractual obligations and the legal implications must be addressed.

9.2.3 Conceptual design

Conceptual design is the generation of solutions to meet the specified requirements. Conceptual design can represent the sum of all subsystems and component parts which go on to make up the whole system. Ion and Smith (1996) describe conceptual design as an iterative process comprising a series of generative and evaluative stages which converge to the preferred solution. At each stage of iteration the concepts are defined in greater detail, allowing more thorough evaluation.

It is important to generate as many concepts and ideas as possible or economically expedient. There is a temptation to accept the first promising concept and proceed towards detailed design and the final product. This should be resisted as such results can invariably be bettered. It is worth noting that sooner or later your design will have to compete against those from other manufacturers, so the generation of developed concepts is prudent.

According to McGrath (1984), concepts are often most effectively generated by working individually and then coming together with other members of the design team at a later stage to evaluate the collective concepts. The strengths and weaknesses of the concept should be identified and one of the concepts selected and further developed, or the strengths of several of the concepts combined, or the process repeated, to generate further new concepts. Several techniques are in common use to aid idea and concept generation. These include boundary shifting, brainstorming and synectics, analogies, function trees, morphological analysis and software tools. Boundary shifting involves challenging the constraints defined in the PDS to identify whether they are necessary. For example, the PDS may define that steel should be used for a component. Boundary shifting would challenge this specification to see whether it is appropriate and, if not, other materials could be considered. Brainstorming involves a multidisciplinary group

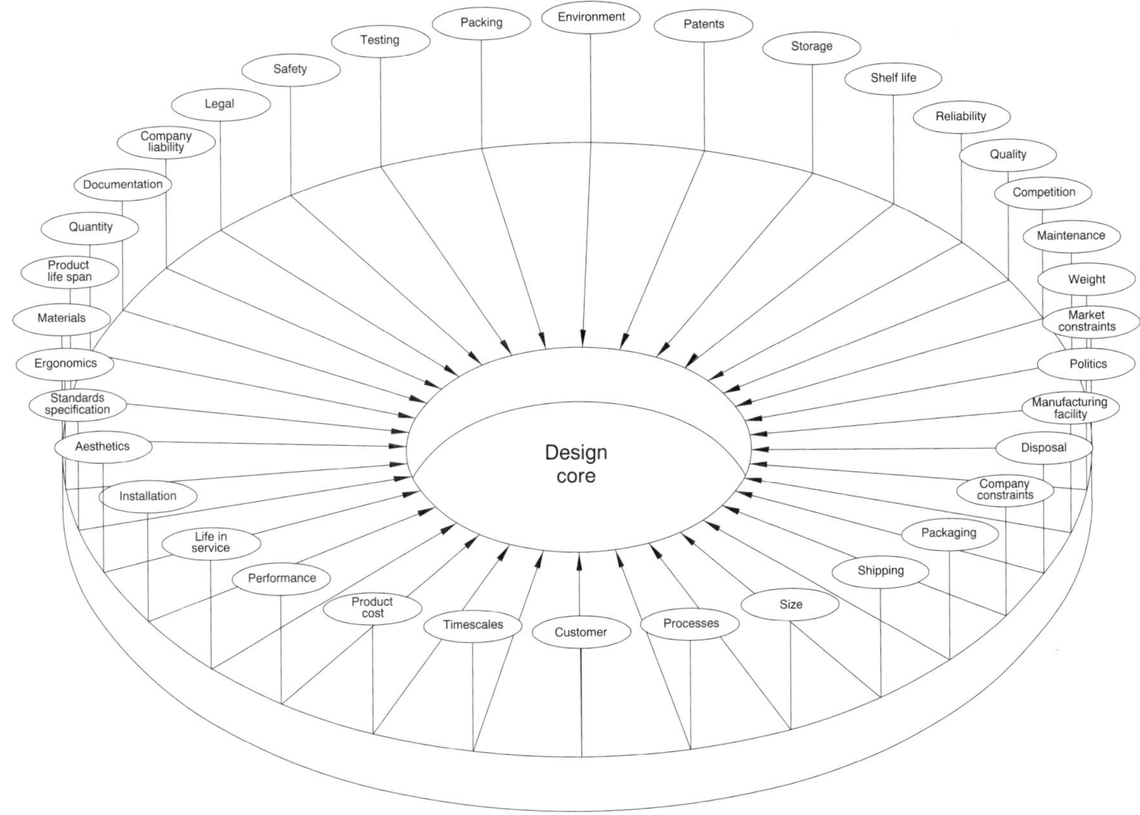

Fig. 9.4 Information content of a typical product design specification (after Pugh, 1990).

Date: .... Product: .... Issue: ....

	Parameters			
	Competition best	Current model (ours)	This design (intent)	World class (target)
Performance Description				
Safety Description				

Fig. 9.5 Format for the documentation of the product design specification (after Pugh, 1990).

Table 9.1 Evaluation tabular matrix for a pallet moving device

		Concepts							
		Hydraulic mechanism		Cable hoist		Linkage mechanism		Rack and pinion	
Selection criteria	Weight (%)	Rating	Weighted score	Rating	Weighted score	Rating	Weighted score	Rating	Weighted score
Safety	20	2	0.4	3	0.6	8	0.8	2	0.4
Positional accuracy	35	6	2.1	5	1.75	7	2.45	6	2.1
Ease of control	10	6	0.6	5	0.5	7	0.7	6	0.6
Ease of manufacture	20	7	1.4	8	1.6	5	1.0	6	1.2
Durability	15	7	1.05	6	0.9	5	0.75	6	0.9
Total score			5.55		5.35		5.7		5.2
Rank			2		3		1		4
Continue?			No		No		Develop		No

meeting together to propose and generate ideas to solve the stated problem. The emphasis within brainstorming is on quantity rather than quality of ideas and criticism of another person's idea is strictly forbidden. Synectics is a sophisticated form of brainstorming aimed at stimulating the thinking process (Chaplin, 1989). Conceptual design by analogy involves looking for solutions to equivalent problems. For example, analogies with solutions found in the natural world may provide insight to the problem in hand (French, 1994). The function tree method (Cross, 1994) involves decomposing the function of a product into different subfunctions. Using the function tree method, conceptual design takes place at the subsystem or component level. For example, it may not be possible to produce new concepts for the power plant of an automobile, but it may be possible to generate new concepts for the subsystems or for individual components. Ulrich and Eppinger (1995) propose the scheme illustrated in Fig. 9.6 for concept generation.

Having generated ideas and concepts, the next step is to evaluate them and select the best concept. One method of evaluation is the use of tabular matrices, as illustrated in Table 9.1. These consist of a series of criteria against which the concepts must be marked. The importance of the criteria can be weighted if appropriate and the most suitable concept is identified as the one with the highest overall mark. This method provides a structured technique of evaluation and makes it difficult for individuals within a design team to push their own ideas for irrational reasons.

Having generated the conceptual solutions the next step is to express these so that they are communicable to all involved in the total design process. In practice this may take the form of a drawn scheme or 3D model, either physical or computer generated. By the time the general scheme has been completed and calculations undertaken to determine the solution's compatibility to the PDS, the basis will have been established for the detailed design phase to commence.

In order to illustrate the process of concept generation a case study for a commercial nailing machine is considered. This case study reported by Ulrich and Eppinger (1995) concerns the commission of a new design for a hand-held nailer for the high-quality consumer market. The overall approach to this challenge would be to perform a marketing exercise to determine whether there is a sales opportunity for such a device, to define the product design specification and then to undertake conceptual and detailed design prior to manufacture and retailing. Here the conceptual phase only will be considered in isolation, to illustrate attributes of the conceptual design process using the scheme proposed in Fig. 9.6.

As a first step it is necessary to develop a general understanding of the problem. It can be helpful to break a large problem into smaller subproblems. As part of the clarification for this product the following information should be clarified with the management commissioning the design:

- The device will use 50 mm to 75 mm nails.
- It will be used for nailing wood.
- The nails should be fired in rapid succession (one nail per second).
- The nailer should be able to access tight working spaces.

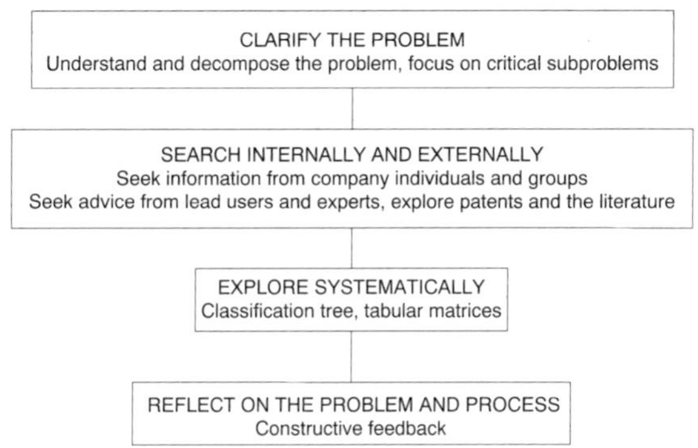

Fig. 9.6 Scheme to aid the generation of conceptual solutions (after Ulrich, 1995).

- It should be lightweight (less than 4 kg).
- There should be no noticeable delay after initialising the machine.
- The nail magazine should be common to other existing tools.

Decomposing the problem into smaller subproblems could result in considering the energy source and conversion, the nail storage, isolation and application and the input signal and triggering process each in isolation. Information gathering may include interviewing lead users of the product, seeking advice from expert consultants, a patent search, a literature survey and looking at competitor's products (competition bench marking). The knowledge and creativity of the design team should then be exploited to generate solutions. As a result of the external and internal information gathering and the design team work, many conceptual solutions may have been developed to the individual subproblems. For example, possible methods of applying the nail are illustrated in Fig. 9.7. Figure 9.8 shows an increased level of embodiment (greater detail) to some of these methods where the method of application has been proposed. The process of conceptual design in this case has consisted of splitting up the design into smaller subproblems, searching for ideas and then giving the ideas increased levels of detail. The next stage is to evaluate the concepts and select one to proceed with. Figure 9.9 shows a refined concept for the body of the nailer which would encase the nailing mechanism, hold the nail storage cartridge and provide reasonable access for nailing applications. Figure 9.10 shows a high-quality nailing machine manufactured by Senko.

It is worth noting the importance of the conceptual design stage. It has been estimated that by the time the conceptual design phase has been completed, approximately 70 per cent of the total cost of the design has been committed (Institution of Production Engineers, 1984). In addition, time and applied effort spent at the conceptual stage can avoid serious mistakes. Although the Sinclair C5 three-wheeler vehicle is now an icon of 1980s design it was actually a commercial flop. It involved a significant quantity of competent detailed design but the conceptual design was flawed. Sir Clive Sinclair admitted in hindsight that the driver was too close to the ground and too exposed to the elements and that the product did not meet the aspirations and needs of people at that time.

9.2.4 Detailed design

Conceptual design is concerned with finding solutions which meet the PDS. As you converge

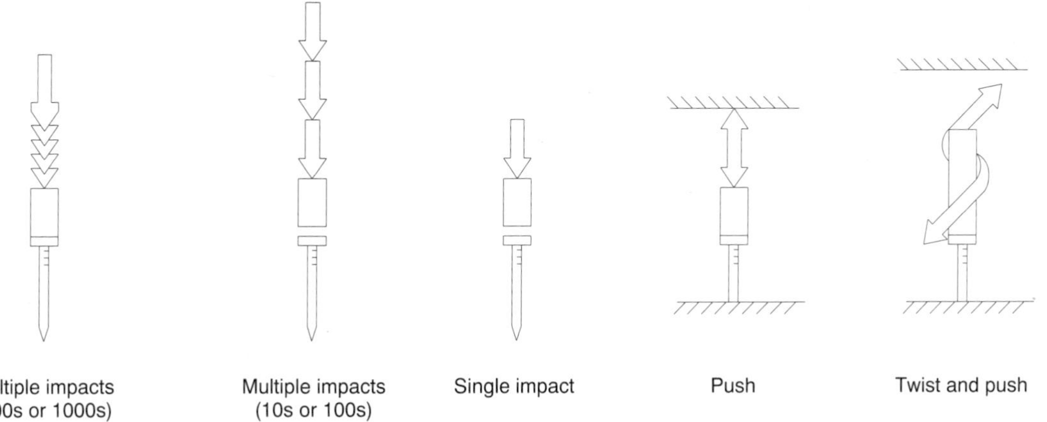

Fig. 9.7 Conceptual solutions for the nailing machine (after Ulrich, 1995).

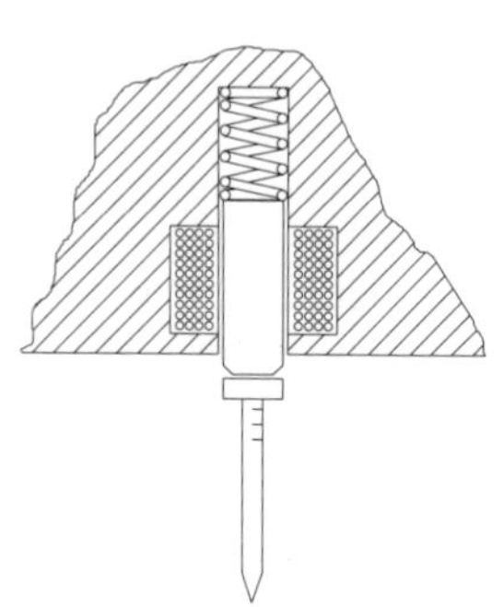
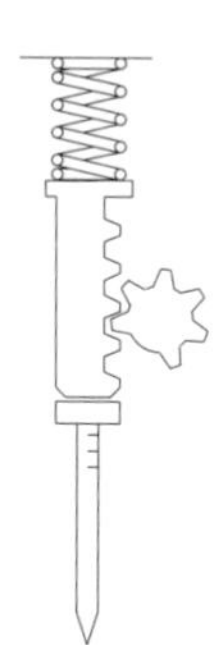
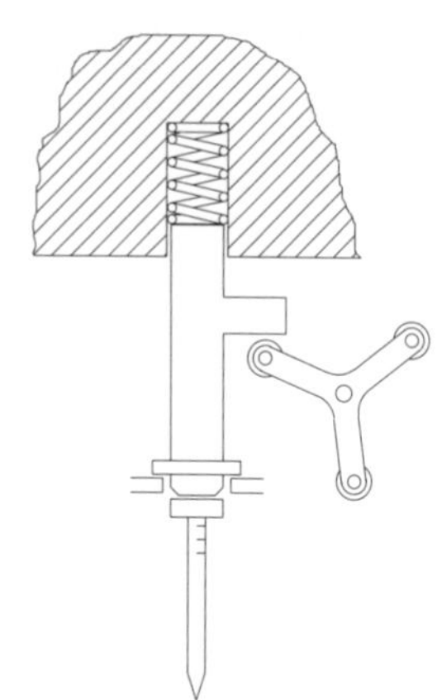

Fig. 9.8 Conceptual mechanisms for a nailing machine (after Ulrich, 1995).

towards selection of one concept you will need to expand the concept in greater detail to give a better understanding and assure engineering compatibility. The design core stages are highly iterative and interactive in practice. The phases are not necessarily distinct and run into each other and overlap. Detailed design is concerned with the technical design of subsystems and components which make up the concept, product overall system and involves defining the layout, form, dimensions, tolerances and surface finishes, materials and machining process of the product components. The product will be made up of components which to some degree will have been defined at the conceptual design phase but the final make-up of these may vary considerably from this following the detailed design phase. Ion and McCracken (1993) describe the detailed design phase as the final stage in the translation of a proposal into a fully detailed and tested product description which can be manufactured. Detailed design can involve the division of a product or process into subsystems, unit assemblies and component parts, analysis and validation and refinement of the concept and its subelements, the design and selection of components, prototyping and test evaluation and preparation of the manufacturing information such as drawings, computer models and machining coordinates. Common sense dictates that poor detailed design can wreck a good concept and excellent detailed design can turn a marginal concept into a marketing success.

Detailed design requires knowledge of materials, analysis techniques, manufacture quantities, life

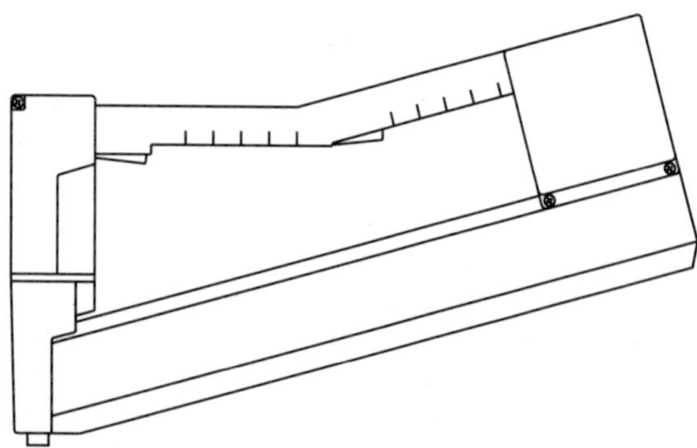

Fig. 9.9 Refined conceptual solution (Ulrich, 1995).

Fig. 9.10 Nailing machine. (Photograph courtesy of Senko.)

prediction, aesthetic appeal etc. The principal emphases in component design are: local performance, local environment, local constraints.

During the design of a component the context of the component is critical to its design. The context can be defined or quantified by the component design specification, or CDS. If issues are neglected these will contribute towards shortened life, lower quality and customer dissatisfaction. Typical component design specification elements are:

- product performance
- company constraints
- local environment
- customer requirements
- testing
- quality
- maintenance
- weight
- manufacture.
- standards
- component cost
- ergonomics
- safety and product liability
- aesthetics
- quantity
- materials

The detailed design phase relates directly to the manufacturing phase as the decisions taken have implications for the manufacturing process.

9.2.5 Manufacture

Manufacture considerations must be accounted for at all stages in a design programme, from the market investigation phase and during the specification, conceptual design and detailed design. The specification phase has particular implications for manufacture as aspects such as quality and reliability, weight, finish, quantity, cost and size may be specified. There must be materials available for the design and manufacture and assembly must be technically and economically possible. The modern approach to material selection should include consideration of metals, polymers, ceramics or composites for any component. In the preparation of a detailed manufacturing drawing considerations should include (Corbett, 1989):

Can a less expensive material be selected?
Can the quantity of material used for a product be reduced?
Is a stock material available from a supplier which does not need machining?
Must all the surfaces be machined?
What tolerances are necessary?
Can the components be held for manufacture?
Is assembly possible?

Various strategies for consideration of manufacture throughout the design process exist, such as design for manufacture (DFM), design for assembly (DFA) and concurrent engineering (McMahon and Browne (1998)). The principal aims of DFM are to minimise component and assembly costs, to minimise development cycles and enable higher quality products to be generated. DFM has two key constituent elements: design for assembly (DFA) and design for piece-part producibility (DFP). The essence of DFA is the selection of the assembly process, for example manual, high speed or robot assembly, and the subsequent design of the product for the chosen assembly process. DFP is concerned with the production of the individual constituent components. Many products are subject to cost reduction at some later stage in their commercial life. A key aim of DFM is to get the costs and quality of manufacture right initially.

Some companies operate with distinct product design and manufacturing operations. Others have integrated design and manufacture teams from the outset of a project. Concurrent engineering is a systematic approach to the integrated simultaneous design and development of both the components and their associated process such as manufacture and testing.

9.2.6 Marketing

During the marketing or sales phase of the design activity model the product is launched into the market place. The marketing technique will depend on whether selling a one-off product, a batch-produced product or a mass-produced product. Sales marketing will involve advertising, distribution, servicing and maintenance. Of direct relevance to the design process is feedback of information to the design team concerning customer reaction to the product: features that are liked or disliked, any component failures and requirements for redesign, any servicing or maintenance issues.

9.2.7 The total design process

As outlined, the total design process requires overall system and detailed constituent component consideration, with an interdisciplinary input of skills from marketing to system and component analysis. The total design process is shown again in Fig. 9.11, illustrating the design core, the detailed considerations at each level, which may be prioritised, and also the disciplinary skills required. As mentioned previously the total design process is iterative, with tasks being repeated until acceptable or optimal solutions have been generated.

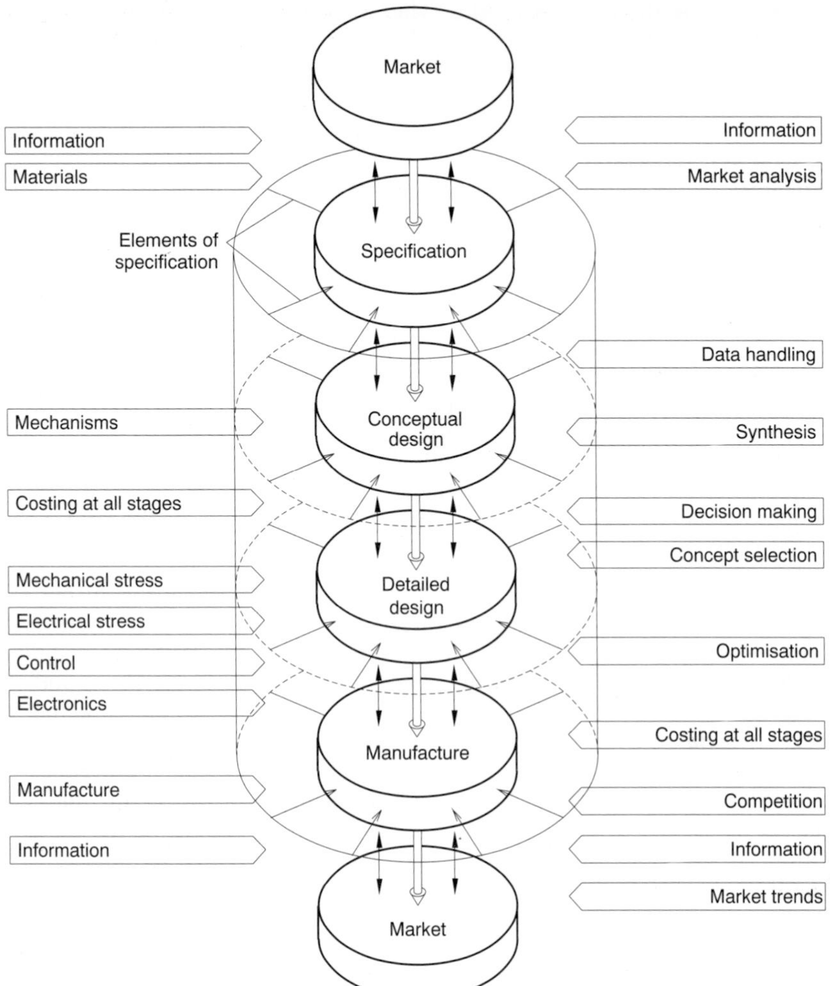

Fig. 9.11 The total design process, showing the core, steps and associated inputs (after Pugh, 1990).

9.3 Design tools

Designers use various analysis and management tools, such as optimisation, computer-aided design, project management and decision-making strategies. The need for competitive products and the development of complex systems with multiple requirements demand that products be optimised. Some of the methods available are outlined in Section 9.3.1. A typical engineering project may involve the performance of hundreds of tasks for its successful completion. Two methods, CPM and PERT, are currently in widespread use for the management of design projects. These are discussed in Section 9.3.3. As mentioned in Section 9.2.3, the evaluation of concepts to select one option is critical to the design process. Decision making is an important engineering skill and various methods are available for its quantification; some of these are outlined in Section 9.3.4.

9.3.1 Optimisation

For a given design requirement many possible solutions may exist. The design task is to seek the best trade-off or compromise between often conflicting requirements and constraints. The process of determining the best solution is called optimisation. There are many techniques available to the engineer including differentiation, Lagrange multipliers and linear programming.

Differentiation For a multiple independent variable function, take partial derivatives with respect to each variable and equate the resulting equations to zero; e.g. for the function $f(x_1, x_2, x_3, \ldots, x_n)$, taking derivatives with respect to each variable in turn gives

$$\frac{\partial f(x_1, x_2, x_3, \ldots, x_n)}{\partial x_1} = 0 \qquad (9.1)$$

$$\frac{\partial f(x_1, x_2, x_3, \ldots, x_n)}{\partial x_2} = 0 \qquad (9.2)$$

$$\vdots$$

$$\frac{\partial f(x_1, x_2, x_3, \ldots, x_n)}{\partial x_n} = 0 \qquad (9.3)$$

The next step is to solve the resulting equations simultaneously.

Example The revenue from selling x components is given by $r(x) = 9x$ and the cost of production and marketing is given by $c(x) = x^3 - 6x^2 + 15x$. What, if any, is the level of production which maximises profit?

Solution The profit is given by $p(x) = r(x) - c(x)$:

$$p = 9x - x^3 + 6x^2 - 15x$$

$$\frac{\partial p}{\partial x} = -3x^2 + 12x - 6$$

Setting $dp/dx = 0$ gives

$$-3x^2 + 12x - 6 = 0 \quad \text{or} \quad x^2 - 4x + 2 = 0$$

Solving for x gives

$$x = 2 \pm \sqrt{2}$$

The solution represents a local maximum given by $2 + \sqrt{2}$ and a minimum given by $2 - \sqrt{2}$; i.e. the profit is a maximum when $x = 2 + \sqrt{2}$.

Lagrange multipliers To find the stationary points of the function $u = f(x, y, z)$ subject to the constraint $\phi(x, y, z) = 0$, determine the differentials:

$$\frac{\partial u}{\partial x} + \lambda \frac{\partial \phi}{\partial x} = 0$$

$$\frac{\partial u}{\partial y} + \lambda \frac{\partial \phi}{\partial y} = 0 \qquad (9.4)$$

$$\frac{\partial u}{\partial z} + \lambda \frac{\partial \phi}{\partial z} = 0$$

The solution of these equations will give values for x, y and z.

Example Determine the total height and diameter of a storage tank, comprising a hemispherical top and cylindrical body, if the container holds $500\,\text{m}^3$ of liquid and the surface area is to be minimised.

Solution The surface area is given by

$$A = 3\pi r^2 + 2\pi hr$$

The volume of the tank is given by

$$V = \pi r^2 h + (2/3)\pi r^3 = 500$$

$$\pi r^2 h + (2/3)\pi r^3 - 500 = 0 \quad \text{(constraint equation)}$$

We need to find

$$\frac{\partial A}{\partial r} + \lambda \frac{\partial V}{\partial r} = 0 \quad \text{and} \quad \frac{\partial A}{\partial h} + \lambda \frac{\partial V}{\partial h} = 0$$

$$\frac{\partial A}{\partial r} = 6\pi r + 2\pi h \qquad \frac{\partial V}{\partial r} = 2\pi hr + 2\pi r^2$$

$$\frac{\partial A}{\partial h} = 2\pi r \qquad \frac{\partial V}{\partial h} = \pi r^2$$

$$6\pi r + 2\pi h + \lambda(2\pi hr + 2\pi r^2) = 0 \quad 2\pi r + \lambda \pi r^2 = 0$$

$$\lambda = -2/r$$

$$6\pi r + 2\pi h - 4\pi h - 4\pi r = 0$$

$$r = h$$

From the constraint equation

$$\pi r^3 + (2/3)\pi r^3 = 500,$$

$$r = (1500/5\pi)^{0.333} = 4.57\,\text{m}$$

$$\phi = 9.14\,\text{m and height} = h + r = 9.14\,\text{m}.$$

Linear programming Linear programming is a mathematical method of allocating constrained resources to attain an objective such as minimum cost, maximum life or profit. There are three basic steps involved in developing a linear programming model:

1. Define the decision variable.
2. Define the objective function.
3. Define the constraints.

The generalised version of the linear programming model can be expressed as the requirement to minimise or maximise a function such as $M = c_1 y_1 + c_2 y_2 + c_3 y_3 + \cdots + c_m y_m$ (called the objective function), subject to the constraints:

$$b_{11} y_1 + b_{12} y_2 + b_{13} y_3 + \cdots + b_{1m} y_m \quad (\leq, =, \geq) R_1 \qquad (9.5)$$

$$b_{21} y_1 + b_{22} y_2 + b_{23} y_3 + \cdots + b_{2m} y_m \quad (\leq, =, \geq) R_2 \qquad (9.6)$$

$$b_{31}y_1 + b_{32}y_2 + b_{33}y_3 + \cdots + b_{3m}y_m \quad (\leq, =, \geq)R_3 \tag{9.7}$$

$$\vdots$$

$$b_{k1}y_1 + b_{k2}y_2 + b_{k3}y_3 + \cdots + b_{km}y_m \quad (\leq, =, \geq)R_k \tag{9.8}$$

$$y_1, y_2, y_3, \ldots, y_m \geq 0 \tag{9.9}$$

where

m = number of decision variables
k = number of constraints
y_i = the ith decision
b_{ji} = amount of the resource the decision variable y_i consumed per unit of activity for $j = 1, 2, 3, \ldots, k$ and $i = 1, 2, 3, \ldots, m$
R_j = total resource for $j = 1, 2, 3, \ldots, k$
c_i = penalty or contribution per unit of activity for $i = 1, 2, 3, \ldots, m$.

Linear programs can be solved analytically or graphically, although graphical methods are quite complex for problems involving more than two decision variables.

Optimisation has been the focus of intense study and the reader is referred to the texts by Siddall (1972) and Johnson (1980) for a formal introduction.

9.3.2 Computer-aided design

Computer-aided design is the process of solving design problems using computers. This process includes analysis of design, manipulation of design-related information, generation of design concepts and generation and manipulation of graphical and geometrical drawings or solid models. There are many types of computer tools that aid in design, each with certain capabilities and limitations that make it suitable to particular phases or techniques in the design process. Computer software which supports design activities is generally called CAD software. Ideally, CAD software would be used to support all the phases of the design process, from specification, development and conceptual design, to product design. However, this is not yet possible because existing computer software/tools require a refined representation of an object on which to operate and are not good at handling the abstract information generated at the conceptual design phase. Current design software packages are basically evaluation tools. The techniques for generating concepts and products are not yet well-enough understood to be incorporated in commercial software.

Software tools can be subdivided into four categories:

- general purpose analysis tools
- special purpose analysis tools
- drafting/visualisation tools
- expert systems.

General purpose analysis tools are like mathematical word processors which allow calculation of whatever can be modelled in terms of simple equations. The most common type of analysis tool is the spreadsheet – a multidimensional grid used to collect and calculate data. During the design process it is often useful to explore the sensitivity of one or more parameters to variations of another parameter. For problems with just a few parameters this can easily be achieved using a spreadsheet. Numerical equation solvers allow solutions of much more complex equations than spreadsheets. Symbolic equation solvers are also available. These form powerful tools which treat each variable as an object with a known relationship to other variables (it makes no difference which variables are dependent and which are independent).

Special analysis tools include stress and strain analysis tools, kinematic and dynamics analysis tools and fluid and thermal analysis tools. These programs typically model geometry as a grid, mesh or series of elements and solve the appropriate set of equations by approximate numerical methods. Commercial examples include ANSYS, LUSAS, CFD 2000, FLOW3D, STAR-CD and FLUENT.

Computer-aided drafting packages are now standard tools for most engineers. CAD drafting and visualisation tools aid the design process by assisting in the visualisation of the design, improving data organisation and communication, acting as pre-/post-processors for analysis (e.g. Autocad and ANSYS) and manufacture.

An expert system is a set of rules, called a rule base, and an inference engine which, given a set of conditions, goes and looks at the rules and comes up with a response. Expert systems can be programmed with languages such as Prolog or Lisp. These are not procedural languages (like Fortran or Basic) but declarative (i.e. there is no order imposed, they are essentially a set of rules; e.g. if this ... then that ... else ...). Expert systems can be used to assist in the location of complex design network configurations, cost estimation and project planning.

9.3.3 Design management

The management of design projects is vitally important. Several methods have been proposed for effective management, including two which are in widespread use: the critical path method (CPM) and the program evaluation and review technique (PERT) (*see* Malcolm *et al.*, 1959; West and Levy, 1969). The basic theory and terms used in both methods are similar. The general characteristics of a CPM or PERT project are:

1. There are clearly defined activities which once completed will result in the completion of the project.
2. Once started, the activity continues without interruption.
3. The activities are independent.
4. The activities are ordered and follow each other in a specified manner.

The steps involved in constructing a CPM network are:

1. Break the project down into individual activities and identify each activity.
2. Estimate the time required for each activity.
3. Determine the sequence of activities.
4. Construct the CPM network.
5. Determine the critical path of the network.

The steps involved in developing a PERT network are:

1. Break the project down into individual activities and identify each activity.
2. Determine the sequence of activities.
3. Develop the PERT network.
4. Obtain the expected time to perform each activity (equation 9.10).
5. Determine the critical network path.
6. Calculate the variance associated with the estimated time of each activity (equation 9.11).
7. Calculate the probability of accomplishing the design project in the stated time (equation 9.12):

$$T_e = \frac{x + 4y + z}{6} \qquad (9.10)$$

where

T_e = expected time for the activity
x = the optimistic estimate for the activity
y = the most likely estimate for the activity
z = the pessimistic estimate for the activity.

$$\sigma^2 = \left(\frac{z - x}{6}\right)^2 \qquad (9.11)$$

$$P = \frac{T - T_L}{\sqrt{\sum \sigma_{cr}^2}} \qquad (9.12)$$

where

σ_{cr} = standard deviation of the activities on the critical path
T_L = the last activity's earliest expected completion time
T = design project due date.

The network consists of a series of paths used to reach project completion. The critical path is the longest path with respect to time through the PERT or CPM network.

CPM is generally used when the activity times are reasonably predictable and PERT is used when the duration of activities are uncertain, as in, for example, research and development projects.

9.3.4 Decision making

Decision making is a key part of the design process during which the designer endeavours to provide a solution to the design opportunity posed by the customer. The designer generally uses the criteria of function, safety, economy, ease of manufacture and marketability. To achieve these criteria the designer may make product decisions such as: anticipated market, component elements, fabrication methods, evolutionary design or original design, expected maintenance, types of loading, target costs, energy sources, controls, materials, expected life, permissible stresses and distortions.

From this point on the decisions required to establish the solution may be more detailed such as: strength of each element, allowable distortion, government regulations, control requirements, anticipated friction, geometry, reliability of each element, style, governing standards, surface finish, lubrication requirements, tolerances, maintenance requirements, noise limitations, anticipated corrosion, anticipated wear.

To make decisions effectively a rational problem-solving technique is necessary. The first step in problem solving is to provide a statement defining the problem. The essential steps are:

- a need statement
- goals, aims, objectives
- constraints and allowable trade-offs
- definitions of terms and conditions
- criteria for design evaluation.

The steps for making a good design may be summarised as:

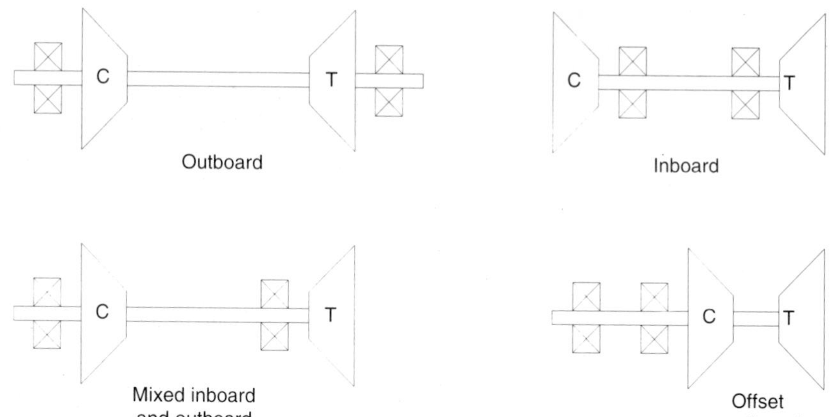

Fig. 9.12 Small gas turbine engine and turbocharger bearing configurations.

- Establish the objectives of the decision to be made.
- Classify objectives by importance, identifying musts, should and wants.
- Develop alternate actions.
- Evaluate alternatives for meeting the objectives.
- Choose the option having the most promising potential for achieving the objectives as the tentative decision.
- Explore future consequences of the tentative decision for adverse effects.
- Control effects of final decision by taking appropriate action, whilst monitoring both the implementation of the final decision and the consequences of the implementation.

Techniques in common use for the quantification of decision making include the suitability, feasibility, acceptability and figure of merit methods.

The suitability, feasibility, acceptability (SFA) method of evaluation can be used to evaluate several proposals. The method is based on determining the suitability, feasibility and acceptability of the proposal using the following procedure, and then evaluating the results (Mischke, 1980):

1. Develop a problem statement.
2. Specify a solution.
3. Is the solution suitable? Does it solve the problem?
4. Is the solution feasible? Can it be implemented with the personnel, time and knowledge available?
5. Is the proposal acceptable?

The figure-of-merit (FOM) method can be used to judge between alternate proposals. Marks are assigned for, say, cost, design factor, safety factor, reliability and time. The marks are combined to produce a total FOM. This is performed for all the solutions and the FOMs compared. The proposal with the highest FOM can then be selected.

9.4 Case studies

The case studies following have been selected to show the wider context of some of the machine components studied. For a range of in-depth case studies the reader is referred to the text by Matthews (1998).

9.4.1 Small gas turbine engine bearing configuration

The four possible basic bearing arrangements for small gas turbine engines and turbochargers are (*see* also Fig. 9.12):

1. outboard
2. inboard
3. mixed inboard and outboard
4. offset outboard.

Outboard bearing installation (*see* Fig. 9.13) provides the minimum load on the bearings due to rotor imbalance. The shaft diameter can be kept small without serious problems occurring at critical speeds. The installation provides good access to the bearings, permitting easy maintenance and bearing replacement. When using rolling element bearings the arrangement is ideal for self-lubricating systems. The design layout is suitable for turbochargers with axial flow turbines and is popular with medium and large turbocharger manufacturers. From the aerodynamic perspective the outboard bearing arrangement requires a long inefficient U-shaped intake to the impeller eye, with webs supporting the bearing housing which obstruct the inlet flow.

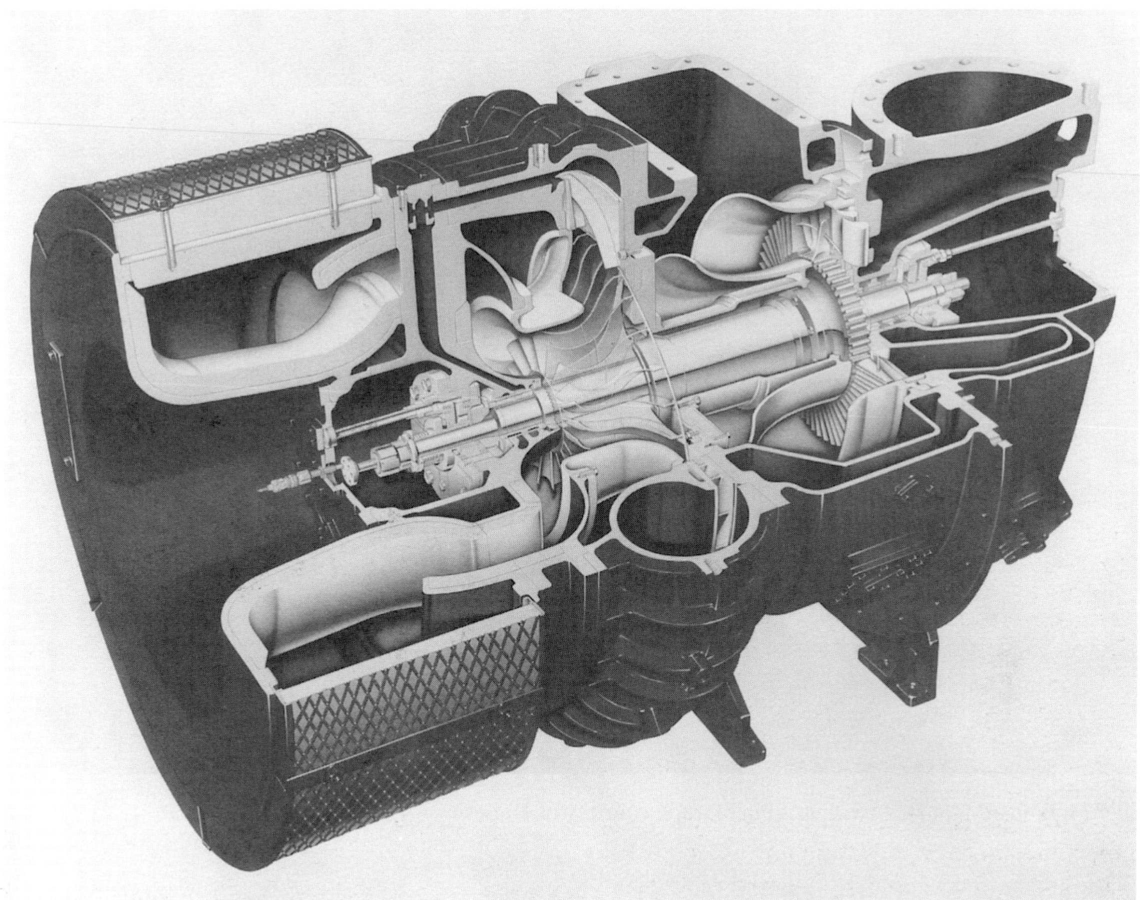

Fig. 9.13 Outboard mounted bearings. (Photograph courtesy of Napier.)

The demand for a simple, light and cheap turbocharger has led to the use of inboard bearings. This is ideally suited for radial turbine turbochargers, as shown in Fig. 9.14, because the arrangement permits direct entry of air into the compressor and turbine. The inboard bearing arrangement reduces the component count, weight and length in comparison to the outboard arrangement. The result is a cheaper product. The disadvantage of the layout is the inherently unstable running of the rotor assembly due to the short distance between the bearings and the weight of the overhanging turbine wheel (e.g. a steel alloy turbine versus an aluminium compressor).

The location of the turbine end bearing between the compressor and turbine provides easy access for the flow to and from the turbine. However, this arrangement has many of the disadvantages of the two arrangements described above and is rarely used.

Offset bearing mounting provides a simple and self-contained bearing housing assembly with good bearing access (*see* Fig. 9.15). The mounting allows use of narrow shafts, resulting in stable and quiet running above the critical speed of the turbocharger assembly. This assembly is only suitable for centrifugal compressor–radial turbine arrangements, and leads to a complex impeller–turbine unit which presents problems of heat transfer from the turbine to the centrifugal impeller.

Both small and large turbochargers have been designed with both sliding and rolling element bearings, except for automotive turbochargers where cost minimisation demands use of sliding bearings. The choice of bearing type is dependent on the application and user's preference. Rolling bearing merits include:

1. low frictional loss in the bearings at start-up, and at low-speed operation;

Fig. 9.14 Inboard mounted bearings. (Photograph courtesy of Holset.)

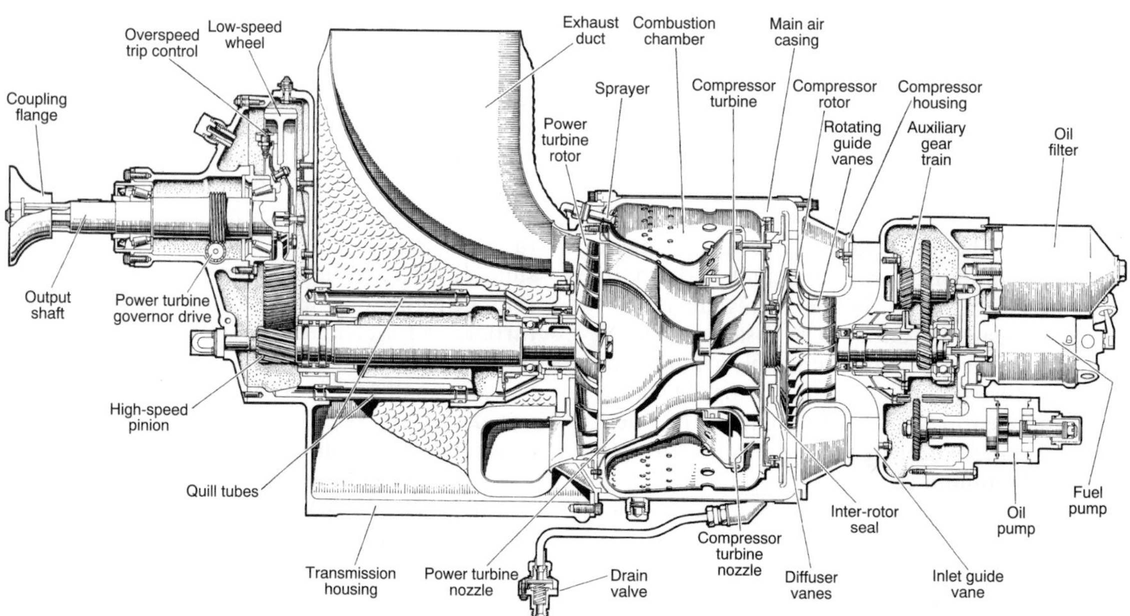

Fig. 9.15 Offset mounted bearings. (Courtesy of Rover.)

2. ability to withstand heavy overload for short periods;
3. insensitivity to lubrication starvation for short periods;
4. self-contained internal lubrication system.

Journal bearings give almost an indefinite life, are capable of taking larger out-of-balance loads and are insensitive to external shocks and vibrations. However, they require high oil pressure and large oil flow rates for lubrication and cooling, necessitating an independent oil supply including a high pressure oil pump, filter and cooler.

For small turbochargers, plain bearings in the form of rotating floating bushes are generally used. The relative speed between the rotating bearing surfaces are kept low to reduce friction losses. The extra fluid film provides a fractional frequency whirl which is better damped; it provides better damping of the rotor critical frequency and larger effective bearing clearance, permitting larger out of balance.

9.4.2 Shaft design

The shaft shown in Fig. 9.16 is to be designed taking into account strength, critical speed and rigidity. 8 kW of power at 900 rpm is supplied to the shaft by means of a belt drive and transmitted through a gear. The shaft is supported on deep groove ball bearings. The shaft is manufactured using 817M40 hot-rolled alloy steel. The pitch circle diameter of the 20° pressure angle gear is 250 mm and the pulley diameter is 250 mm. The masses of the gear and pulley are 8 kg and 10 kg respectively. The ratio of belt tensions is 2.5. Profiled keys are used to transmit torque through the gear and pulley. A nominal reliability of 90 per cent is required. The slope at the bearings must not exceed 0.05° and the deflection at the gear must not exceed 0.025 mm for reliability and noise reasons. The operating speed of the shaft should not exceed 60 per cent of the lowest critical speed.

The specifications for this example are the same as used to illustrate the application of the ASME equation for the design of transmission shafting in Chapter 3. From that example, the maximum torque and bending moment were found to be 158.5 N·m and 84.9 N·m respectively, and the resulting required shaft diameter was approximately 27 mm. The ASME equation, however, does not consider deflection, slope or critical speed requirements, and these must be addressed independently.

The deflection at the gear can be determined using Macaulay's method and the resultant deflection is found to be

$$y_{\text{resultant}} = \frac{0.2654}{EI}$$

Substituting for I and rearranging for the shaft diameter required to limit the deflection to 0.025 mm

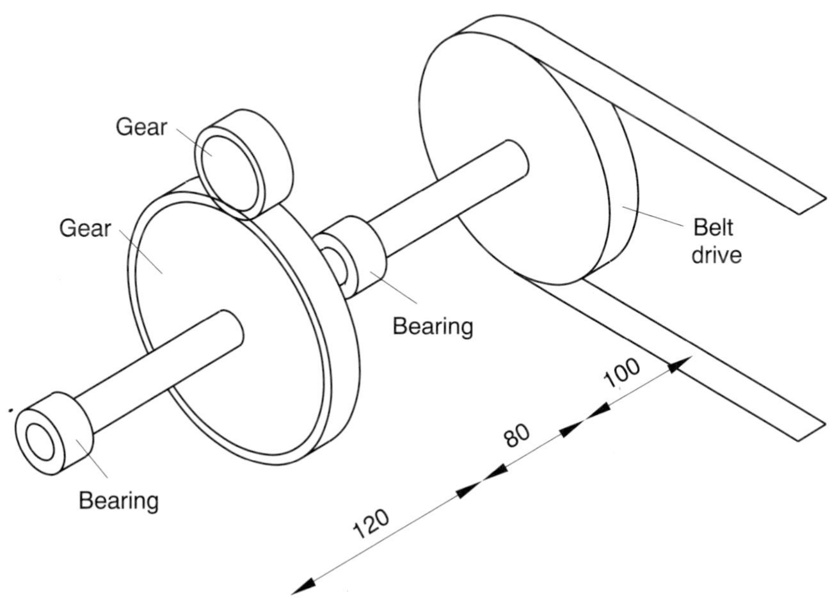

Fig. 9.16 Power transmission shaft.

gives

$$D = \left(\frac{64 \times 0.2654}{\pi \times 200 \times 10^9 \times 0.025 \times 10^{-3}} \right)^{0.25}$$

$$= 0.032\,\text{m}$$

Similarly, the resultant slope at the left and right hand bearings can be found:

$$\left(\frac{dy}{dx} \right)_{\text{resultant left bearing}} = \frac{3.261}{EI}$$

$$\left(\frac{dy}{dx} \right)_{\text{resultant right bearing}} = \frac{8.127}{EI}$$

The diameter required to limit the slope to 0.05° for the right hand bearing (greater slope) is

$$D = \left(\frac{64 \times 8.127}{\pi \times 200 \times 10^9 \times 8.727 \times 10^{-4}} \right)^{0.25}$$

$$= 0.031\,\text{m}$$

The deflection of the shaft due to the gear and pulley weights at the gear and pulley locations respectively are

$$y_{\text{gear}} = -\frac{0.01306}{EI} \quad \text{and} \quad y_{\text{pulley}} = \frac{0.07803}{EI}$$

Using absolute values for these deflections and solving the Rayleigh–Ritz equation so that the critical frequency is $\geq 900/0.6 = 1500$ rpm gives $D = 0.012$ m.

The necessary shaft diameters for each of the design criteria are summarised in Table 9.2. As can be seen in this case, the constraint on gear deflection is dominant and the shaft should be manufactured with a diameter greater than 32 mm.

Table 9.2 Calculated shaft diameters

	Diameter (mm)
ASME equation	27
Gear deflection	32
Slope at bearings	31
Critical speed	12

9.4.3 Double reduction gear box

Consider the design of a double reduction gear box illustrated in Fig. 9.17. This consists of four gears, six bearings, three shafts and the housing. The arrangement of the gears, the location of the bearings and the general configuration of the housing are all design decisions. The design process cannot readily proceed until these decisions have been made. In the schematic of Fig. 9.17 the integration of the individual components has begun. When the overall design is conceptualised the design of the individual elements can proceed.

For the gear pairs the designer must specify the number of teeth in the gears, the module, the pitch diameters, the face width, the material and any heat treatment. These specifications depend on consideration of the strength and wear on the gear teeth and the kinematic requirements. The gears must be mounted on the shafts in a manner that ensures location of the gears, adequate torque transmission and safe shaft design.

Having designed the gear pairs, the shaft design should be considered. The shafts are loaded in bending and torsion because of the forces acting at the gear teeth. The shaft design must consider strength and rigidity, and permit mounting of the

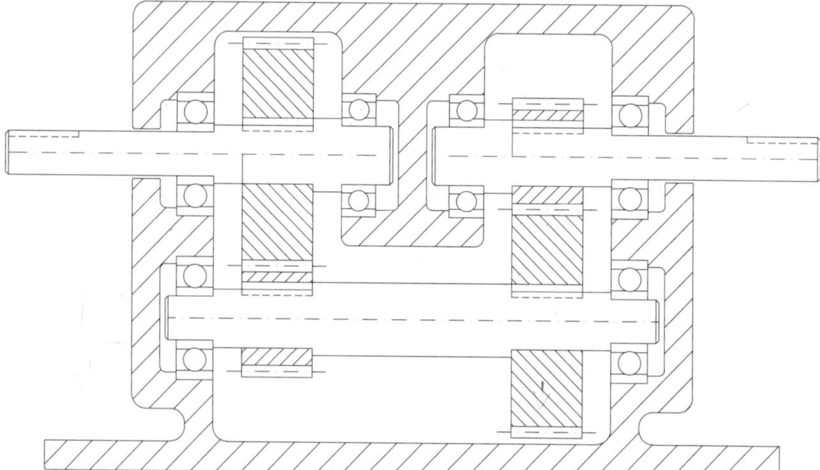

Fig. 9.17 Double reduction gear box design example.

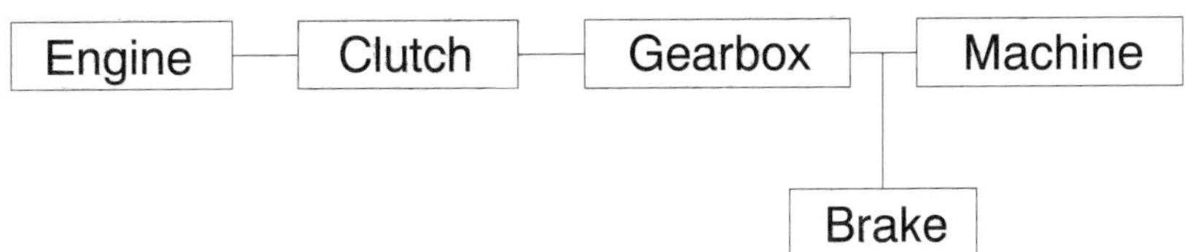

Fig. 9.18 Gear transmission drive.

gears and bearings. Shafts of varying diameter can be used to provide shoulders against which to seat the gears and bearings. Keyseats can be used to locate the gears and transmit the torque. The diameter of the shaft at the key determines the key's allowable size (width, height). The torque that must be transmitted is used in strength calculations to specify the key length and material.

If rolling contact bearings are to be used, these can be selected from commercially available stock bearing catalogues. The magnitude of the loads on the bearings, both axial and radial, must be determined. By considering the rotational speed and reasonable design life, the bearing can be selected. On the basis of the shaft analysis, the designer could specify the minimum allowable diameter at each bearing seat location to ensure safe stress levels.

Once the working components have been selected and designed, the housing design can proceed. This should be both creative and practicable. Provision should be made to mount the bearings accurately and transmit the bearing loads safely through the case to the structure on which the speed reducer is mounted. Consideration of how the various elements are to be assembled within the housing must be made. Provision for gear and bearing lubrication in terms of lubricant circulation and containment should also be considered. Decisions concerning the housing material (whether it should be a casting or of welded construction etc.) form the next step. Maintenance, ease of assembly, uniformity of bolts and fixtures, need for special tools etc., and modularity to allow the component to be fixed to as many power sources and loads as possible, are additional considerations. The design of this kind of system may well require several iterations, both for the individual components and for the overall assembly.

9.4.4 Power transmission

The two examples given below are summative of the content of this text in that they utilise variants of the machine elements presented.

Example A transmission drive is required to transmit 50 kW from a diesel engine running at 3000 rpm, to a low-speed machine running at 40 rpm in an oily environment. The application requires no load on the engine at start up and provision to reduce the machine speed suddenly. Select and scheme out the basic elements of the drive.

Solution The requirement is for a significant speed reduction ratio operating within an oily environment. This could be achieved in a number of ways, one of which would be to use a reduction gear box with, say, three gear sets, another to use a double reduction gear box with a final chain drive. The oily environment precludes the use of a belt drive. The requirements for an unloaded start-up and sudden speed reduction demand the use of a clutch and brake respectively. The two proposed solutions are schemed out in Figs 9.18 and 9.19.

Example A drive is required to transmit 40 kW from an electric motor running at 1490 rpm, to an agricultural machine running at 130 rpm. The motor must start up unloaded and there is a requirement for a panic stop for the machine. Select an appropriate drive system.

Solution The speed reduction ratio here could be achieved by a double reduction belt drive. The requirements for an unloaded speed up can once again be achieved by use of a clutch and the requirement for panic stop capability by a self-energising brake. This possible solution is illustrated in Fig. 9.20.

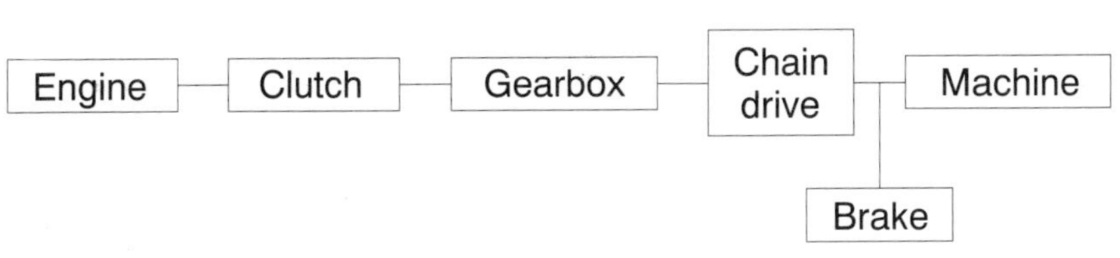

Fig. 9.19 Gear and chain transmission drive.

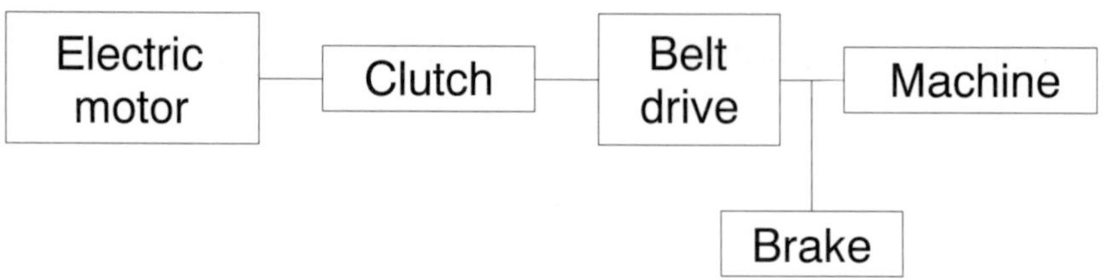

Fig. 9.20 Belt transmission drive.

9.5 A guide to the design literature and sources of information

The sources of information for design are varied and extensive and include books, the general media, patents, standards, journals, manufacturers' brochures and catalogues and electronic databases and the web.

The textbooks relevant to mechanical engineering design broadly split into those on general design and those covering machine design. As an introduction to the broad design and creation process, *The Material of Invention* by Enzio Manzini is excellent. For specific information on the overall design process within an engineering context, *Total Design* by Pugh (1990) and *Product Design and Development* by Ulrich and Eppinger (1995) are primary texts. Additional texts of value include those by Pahl and Beitz (1988) for an overview of the design process, by French (1985, 1994) for an introduction to conceptual design and the lessons to be gained from nature, and Dhillon (1996) for an overview of up-to-date design techniques.

There have been many books on machine design and mechanical component design, varying from engineering reference handbooks and encyclopaedias to introductory texts. One of the most comprehensive reference books is *The Standard Handbook of Machine Design*, edited by Shigley and Mischke (1986). For an introduction to component and machine design the texts by Norton (1996), Juvinall and Marshek (1991), Hurst (1994) and Mott (1992) are excellent.

Valuable sources of specific information for mechanical components are the manufacturers' brochures, catalogues and design guides. Obtaining these can appear awesome to new engineers, but companies are invariably willing to post these out following a telephone call. The trick is finding out the names, addresses and contact numbers. Excellent sources of this information are the business directories such as *Kelly's Business Directory*, *Who's Who in Engineering*, *Dial Engineering* etc.

There are many academic and industrial journals on mechanical design; university and the professional institution libraries are a good place to access these. Specific articles can also be obtained through national lending libraries such as the British Library Document Supply Centre, Boston Spa, Wetherby LS23 7BQ, UK.

Professional engineers must operate according to statutory standards. These can be obtained from national standard organisations such as the British Standards Institution (BSI), DIN, ISO, AGMA and SAE. Some standards are now available on CD-ROM at libraries and in company databases.

Electronic databases and communication are still in an evolutionary stage, but these can be good sources of information. Electronic information on design ranges from the CD-ROM encyclopaedias, some allowing interactive calculations, to proprietary databases and technical and general interest engineering articles on the web.

9.6 Concluding remarks

This text has outlined the total design process, highlighting the importance of considering each area. The bulk of the book has been concerned with one particular aspect of the total design process, detailed design. This has been deliberate as it is hoped that you will have developed some skills as a result and can approach a design with various 'building blocks' at your disposal. Considered in isolation these building blocks are not much use, but they can be used in conjunction to create many machines and mechanisms.

Mechanical engineering is a fulfilling profession, allowing the design and creation of new and better products. It is hoped that the information presented here will encourage you to apply and develop your design skills.

References and sources of information

Books and papers

Adams, J.L. 1979: *Conceptual blockbusting: a guide to better ideas.* Norton.

Chaplin, C.R. 1989: *Creativity in engineering design – the education function.* The Fellowship of Engineering.

Corbett, J. 1989: *Manufacturing phase.* SEED, Sharing Experience in Engineering Design.

Cross, N. 1994: *Engineering design methods. Strategies for product design*, 2nd edition. Wiley.
Design Information Group, 1995: *Multimedia handbook for engineering design*. University of Bristol.
Dhillon, B.S. 1996: *Engineering design*. Irwin.
Dyson, J. 1997: *Against the odds. An autobiography*. Orion Business Books.
French, M.J. 1985: *Conceptual design for engineers*. Springer-Verlag.
French, M.J. 1994: *Invention and evolution: design in nature and engineering*. CUP.
Gaither, N. 1975: The adoption of operations research techniques by manufacturing organisations. *Decision Sciences* **6**, 794–814.
Hamilton, P.H. 1989: *Costing in design*. SEED, Sharing Experience in Engineering Design.
Hurst, K. (ed.) 1994: *Rotary power transmission design*. McGraw Hill.
Institution of Production Engineers 1984: *A guide to design for production*.
Ion, B. and McCracken, B. 1993: *The detail design phase*. SEED, Sharing Experience in Engineering Design.
Ion, B. and Smith, D. 1996: *The conceptual design phase*. SEED, Sharing Experience in Engineering Design.
Johnson, R.C. 1980: *Optimum design of mechanical elements*. Wiley.
Juvinall, R.C. and Marshek, K.M. 1991: *Fundamentals of machine component design*. Wiley.
Macdonald, A.S. 1992: *Aesthetics in design*. SEED, Sharing Experience in Engineering Design.
Malcolm, D.G., Roseboom, J.H., Clark, C.E. and Fazar, W. 1959: Application of a technique for research and development program evaluation. *Operations Research* **7**, 646–69.
Manzini, E. 1989: *The material of invention*. The Design Council.
Matthews, C. 1998: *Case studies in engineering design*. Arnold.
McGrath, J.E. 1984: *Groups: interaction and performance*. Prentice Hall.
McMahon, C., Browne, J. 1998: *CADCAM principles, practice and manufacturing management*, 2nd edition. Addison Wesley.
Mischke, C.R. 1980: *Mathematical model building*. Iowa State University Press.
Mott, R.L. 1992: *Machine elements in mechanical design*. Merrill.
Norton, R.L. 1996: *Machine design, an integrated approach*. Prentice Hall.
Pahl, G. and Beitz, W. 1988: *Engineering design. A systematic approach*. Design Council.
Pugh, S. 1986: *Specification phase*. SEED, Sharing Experience in Engineering Design.
Pugh, S. 1990: *Total design*. Addison Wesley.
SEED 1985: Curriculum for design. Engineering undergraduate courses.
Shigley, J.E. and Mischke, C.R. (eds) 1986: *Standard handbook of machine design*. McGraw Hill.
Siddall, J.N. 1972: *Analytical decision making in engineering design*. Prentice Hall.
Ulrich, K.T. and Eppinger, S.D. 1995: *Product design and development*. McGraw Hill.
Watson, N. and Janota, M.S. 1982: *Turbocharging the internal combustion engine*. Macmillan.
West, J.D. and Levy, F.K. 1969: *A management guide to PERT/CPM*. Prentice Hall.
Whitfield, A. and Baines, N.C. 1990: *Design of radial turbomachines*. Longman.

Standards

BS 5750: Part 1: 1987. Quality systems. Part 1. Specification for design/development, production, installation and servicing. BSI.
BS 7000: Part 10: 1995. Design management systems. Glossary of terms used in design management. BSI.
BS 7000: Part 2: 1997. Design management systems. Guide to managing the design of manufactured components. BSI.

Web sites

http:/www.astm.com/index.html (American Society for the Testing of Materials)
http://www.bsi.org.uk/bsi/ (British Standards Institute)
http://www.eevl.ac.uk (Edinburgh Engineering Virtual Library)
http://www.iso.ch/ (International Organisation for Standardisation)
http://www.sae.org (Society of Automotive Engineers)
http://www.strath.ac.uk/~clds13/SEED/seed.htm (Sharing Experience in Engineering Design)

Design worksheet

1. Describe the total design process and explain why it is necessary.
2. What information should occur on a product design specification?
3. What criteria would be used in decision making during the design process?
4. Describe the key aspects of the CPM and PERT management techniques and give examples of their application.
5. What role do computers play in design?
6. Discuss in about 400 words, with sketches, the design decisions that would be considered in the design of a double reduction gear box.

Appendix A The designation of steels

The coding system used for steels in the UK is given in British Standard 970. The standard uses a six-digit designation, for example 070M20. The first three digits of the designation denote the family of steels to which the alloy belongs:

- 000–199 Carbon and carbon–manganese steels. The first three figures indicate 100 times the mean manganese content.
- 200–240 Free cutting steels where the second and third digits represent 100 times the minimum or mean sulphur content.
- 250 Silicon–manganese spring steels.
- 300–399 Stainless, heat resisting and valve steels.
- 500–999 Alloy steels in groups of ten or multiples of groups of ten according to alloy type.

The fourth digit of the designation is a letter (A, H, M or S):

- A Denotes that the steel will be supplied to close limits of chemical composition.
- H Denotes a hardenability requirement for the material.
- M Denotes specified mechanical properties.
- S Denotes a stainless steel.

The fifth and sixth digit of the designation represent 100 times the mean carbon content of the steel. For example 080A40 denotes a carbon–manganese steel, supplied to close limits of chemical composition, containing 0.7–0.9 per cent manganese and 0.4 per cent carbon.

Tables A.1 and A.2 list some steel types and values of typical material properties.

The choice of steel for a particular application can sometimes be a bewildering experience. Within the current British Standard for steels (BS 970) alone there are several hundred steel specifications. In practice relatively few steels are used for the majority of applications and some of the popular specifications are listed below.

Steels can be divided into seven principal groupings.

1. **Low-carbon free cutting steels**. These are the most popular type of steel for the production of turned components where machineability and surface finish are important. Applications include automotive and general engineering. The principal specification is 230M07.
2. **Low-carbon steels or mild steels**. These are used for lightly stressed components, welding, bending, forming and general engineering applications. Some of the popular specifications are 040A10, 045M10, 070M20, 080A15 and 080M15.
3. **Carbon and carbon–manganese case hardening steels**. These steels are suitable for components that require a wear resisting surface and tough core. Specifications include 045A10, 045M10, 080M15, 210M15, 214M15.
4. **Medium-carbon and carbon–manganese steels**. These offer greater strength than mild steels and respond to heat treatment. Tensile strengths in the range 700–1000 MPa can be attained. Applications include gears, racks, pinions, shafts, rollers, bolts and nuts. Specifications include 080M30, 080M40, 080A42, 080M50, 070M55, 150M36.
5. **Alloy case-hardening steels**. These are used when a hard wear resisting surface is required, but because of the alloying elements, superior mechanical properties can be attained in comparison with carbon and carbon–manganese case hardening steels. Typical applications include gears, cams, rolled and transmission components. Types include 635M15, 655M13, 665M17, 805M20 and 832M13.
6. **Alloy direct hardening steels**. These steels include alloying elements such as Ni, Cr, Mo and V and are used for applications where high strength and shock resistance are important. Types include 605M36, 708M40, 709M40, 817M40 and 826M40.

Table A.1 Guide to the ultimate tensile and yield strengths for selected steels

BS 970 designation	σ_{uts} (MN/m^2)	σ_y (MN/m^2)	Old EN code
230M07	475		1A
080M15	450[a]	330[a]	32C
070M20	560[a]	440[a]	3B
210M15	490		32M
214M15	720		202
080M30	620[a]	480[a]	6
080M40	660[a]	530[a]	8
635M15	770_{min}		351
655M13	1000_{min}		36
665M17	770_{min}		34
605M36	700–850[b]	540[b]	16
709M40	700–1225[b]	540–955[b]	19
817M40	850–1000[b]	700[b]	24
410S21	700–850[c]	525[c]	56A
431S29	850–1000	680	57
403S17	420_{min}	280	
430S17	430_{min}	280	
304S15	480		58E
316S11	490		
303S31	510		58M
303S42	510		58AM

[a] Hot rolled and cold drawn.
[b] Hardened and tempered and cold drawn.
[c] Tempered.

Table A.2 General physical properties

Steel	ρ (kg/m^3)	E (GPa)	G (GPa)	ν	α (K^{-1})
080A15	7859	211.665	82.046	0.291	11.7×10^{-6}
080M40	7844	211.583	82.625	0.287	11.2×10^{-6}
708M40	7830	210.810	83.398	0.279	12.3×10^{-6}
303S–	7930	193.050	–	–	17.2×10^{-6}

7. **Stainless steels**. There are three types:
   (a) Martensitic stainless steels, which can be hardened and tempered to give tensile strengths in the range 550–1000 MN/m^2. Applications include fasteners, valves, shafts, spindles, cutlery and surgical instruments. Specifications include 410S21, 420S29, 420S45, 431S29, 416S21, 416S41, 416S37 and 441S49.
   (b) Ferritic stainless steels, which are common in strip and sheet form. Applications include domestic and automotive trim, catering equipment and exhaust systems. They have good ductility and are easily formed. Specifications include 403S17 and 430S17.
   (c) Austenitic stainless steels, which offer the highest resistance to corrosion. Applications include the food, chemical, gas and oil industries as well as medical equipment and domestic appliances. Specifications include 302S31, 304S15, 316S11, 316S31, 320S31, 321S31, 303S31, 325S31, 303S42 and 326S36.

References and sources of information

British Standards Institute. BS 970: Part 1: 1996. Specification for wrought steels for mechanical and allied engineering purposes. General inspection and testing procedures and specific requirements for carbon, carbon manganese, alloy and stainless steels.

Macreadys 1995: *Standard stock range of quality steels and specifications*. Glynwed Steels Ltd.

Appendix B Worksheet solutions

Chapter 2: Bearings

Question 1
$W = 2000\,\text{N}$, $N = 20\,\text{rpm}$. Select $L = D = 0.05\,\text{m}$.

$$P = \frac{W}{LD} = \frac{2000}{0.05 \times 0.05} = 0.8\,\text{MN/m}^2$$

$$V = 20 \times \frac{2\pi}{60} \times \frac{0.05}{2} = 0.05235\,\text{m/s}$$

$$PV = 0.04188\,(\text{MN/m}^2)(\text{m/s})$$

Taking a safety factor of 2, carbon graphite bearings would be suitable as they can be used for continuous operation.

Question 2
The light load and low speed, together with the requirement that the bearings should be cheap, clean and maintenance free suggest that dry rubbing bearings be considered.

The length of A4 paper is 297 mm. The velocity of the roller surface is given by

$$V_{\text{roller}} = \frac{30 \times 0.297}{60} = 0.1485\,\text{m/s}$$

The angular velocity of the rollers is

$$0.1485/0.01 = 14.85\,\text{rad/s}$$

The velocity of the bearing surface is therefore

$$V = 14.85 \times 0.005 = 0.07425\,\text{m/s}$$

An L/D ratio of 0.5 could be suitable, giving a bearing length of 5 mm. The load capacity is

$$P = \frac{W}{LD} = \frac{20}{0.005 \times 0.01} = 0.4\,\text{MPa}$$

$$PV = 0.4 \times 0.07425 = 0.0297\,(\text{MN/m}^2)(\text{m/s})$$

For this low value of PV the cheapest bearing materials, thermoplastics, are adequate.

Question 3
The primary data are $D = 50.0\,\text{mm}$, $L = 25\,\text{mm}$, $W = 2500\,\text{N}$, $N = 3000\,\text{rpm}$, $c = 0.04\,\text{mm}$, SAE 10, $T_1 = 45°\text{C}$.

Guess a value for the lubricant temperature rise ΔT across the bearing to be, say, $\Delta T = 20°\text{C}$.

$$T_{\text{av}} = T_1 + \frac{\Delta T}{2} = 45 + \frac{20}{2} = 55°\text{C}$$

From Fig. 2.11 for SAE 10 at 55°C, $\mu = 0.017\,\text{Pa}\cdot\text{s}$.

$$N_s = 3000/60 = 50\,\text{rps},\; L/D = 25/50 = 0.5$$

$$P = \frac{W}{LD} = \frac{2500}{0.025 \times 0.05} = 2 \times 10^6\,\text{N/m}^2$$

$$S = \left(\frac{r}{c}\right)^2 \frac{\mu N_s}{P} = \left(\frac{25 \times 10^{-3}}{0.04 \times 10^{-3}}\right)^2 \frac{0.017 \times 50}{2 \times 10^6} = 0.1660$$

From Fig. 2.15 with $S = 0.1660$ and $L/D = 0.5$, $(r/c)f = 4.8$.
From Fig. 2.16 with $S = 0.1660$ and $L/D = 0.5$, $Q/(rcN_sL) = 5.15$.
From Fig. 2.17 with $S = 0.1660$ and $L/D = 0.5$, $Q_s/Q = 0.816$.

The value of the temperature rise of the lubricant can now be calculated using equation 2.8:

$$\Delta T = \frac{8.3 \times 10^{-6} P}{1 - 0.5Q_s/Q} \times \frac{(r/c)f}{Q/(rcN_sL)} = \frac{8.3 \times 10^{-6} \times 2 \times 10^6}{1 - (0.5 \times 0.816)} \times \frac{4.8}{5.15} = 26.13°\text{C}$$

Using $\Delta T = 26.13°$C to calculate T_{av} gives

$$T_{av} = 45 + \frac{26.13}{2} = 58.07°\text{C}$$

Repeating the procedure using the new value for T_{av} gives

$\mu = 0.015\,\text{Pa} \cdot \text{s}$

$S = 0.1465$

From Fig. 2.15 with $S = 0.1465$ and $L/D = 0.5$, $(r/c)f = 4.5$.
From Fig. 2.16 with $S = 0.1465$ and $L/D = 0.5$, $Q/(rcN_sL) = 5.22$.
From Fig. 2.17 with $S = 0.1465$ and $L/D = 0.5$, $Q_s/Q = 0.83$.

$$\Delta T = \frac{8.3 \times 10^{-6} \times 2 \times 10^6}{1 - (0.5 \times 0.83)} \times \frac{4.5}{5.22} = 24.46°\text{C}$$

$$T_{av} = 45 + \frac{24.46}{2} = 57.23°\text{C}$$

This value for T_{av} is close to the previous calculated value, suggesting that the solution has converged. For $\Delta T \approx 24.5°$C, $T_{av} = 57.2°$C, $\mu = 0.015\,\text{Pa} \cdot \text{s}$ and $S = 0.1465$.

The other parameters can now be found:

$$Q = rcN_sL \times 5.22 = 25 \times 0.04 \times 50 \times 25 \times 5.22 = 6525\,\text{mm}^3/\text{s}$$

$$f = 4.5 \times (c/r) = 4.5 \times (0.04/25) = 0.0072$$

$$\text{Torque} = fWr = 0.0072 \times 2500 \times 0.025 = 0.45\,\text{N} \cdot \text{m}$$

$$\text{Power} = 2\pi \times \text{Torque} \times N_s = 141.4\,\text{W}$$

Question 4

$D = 25\,\text{mm}$, $L = 25\,\text{mm}$, $W = 1450\,\text{N}$, $N = 2500\,\text{rpm}$, $N_s = 2500/60\,\text{rps} = 41.67\,\text{rps}$, SAE 10, $T_1 = 50°$C. $L/D = 1$.

$$P = \frac{W}{LD} = \frac{1450}{0.025 \times 0.025} = 2.32 \times 10^6\,\text{N/m}^2$$

(a) $c = 0.025\,\text{mm}$:

$$S = \left(\frac{r}{c}\right)^2 \frac{\mu N_s}{P} = \left(\frac{12.5 \times 10^{-3}}{0.025 \times 10^{-3}}\right)^2 \frac{\mu 41.67}{2.32 \times 10^6} = 4.49\mu$$

The solution for the iterative procedure, using the Raymondi and Boyd charts (Figs 2.15–2.17) to find the average temperature rise of the lubricant across the bearing, is given in tabular form below. The first estimate for the temperature rise is taken as $\Delta T = 10°$C.

Temperature rise, ΔT (°C)	10	16.13	15.36
Average lubricant temperature, T_{av} (°C)	55	58.07	57.68
Average lubricant viscosity, μ (Pa · s)	0.017	0.015	Converged
Sommerfield number, S	0.07633	0.06735	
Coefficient of friction variable, $(r/c)f$	2.3	2.2	
Flow variable, Q/rcN_sL	4.45	4.52	
Side flow to total flow ratio, Q_s/Q	0.766	0.78	

$\Delta T \approx 15.4°\text{C}$

$Q = rcN_sL \times 4.52 = 12.5 \times 0.025 \times 41.67 \times 25 \times 4.52 = 1471\ \text{mm}^3/\text{s}$

$f = 2.2 \times (c/r) = 2.2 \times 0.025/12.5 = 0.0044$

$T = fWr = 0.0044 \times 1450 \times 0.0125 = 0.07975\ \text{N} \cdot \text{m}$

Heat generated $= 2\pi N_s \times$ Torque $= 20.88$ W

With $S = 0.06735$ and $L/D = 1$, the design selected is outside the optimum operating zone for minimum friction and optimum load capacity indicated in Fig. 2.14.

(b) $c = 0.015$ mm:

$$S = \left(\frac{r}{c}\right)^2 \frac{\mu N_s}{P} = \left(\frac{12.5 \times 10^{-3}}{0.015 \times 10^{-3}}\right)^2 \frac{\mu 41.67}{2.32 \times 10^6} = 12.47\mu$$

Assuming $\Delta T = 10°\text{C}$:

Temperature rise, ΔT (°C)	10	27.01	25.17	25.85
Average lubricant temperature, T_{av} (°C)	55	63.51	62.58	62.93
Average lubricant viscosity, μ (Pa · s)	0.017	0.0125	0.013	Converged
Sommerfield number, S	0.2120	0.1559	0.1621	
Coefficient of friction variable, $(r/c)f$	4.8	3.8	3.9	
Flow variable, Q/rcN_sL	4.08	4.22	4.21	
Side flow to total flow ratio, Q_s/Q	0.55	0.622	0.62	

$\Delta T \approx 25.9°\text{C}$

$Q = rcN_sL \times 4.21 = 12.5 \times 0.015 \times 41.67 \times 25 \times 4.21 = 822.3\ \text{mm}^3/\text{s}$

$f = 3.9 \times (c/r) = 3.9 \times 0.015/12.5 = 0.00468$

$T = fWr = 0.00468 \times 1450 \times 0.0125 = 0.08483\ \text{N} \cdot \text{m}$

Heat generated $= 2\pi N_s \times$ Torque $= 22.21$ W

With $S = 0.1621$ and $L/D = 1$, the design selected is inside the optimum operating zone for minimum friction and optimum load capacity indicated in Fig. 2.14. Therefore, this clearance ($c = 0.015$ mm) should be selected.

Question 5

There is no single answer to this question and a range of possible good solutions exist. The answer given here is one such possibility.

From Fig. 2.8 for $D = 40$ mm and a speed of 1000 rpm a suitable diametral clearance is 50 μm. Taking $2c = 0.05$ mm, $c = 0.025$ mm.

Normally the L/D ratio is in the range of 0.5 to 1.5. Taking $L/D = 0.5$ gives $L = 20$ mm. This gives a load capacity of

$$P = \frac{W}{LD} = \frac{2200}{0.02 \times 0.04} = 2.75 \times 10^6\ \text{N/m}^2$$

$$S=\left(\frac{r}{c}\right)^2\frac{\mu N_s}{P}=\left(\frac{20\times10^{-3}}{0.025\times10^{-3}}\right)^2\frac{\mu 16.67}{2.75\times10^6}=3.879\mu$$

$T_1 = 60°\text{C}$, SAE 20. Taking $\Delta T = 15°\text{C}$ as the first estimate:

Temperature rise, ΔT (°C)	15	18.14	18.17
Average lubricant temperature, T_{av} (°C)	67.5	69.07	69.09
Average lubricant viscosity, μ (Pa · s)	0.0158	0.015	Converged
Sommerfield number, S	0.06129	0.05819	
Coefficient of friction variable, $(r/c)f$	2.4	2.4	
Flow variable, Q/rcN_sL	5.55	5.55	
Side flow to total flow ratio, Q_s/Q	0.912	0.914	

$\Delta T \approx 18.2°\text{C}$

With $S = 0.05819$ and $L/D = 0.5$, the design selected is just inside the optimum operating zone for minimum friction and optimum load capacity indicated in Fig. 2.14.

Question 6

$W = 1600\,\text{N}$, $N = 5000\,\text{rpm}$, $N_s = 5000/60 = 83.33\,\text{rps}$, $D = 40\,\text{mm}$, $L = 20\,\text{mm}$, $L/D = 20/40 = 0.5$, $T_1 = 60°\text{C}$, SAE 10.

$$P=\frac{W}{LD}=\frac{1600}{0.02\times0.04}=2\times10^6\,\text{N/m}^2$$

(a) $c = 0.05\,\text{mm}$:

$$S=\left(\frac{r}{c}\right)^2\frac{\mu N_s}{P}=\left(\frac{20\times10^{-3}}{0.05\times10^{-3}}\right)^2\frac{\mu 83.33}{2\times10^6}=6.666\mu$$

Taking $\Delta T = 10°\text{C}$ as the first estimate:

Temperature rise, ΔT (°C)	10	15.89	17.13
Average lubricant temperature, T_{av} (°C)	65	67.94	68.56
Average lubricant viscosity, μ (Pa · s)	0.0116	0.0103	Converged
Sommerfield number, S	0.07733	0.06866	
Coefficient of friction variable, $(r/c)f$	2.9	2.7	
Flow variable, Q/rcN_sL	5.47	5.52	
Side flow to total flow ratio, Q_s/Q	0.892	0.898	

$\Delta T \approx 17.1°\text{C}$

$Q = rcN_sL \times 5.52 = 20 \times 0.05 \times 83.33 \times 20 \times 5.52 = 9200\,\text{mm}^3/\text{s}$

$f = 2.7 \times (c/r) = 2.7 \times 0.05/20 = 0.00675$

$h_0/c = 0.16$

$h_0 = 0.05 \times 0.16 = 0.008\,\text{mm}$

From Fig. 2.14, with $S = 0.06866$ and $L/D = 0.5$, the design is within the optimum zone for minimum friction and maximum load.

(b) $c = 0.07\,\text{mm}$:

$$S=\left(\frac{r}{c}\right)^2\frac{\mu N_s}{P}=\left(\frac{20\times10^{-3}}{0.07\times10^{-3}}\right)^2\frac{\mu 83.33}{2\times10^6}=3.401\mu$$

Taking $\Delta T = 10°C$ as the first estimate:

Temperature rise, ΔT (°C)	10	9.921
Average lubricant temperature, T_{av} (°C)	65	64.96
Average lubricant viscosity, μ (Pa · s)	0.0116	Converged
Sommerfield number, S	0.03945	
Coefficient of friction variable, $(r/c)f$	1.8	
Flow variable, Q/rcN_sL	5.64	
Side flow to total flow ratio, Q_s/Q	0.932	

$\Delta T \approx 9.9°C$

$Q = rcN_sL \times 5.64 = 20 \times 0.07 \times 83.33 \times 20 \times 5.64 = 13\,160\,\text{mm}^3/\text{s}$

$f = 1.8 \times (c/r) = 1.8 \times 0.07/20 = 0.0063$

$h_0/c = 0.114$

$h_0 = 0.114 \times 0.07 = 0.00798\,\text{mm}$

From Fig. 2.14, with $S = 0.03945$ and $L/D = 0.5$, the design is on the edge of the optimum zone for minimum friction and maximum load. As a result either design would be acceptable, but the $c = 0.05$ mm design gives a lower flow requirement and is more 'safely' within the optimum zone.

Question 7

The primary design data are: $D = 30\,\text{mm}$, $L = 15\,\text{mm}$, $L/D = 0.5$, $W = 1000\,\text{N}$, $N = 6000\,\text{rpm}$, $c = 0.02\,\text{mm}$, SAE 10, $T_1 = 50°C$.

$$P = \frac{W}{LD} = \frac{1000}{0.015 \times 0.03} = 2.222\,\text{MPa}$$

Taking $\Delta T = 40°C$ as the first estimate:

Temperature rise, ΔT (°C)	40	39.32
Average lubricant temperature, T_{av} (°C)	70	69.66
Average lubricant viscosity, μ (Pa · s)	0.0097	Converged
Sommerfield number, S	0.2455	
ε	0.6533	
I_s	0.1615	
I_c	−0.1668	
Flow variable, Q/rcN_sL	5.032	
Q_π/rcN_sL	1.203	
Side flow to total flow ratio, Q_s/Q	0.7610	
Coefficient of friction variable, $(r/c)f$	6.646	

$\Delta T \approx 39.3°C$

The values for Q, h_0, the torque and the heat generated in the bearing can now be determined:

$Q = 5.032 \times 15 \times 0.02 \times 100 \times 15 = 2264\,\text{mm}^3/\text{s}$

$h_0 = c(1 - \varepsilon) = 0.0069\,\text{mm}$

Torque $= fWr = 6.646(0.02/15) \times 1000 \times 0.015 = 0.1329\,\text{N} \cdot \text{m}$

Power $= 83.52\,\text{W}$

Question 8
$L/D = 0.75$, $P = 1.562\,\text{MN/m}^2$.

Temperature rise, ΔT (°C)	20	19.16
Average lubricant temperature, T_{av} (°C)	60	59.58
Average lubricant viscosity, μ (Pa · s)	0.032	Converged
Sommerfield number, S	0.1748	
ε	0.6006	
I_s	0.2352	
I_c	−0.1928	
Flow variable, Q/rcN_sL	4.715	
Q_π/rcN_sL	1.440	
Side flow to total flow ratio, Q_s/Q	0.6946	
Coefficient of friction variable, $(r/c)f$	4.547	

$\Delta T \approx 19.2°\text{C}$

$h_0 = 0.05 \times 10^{-3}(1 - 0.6006) = 19.97\,\mu\text{m} \approx 20.0\,\mu\text{m}$

$Q = 4.715 \times 40 \times 0.05 \times (800/60) \times 60 = 7544\,\text{mm}^3/\text{s}$

$\text{Torque} = fWr = 4.547(0.05/40) \times 7500 \times 0.04 = 1.705\,\text{N} \cdot \text{m}$

$\text{Heat generated} = 2\pi N_s \times \text{Torque} = 142.8\,\text{W}$

Question 9
There is no unique solution to this question. The answer given here is one possibility.
Resolving vertically:

$R_1 + R_2 = 1500 + 200 = 1700\,\text{N}$

Taking moments around the left hand bearing:

$1500 \times 0.25 + 200 \times 0.5 - R_2 \times 0.75 = 0$

$R_2 = 633.3\,\text{N} \quad R_1 = 1066.7\,\text{N}$

Applying a safety factor of 2 gives the maximum load on either bearing as approximately $W = 2000\,\text{N}$. This load will be used to design the bearings.

An SAE 30 oil is selected and it is assumed that the lubrication system can deliver the lubricant at 50°C. From Fig. 2.8 for $D = 50\,\text{mm}$ and 300 rpm a suitable clearance is $c = 0.025\,\text{mm}$. The L/D ratio is arbitrarily set as 0.5. So $L = 25\,\text{mm}$.

Analysing this design proposal gives

$\Delta T \approx 14.5°\text{C}$

$Q = 424.9\,\text{mm}^3/\text{s}$

$\text{Torque} = 0.169\,\text{N} \cdot \text{m}$

$\text{Heat dissipated in the bearing} = 5.3\,\text{W}$

$h_0 = 5.1\,\mu\text{m}$

The resulting Sommerfield number, $S = 0.0968$, and L/D ratio of 0.5 is within the optimum zone of Fig. 2.14 for the minimum friction and maximum load.

Question 10

$C = P_d L_d^{1/k} = 14\,200 \times (3800 \times 60 \times 925/10^6)^{3/10} = 70.7\,\text{kN}$

Question 11

The requirement is for a floating bearing carrying only radial loading, so a cylindrical roller bearing is appropriate.

The total number of revolutions in life is

$7500 \times 3000 \times 60 = 1350$ million. $L = 1350$

The load is radial. $P = 9000$ N.

The required basic dynamic load rating is

$$C = PL^{1/3.33} = 9000 \times 1350^{1/3.33} = 78\,224\,\text{N}$$

From the limited bearings, catalogue listing in a suitable bearing is NU310E.

$d = 50$ mm, $D = 110$ mm, $B = 27$ mm, $C = 111\,000$ N, $C_0 = 112\,000$ N. Speed limit using grease is 5000 rpm, using oil 6000 rpm.

Question 12

$30\,\text{s} + 45\,\text{s} + 15\,\text{s} = 90\,\text{s}$

By proportion: 1333.3 hours + 2000 hours + 666.6 hours = 4000 hours.

$$F_m = \left(\frac{95^3(1333.3 \times 60 \times 2500) + 75^3(2000 \times 60 \times 3500) + 115^3(666.6 \times 60 \times 3100)}{1333.3 \times 60 \times 2500 + 2000 \times 60 \times 3500 + 666.6 \times 60 \times 3100} \right)^{1/3}$$

$$= \left(\frac{5.3725 \times 10^{14}}{7.434 \times 10^8} \right)^{1/3} = 89.7\,\text{N}$$

Question 13

There is no unique solution to this question. The answer given below is one possibility.

The installation requires low-friction self-contained bearings which can operate for sustained time periods between servicing. Grease lubricated bearings would be suitable.

The load at each idler is obtained from the idler separation L, belt speed V, belt mass per unit length, m, and belt loading W.

$$F = L\left(\frac{W}{V} + m\right)g = 1 \times \left(\frac{3000 \times 10^3/3600}{3} + 40\right) \times 9.81 = 3117.4\,\text{N}$$

The angular velocity of the rollers is

$$\omega = V/r = 3/0.05 = 60\,\text{rad/s}$$

The rotational speed of the rollers is therefore $60 \times 60/2\pi = 573.0$ rpm.

The centre idler supports 60 per cent of the load giving a bearing load of $(3117.4 \times 0.6/2) = 935.2$ N.

As the axial loads are small, deep groove ball bearings would be appropriate.

Using $L = (C/P)^k$ to determine the basic dynamic load rating for 10 years life gives

$$C = PL^{1/3} = 935.2 \times (10 \times 365 \times 24 \times 60 \times 573)^{1/3} = 13505\,\text{N}$$

A suitable bearing could be a 6403 deep groove ball bearing: $d = 17$ mm, $D = 62$ mm, $B = 17$ mm, $C = 23\,000$ N, $C_0 = 10\,800$ N, speed limit using grease is 12 000 rpm.

The bearings in the side idlers support both axial and radial loads, so it is necessary to use $P = VXR + YT$ (equation 2.25) to calculate an equivalent loading. The factors X and Y must be found from specific bearing manufacturers' catalogues.

Chapter 3: Shafts

Question 1

$$-R_1x + W_1[x - L_1] + W_2[x - (L_1 + L_2)] + W_3[x - (L_1 + L_2 + L_3)]$$

$$+ W_4[x - (L_1 + L_2 + L_3 + L_4)] = EI\frac{d^2y}{dx^2}$$

$$-\frac{R_1x^3}{6} + \frac{W_1}{6}[x - L_1]^3 + \frac{W_2}{6}[x - (L_1 + L_2)]^3 + \frac{W_3}{6}[x - (L_1 + L_2 + L_3)]^3$$

$$+ \frac{W_4}{6}[x - (L_1 + L_2 + L_3 + L_4)]^3 + Ax + B = EIy$$

Boundary conditions: $y = 0$ at $x = 0$, $y = 0$ at $x = L_1 + L_2 + L_3 + L_4 + L_5 = L$. Hence $B = 0$.

$$A = \frac{1}{L}\left(\frac{R_1L^3}{6} - \frac{W_1}{6}(L - L_1)^3 - \frac{W_2}{6}(L_3 + L_4 + L_5)^3 - \frac{W_3}{6}(L_4 + L_5)^3 - \frac{W_4}{6}L_5^3\right)$$

$R_1 + R_2 = W_1 + W_2 + W_3 + W_4$

Moments about R_1: $W_1L_1 + W_2(L_1 + L_2) + W_3(L_1 + L_2 + L_3) + W_4(L_1 + L_2 + L_3 + L_4) - R_2L = 0$.

Hence:

$A = 2.20725$, $R_2 = 353.16\,\text{N}$, $R_1 = 353.16\,\text{N}$, $I = 2.67 \times 10^{-7}\,\text{m}^4$, $E = 200 \times 10^9\,\text{N/m}^2$

$y_{w1} = 1.9286726 \times 10^{-6}\,\text{m}$

$y_{w2} = 3.0996525 \times 10^{-6}\,\text{m}$

$y_{w3} = 3.0996525 \times 10^{-6}\,\text{m}$

$y_{w4} = 1.9286726 \times 10^{-6}\,\text{m}$

$\omega = 1923.8\,\text{rad/s} = 18\,371\,\text{rpm}$

Question 2

$W_1 = 9.81 \times 22.63 = 222\,\text{N}$, $W_2 = 9.81 \times 13.56 = 133\,\text{N}$.

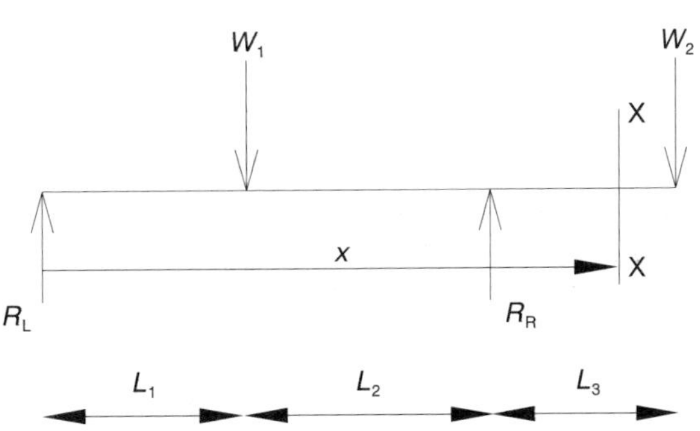

Moments about the left hand bearing: $222 \times 0.254 - R_R \times 0.508 + 133 \times 0.762 = 0$.
$R_R = 310.5\,\text{N}$.

Resolving vertically $R_L + R_R = 222 + 133$.
$R_L = 44.5\,\text{N}$.

Using Macaulay's method to determine the slope and deflection:

$$M_{XX} = -R_L x + W_1[x - L_1] - R_R[x - (L_1 + L_2)]$$

$$EI\frac{dy}{dx} = -R_L\frac{x^2}{2} + \frac{W_1}{2}[x - L_1]^2 - \frac{R_R}{2}[x - (L_1 + L_2)]^2 + A$$

$$EIy = -R_L\frac{x^3}{6} + \frac{W_1}{6}[x - L_1]^3 - \frac{R_R}{6}[x - (L_1 + L_2)]^3 + Ax + B$$

Boundary conditions: at $x = 0$, $y = 0$ (left hand bearing), so from the above equation, $B = 0$. At $x = L_1 + L_2$, $y = 0$ (right hand bearing).

$$0 = -R_L\frac{(L_1 + L_2)^3}{6} + \frac{W_1}{6}L_2^3 + A(L_1 + L_2)$$

$$A = \frac{R_L\dfrac{(L_1 + L_2)^3}{6} - \dfrac{W_1}{6}L_2^3}{L_1 + L_2}$$

$$A = \frac{\dfrac{44.5(0.508)^3}{6} - \dfrac{222(0.254)^3}{6}}{0.508} = 0.7204$$

y_1 = deflection at location of load W_1.
y_2 = deflection at location of load W_2.

$$y_1 = \frac{1}{EI}\left(-\frac{44.5 \times 0.254^3}{6} + 0.7204 \times 0.254\right) = \frac{0.06144}{EI}$$

$$y_2 = \frac{1}{EI}\left(-\frac{44.5 \times 0.762^3}{6} + \frac{222}{6}(0.762 - 0.254)^3 - \frac{310.5}{6}(0.254)^3 + 0.7204 \times 0.762\right) = \frac{1.270}{EI}$$

Let the first critical speed be

$$12\,000\,\text{rpm} = 12\,000 \times 2\pi/60\,\text{rad/s} = 1257\,\text{rad/s}$$

$$1257 = \sqrt{\frac{9.81\left(222 \times \dfrac{0.01644}{EI} + 133 \times \dfrac{1.270}{EI}\right)}{222\left(\dfrac{0.01644}{EI}\right)^2 + 133\left(\dfrac{1.270}{EI}\right)^2}} = \sqrt{EI \times 9.81 \times \frac{172.5}{214.6}} = \sqrt{EI \times 7.89}$$

$$1257^2 = EI \times 7.89 = \frac{\pi d^4}{64} \times 207 \times 10^9 \times 7.89$$

$$d = (1.971 \times 10^{-5})^{0.25} = 0.06663\,\text{m}$$

$d \approx 67$ mm. Take $d = 70$ mm as the nearest standard size.

Question 3
Macaulay's method:

$$-R_1[x - L_1] - R_2[x - (L_1 + L_2)] + W_1 x = EI\frac{d^2y}{dx^2}$$

$$-\frac{R_1}{2}[x - L_1]^2 - \frac{R_2}{2}[x - (L_1 + L_2)]^2 + \frac{W_1 x^2}{2} + A = EI\frac{dy}{dx}$$

$$-\frac{R_1}{6}[x - L_1]^3 - \frac{R_2}{6}[x - (L_1 + L_2)]^3 + \frac{W_1 x^3}{6} + Ax + B = EIy$$

Boundary conditions: $y = 0$ at $x = L_1$, $y = 0$ at $x = L_1 + L_2$. Substituting into the above equation and ignoring terms in square brackets when they go negative:

$$\frac{W_1 L_1^3}{6} + AL_1 + B = 0 \qquad \text{(B.2.1)}$$

$$-\frac{R_1}{6}L_2^3 + \frac{W_1(L_1 + L_2)^3}{6} + A(L_1 + L_2) + B = 0 \qquad \text{(B.2.2)}$$

Solving equations B.2.1 and B.2.2 for A and B.

Equation B.2.2 minus equation B.2.1:

$$-\frac{R_1}{6}L_2^3 + \frac{W_1}{6}((L_1 + L_2)^3) + A(L_1 + L_2) - \frac{W_1 L_1^3}{6} - AL_1 = 0$$

$$A = -\frac{1}{L_2}\left(-\frac{R_1 L_2^3}{6} + \frac{W_1}{6}(L_1 + L_2)^3 - \frac{W_1 L_1^3}{6}\right)$$

$$B = -AL_1 - \frac{W_1 L_1^3}{6}$$

Resolving:

$$R_1 + R_2 = W_1 + W_2$$

Moments about bearing 1:

$$W_2(L_2 + L_3) - R_2 L_2 - W_1 L_1 = 0$$

$$R_2 = \frac{1}{L_2}(W_2(L_2 + L_3) - W_1 L_1)$$

$$R_1 = W_1 + W_2 - R_2$$

$W_1 = 1\,\text{N}$, $W_2 = 0\,\text{N}$, $L_1 = 0.055\,\text{m}$, $L_2 = 0.12\,\text{m}$, $L_3 = 0.04\,\text{m}$, $L_T = 0.215\,\text{m}$.

$$R_2 = \frac{1}{0.12}(-1 \times 0.055) = -0.4583 \quad R_1 = 1.4583$$

$A = -3.7125 \times 10^{-3}$, $B = 1.7646 \times 10^{-4}$.

$E = 200 \times 10^9\,\text{N/m}^2$.

$I = \pi d^4/64 = 7.854 \times 10^{-9}\,\text{m}^4$.

a_{11} = deflection at mass 1 due to 1 N at mass location 1.

a_{21} = deflection at mass 2 due to 1 N at mass location 1.

$$y_{x=0} = a_{11} = B/EI$$

$$y_{x=0} = a_{11} = 1.123368 \times 10^{-7}\,\text{m}$$

$$a_{21} = \frac{1}{EI}(-9.9555 \times 10^{-4} + 4.8889 \times 10^{-6} + 1.6564 \times 10^{-3} - 7.9819 \times 10^{-4} + 1.7646 \times 10^{-4})$$

$$a_{21} = a_{12} = 2.8018 \times 10^{-8}\,\text{m}$$

$W_2 = 1\,\text{N}$, $W_1 = 0$: $R_2 = 1.333\,\text{N}$, $R_1 = -0.333\,\text{N}$, $A = -8 \times 10^{-4}$, $B = 4.4 \times 10^{-5}$.

$$y_{x=L_1+L_2+L_3} = \frac{1}{EI}(2.2756 \times 10^{-4} - 1.4222 \times 10^{-5} - 1.72 \times 10^{-4} + 4.4 \times 10^{-5}) = 5.433 \times 10^{-8} = a_{22}$$

$$\begin{vmatrix} (a_{11}m_1 - 1/\omega^2) & a_{12}m_2 \\ a_{21}m_1 & (a_{22}m_2 - 1/\omega^2) \end{vmatrix} = 0$$

$$\left(a_{11}m_1 - \frac{1}{\omega^2}\right)\left(a_{22}m_2 - \frac{1}{\omega^2}\right) - a_{12}m_2a_{21}m_1 = 0$$

$$a_{11}m_1a_{22}m_2 - \frac{a_{11}m_1}{\omega^2} - \frac{a_{22}m_2}{\omega^2} + \frac{1}{\omega^4} - a_{12}m_2a_{21}m_1 = 0$$

$$\frac{1}{\omega^4} - (a_{11}m_1 + a_{22}m_2)\frac{1}{\omega^2} + (a_{11}a_{22} - a_{12}a_{21})m_1m_s = 0$$

Let $p = \omega^2$. Multiply by p^2:

$$(a_{11}a_{22} - a_{12}a_{21})m_1m_2p^2 - (a_{11}m_1 + a_{22}m_2)p + 1 = 0$$

$$3.9872 \times 10^{-14}p^2 - 4.7273 \times 10^{-7}p + 1 = 0$$

$p = 9\,100\,051.6$, $\sqrt{p} = \omega_{c2} = 3017\,\text{rad/s}$.
$p = 2\,756\,138.3$, $\sqrt{p} = \omega_{c1} = 1660\,\text{rad/s}$.

Question 4

Assuming the bearings act as rigid pin supports. In practice the bearings will deflect slightly and the critical frequencies predicted using the assumption of rigid bearing supports will be overestimates.

Using Macaulay's method:

$$-R_1x + W_1[x - L_1] - R_2[x - (L_1 + L_2)] = M_{XX}$$

$$M = EI\frac{d^2y}{dx^2}$$

Integrating to find the slope:

$$EI\frac{dy}{dx} = -R_1\frac{x^2}{2} + \frac{W_1}{2}[x - L_1]^2 - \frac{R_2}{2}[x - (L_1 + L_2)]^2 + A$$

Integrating to find the deflection:

$$EIy = -R_1\frac{x^3}{6} + \frac{W_1}{6}[x - L_1]^3 - \frac{R_2}{6}[x - (L_1 + L_2)]^3 + Ax + B$$

Use boundary conditions to find the constants A and B, assuming zero deflection at the bearing supports. At $x = 0$, $y = 0$. Hence $B = 0$. At $x = L_1 + L_2$, $y = 0$.

$$0 = -\frac{R_1}{6}(L_1 + L_2)^3 + \frac{W_1}{6}L_2^3 + A(L_1 + L_2)$$

$$A = \frac{1}{L_1 + L_2}\left(\frac{R_1}{6}(L_1 + L_2)^3 - \frac{W_1}{6}L_2^3\right)$$

Resolving vertically:

$$W_1 + W_2 = R_1 + R_2$$

Moments about the left hand bearing:

$$W_1L_1 - R_2(L_1 + L_2) + W_2(L_1 + L_2 + L_3) = 0$$

$$R_2 = \frac{1}{L_1 + L_2}[W_1L_1 + W_2(L_1 + L_2 + L_3)]$$

$$R_1 = W_1 + W_2 - R_2$$

Now determine the influence coefficients.

If $W_1 = 1\,\text{N}$ and $W_2 = 0\,\text{N}$, $L_1 = 0.09\,\text{m}$, $L_2 = 0.07\,\text{m}$, $L_3 = 0.08\,\text{m}$, then $A = 1.509375 \times 10^{-3}$.

$R_2 = 0.09/0.16 = 0.5625,\ R_1 = 1 - R_2 = 0.4375\,\text{N}.$

$$I = \frac{\pi d^4}{64} = 7.8539816 \times 10^{-9}\,\text{m}^4$$

Calculating the deflection at $x = L_1$ to determine a_{11}:

$$EIa_{11} = -R_1\frac{L_1^3}{6} + AL_1 = -\frac{0.4375 \times 0.09^3}{6} + 1.509375 \times 10^{-3} \times 0.09 = 8.268875 \times 10^{-5}$$

$$a_{11} = \frac{8.268875 \times 10^{-5}}{200 \times 10^9 \times 7.8139816 \times 10^{-9}} = 5.2640498 \times 10^{-8}\,\text{m}$$

Determining the influence coefficients a_{21}, a_{12} ($a_{21} = a_{12}$ by reciprocity):

$$EIa_{21} = EIa_{12} = -R_1\frac{x^3}{6} + \frac{W_1}{6}(L_2 + L_3)^3 - \frac{R_2}{6}L_3^3 + A(L_1 + L_2 + L_3)$$

$$= -\frac{0.4375 \times 0.24^3}{6} + \frac{1}{6}0.15^3 - \frac{0.5625}{6} \times 0.08^3 + 1.509375 \times 10^{-3} \times 0.24$$

$$= -1.008 \times 10^{-3} + 5.625 \times 10^{-4} - 4.8 \times 10^{-5} + 3.6225 \times 10^{-4}$$

$$= -1.3125 \times 10^{-4}$$

$$a_{12} = a_{21} = -8.3556345 \times 10^{-8}\,\text{m}$$

Substituting values into the frequency equation for a system consisting of two concentrated masses ($m_1 = 40/9.81\,\text{kg}$, $m_2 = 38/9.81\,\text{kg}$):

$$\frac{1}{\omega^4} - 1.1449744 \times 10^{-6}\frac{1}{\omega^2} + 1.1843419 \times 10^{-13} = 0$$

Let $f = \omega^2$ and solve as a quadratic:

$$f_1 = 970884.46$$

$$\sqrt{f_1} = 985.3347\,\text{rad/s}\ (985\,\text{rad/s})$$

$$f_2 = 8696715.7$$

$$\sqrt{f_2} = 2949.0195\,\text{rad/s}\ (2949\,\text{rad/s})$$

The values are an overestimate, as the mass of the shaft has been neglected and zero deflection assumed for the bearings. In addition the nature of the bearing support would need to be considered, as would any torque between the compressor and turbine.

Question 5

Resolving vertically:

$$R_1 + R_2 = W_1 + W_2$$

Moments about R_1:

$$-R_2L_2 + W_2(L_2 + L_3) - W_1L_1 = 0$$

$$R_2 = \frac{W_2(L_2 + L_3) - W_1L_1}{L_2}$$

$$R_1 = W_1 + W_2 - R_2$$

$$W_1x - R_1[x - L_1] - R_2[x - (L_1 + L_2)] = EI\frac{d^2y}{dx^2}$$

$$W_1\frac{x^2}{2}-\frac{R_1}{2}[x-L_1]^2-\frac{R_2}{2}[x-(L_1+L_2)]^2+A=EI\frac{\mathrm{d}y}{\mathrm{d}x}$$

$$W_1\frac{x^3}{6}-\frac{R_1}{6}[x-L_1]^3-\frac{R_2}{6}[x-(L_1+L_2)]^3+Ax+B=EIy$$

Boundary conditions: at $x=L_1$, $y=0$. Also at $x=L_1+L_2$, $y=0$.

$$\frac{W_1L_1^3}{6}+AL_1+B=0$$

$$A=\frac{1}{L_2}\left(\frac{W_1L_1^3}{6}-\frac{W_1}{6}(L_1+L_2)^3+\frac{R_1}{6}L_2^3\right)$$

$$B=-\frac{W_1L_1^3}{6}-AL_1$$

Solving for the influence coefficients:

$W_1=1\,\mathrm{N}\quad W_2=0\,\mathrm{N}$

$L_1=0.07\,\mathrm{m}\quad L_2=0.14\,\mathrm{m}\quad L_3=0.055\,\mathrm{m}$

Hence, $R_2=-0.07/0.14=-0.5\,\mathrm{N}$, $R_1=1-(-0.5)=1.5\,\mathrm{N}$.

$A=-5.717\times10^{-3}\quad B=3.43\times10^{-4}$

$EI=200\times10^9\pi(0.022^4)/64=2299.8\,\mathrm{Nm}^2$

For the deflection at W_1, $x=0$:

$$y=\frac{1}{2299.8}(3.43\times10^{-4})=1.491\times10^{-7}=a_{11}$$

For the deflection at W_2, $x=0.265$:

$y=3.902\times10^{-8}=a_{21}=a_{12}$

$W_1=0\,\mathrm{N}\quad W_2=1\,\mathrm{N}$

$R_2=1.393\,\mathrm{N}\quad R_1=-0.3929\,\mathrm{N}$

$A=-1.283\times10^{-3}\quad B=8.984\times10^{-5}$

For the deflection at W_2, $x=0.265$:

$$y=\frac{1}{22\,299.8}(1.968\times10^{-4})=8.556\times10^{-8}=a_{22}$$

$$\begin{vmatrix}\left(a_{11}m_1-\frac{1}{\omega^2}\right) & a_{12}m_2\\ a_{21}m_1 & \left(a_{22}m_2-\frac{1}{\omega^2}\right)\end{vmatrix}=0$$

$$\left(a_{11}m_1-\frac{1}{\omega^1}\right)\left(a_{22}m_2-\frac{1}{\omega^2}\right)-a_{12}m_2a_{21}m_1=0$$

$$a_{11}m_1a_{22}m_2-\frac{a_{11}m_1}{\omega^2}-\frac{a_{22}m_2}{\omega^2}+\frac{1}{\omega^4}-a_{12}m_2a_{21}m_1=0$$

$$\frac{1}{\omega^4}-(a_{11}m_1+a_{22}m_2)\frac{1}{\omega^2}+(a_{11}a_{22}-a_{12}a_{21})m_1m_2=0$$

Let $p = \omega^2$. Multiply by p^2:

$(a_{11}a_{22} - a_{12}a_{21})m_1m_2\,p^2 - (a_{11}m_1 + a_{22}m_2)p + 1 = 0$

$m_1 = 4.2\,\text{kg}$, $m_2 = 5.5\,\text{kg}$. This gives: $\omega_1 = 1153\,\text{rad/s}$, $\omega_2 = 1703\,\text{rad/s}$.

Question 6

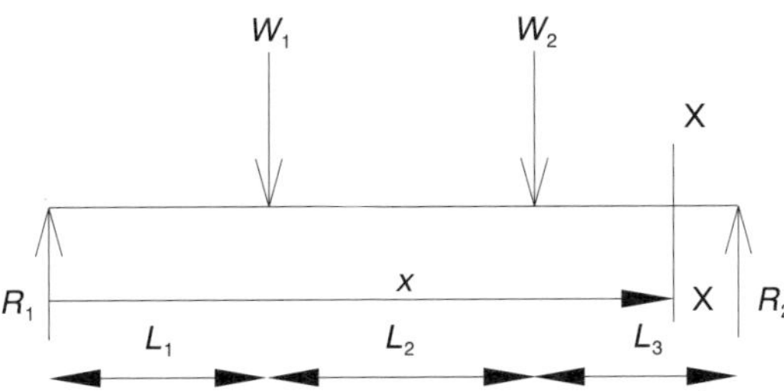

R_1 = reaction at left hand bearing, R_2 = reaction at right bearing.

Clockwise moments about left hand bearing: $W_1L_1 + W_2(L_1 + L_2) - R_2(L_1 + L_2 + L_3) = 0$.

$$R_2 = \frac{W_1L_1 + W_2(L_1 + L_2)}{(L_1 + L_2 + L_3)}$$

Resolving vertically:

$$R_1 = W_1 + W_2 - R_2$$

Defining the moment at XX and using Macaulay's method to determine slope and deflections:

$$M_{\text{XX}} = -R_1x + W_1[x - L_1] + W_2[x - (L_1 + L_2)]$$

$$\left(EI\frac{\mathrm{d}^2y}{\mathrm{d}x^2} = M\right)$$

$$EI\frac{\mathrm{d}y}{\mathrm{d}x} = -R_1\frac{x^2}{2} + \frac{W_1}{2}[x - L_1]^2 + \frac{W_2}{2}[x - (L_1 + L_2)]^2 + A$$

$$EIy = -R_1\frac{x^3}{6} + \frac{W_1}{6}[x - L_1]^3 + \frac{W_2}{6}[x - (L_1 + L_2)]^3 + Ax + B$$

Boundary conditions: at $x = 0$, $y = 0$ (left hand bearing). For these conditions from the above equation, $B = 0$.

At $x = L_1 + L_2 + L_3$, $y = 0$ (right hand bearing), so

$$0 = -R_1\frac{(L_1 + L_2 + L_3)^3}{6} + \frac{W_1}{6}(L_2 + L_3)^3 + \frac{W_2}{6}L_3^3 + A(L_1 + L_2 + L_3)$$

$$A = \frac{R_1\dfrac{(L_1 + L_2 + L_3)^3}{6} - \dfrac{W_1}{6}(L_2 + L_3)^3 - \dfrac{W_2}{6}L_3^3}{L_1 + L_2 + L_3}$$

If $W_1 = 1\,\text{N}$, $W_2 = 0\,\text{N}$, $R_2 = 0.3191\,\text{N}$, $R_1 = 0.6808\,\text{N}$, $A = 0.01345$.

When $x = L_1$,

$$y = a_{11} = \frac{1}{EI}\left(-\frac{R_1L_1^3}{6} + AL_1\right)$$

$$I = \frac{\pi d^4}{64} = \pi\frac{0.04^4}{64} = 1.257 \times 10^{-7}\,\text{m}^4$$

$$a_{11} = \frac{1}{200 \times 10^9 \times 1.257 \times 10^{-7}} \left(-\frac{0.6808 \times 0.15^3}{6} + 0.01345 \times 0.15 \right)$$

$$a_{21} = y_{\text{at } x = L_1 + L_2} = \frac{1}{EI} \left(-\frac{R_1}{6}(L_1 + L_2) + \frac{W_1}{6} L_2^3 + A(L_1 + L_2) \right) = 5.298 \times 10^{-8}\,\text{m}$$

$a_{12} = a_{21}$ by reciprocity.

When $W_1 = 0\,\text{N}$ and $W_2 = 1\,\text{N}$, $R_2 = 0.7021\,\text{N}$, $R_1 = 0.2979\,\text{N}$, $A = 0.009994$. $a_{22} = y_{\text{at } x = L_1 + L_2} = 6.023 \times 10^{-8}\,\text{m}$.

$$\begin{vmatrix} (a_{11}m_1 - 1/\omega^2) & a_{12}m_2 \\ a_{21}m_1 & (a_{22}m_2 - 1/\omega^2) \end{vmatrix} = 0$$

$$a_{11}a_{22}m_1m_2 - (a_{11}m_1 + a_{22}m_2)\frac{1}{\omega^2} + \frac{1}{\omega^4} - a_{12}a_{21}m_1m_2 = 0$$

Multiplying by ω^4, letting $p = \omega^2$ and rearranging gives

$$(a_{11}a_{22} - a_{12}a_{21})m_1m_2p^2 - (a_{11}m_1 + a_{22}m_2)p + 1 = 0$$

Solving for p gives

$$p = \frac{5.770 \times 10^{-6} \pm \sqrt{(5.770 \times 10^{-6})^2 - (4 \times 2.305 \times 10^{-12})}}{2 \times 2.305 \times 10^{-12}}$$

$$p = 2\,315\,271.7, \quad 187\,342.9$$

$$\omega_2 = 1522\,\text{rad/s}, \quad \omega_1 = 432.8\,\text{rad/s}$$

Assumptions include constant diameter, zero deflection at the bearings, concentrated loads and point reactions at the bearings. The bearings would in reality deflect, reducing the values obtained for the critical frequencies.

Question 7

Looking at the shaft from the left:

Section I: $M = R_L x$.
Section II: $M = R_L x$.
Section III: $M = R_R(L - x)$.

$$R_L + R_R = W$$

$$WL_2 - R_R L = 0$$

$$R_R = WL_2/L$$

$$R_L = W\left(1 - \frac{L_2}{L}\right)$$

By definition:

$$U = \int \frac{M^2\,dx}{2EI}$$

$$U_1 = \int_0^{L_1} \frac{R_L^2 x^2}{2EI_1}\,dx = \left[\frac{W^2(1 - L_2/L)^2(x^3/3)}{2EI_1}\right]_0^{L_1} = \frac{W^2(1 - L_2/L)^2(L_1^3/3)}{2EI_1}$$

$$\frac{\partial U_1}{\partial W} = \frac{2W(1 - L_2/L)^2(L_1^3/3)}{2EI_1}$$

$$U_2 = \int_{L_1}^{L_2} \frac{R_L^2 x^2}{2EI_2}\,dx = \left[\frac{W^2}{2EI_2}\left(1-\frac{L_2}{L}\right)^2 \frac{x^3}{3}\right]_{L_1}^{L_2} = \frac{W^2}{2EI_2}\left(1-\frac{L_2}{L}\right)^2\left(\frac{L_2^3}{3}-\frac{L_1^3}{3}\right)$$

$$\frac{\partial U_2}{\partial W} = \frac{2W}{2EI_2}\left(1-\frac{L_2}{L}\right)^2\left(\frac{L_2^3}{3}-\frac{L_1^3}{3}\right)$$

$I_3 = I_2$ (same section)

$$U_3 = \int \frac{W^2(L_2/L)^2}{2EI_3}(L-x)^2\,dx = \left[\frac{W^2(L_2/L)^2(L^2x - x^2L + (x^3/3))}{2EI_3}\right]_{L_2}^{L}$$

$$= \frac{W^2}{2EI_3}\left(\frac{L_2}{L}\right)^2\left(L^3 - L^3 + \frac{L^3}{3} - L^2L_2 + L_2^2L - \frac{L_2^3}{3}\right)$$

$$\frac{\partial U_3}{\partial W} = \frac{W}{EI_3}\left(\frac{L_2}{L}\right)^2\left(\frac{L^3}{3} - L^2L_2 + L_2^2L - \frac{L_2^3}{3}\right)$$

$$\frac{\partial U_1}{\partial W} = 2.4708039 \times 10^{-5}$$

$$\frac{\partial U_2}{\partial W} = 6.6596749 \times 10^{-6}$$

$$\frac{\partial U_3}{\partial W} = 1.0801836 \times 10^{-5}$$

$$\frac{\partial U}{\partial W} = 4.2169551 \times 10^{-5}$$

$\omega_c = 482.3\ \text{rad/s} = 4605.8\ \text{rpm}.$

Question 8

070M20: $\sigma_{uts} = 560$ MPa, $\sigma_y = 440$ MPa.

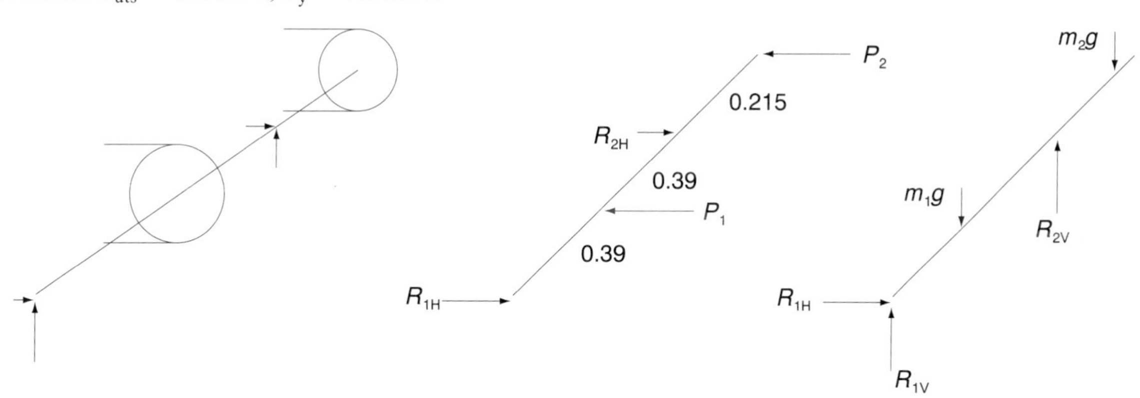

Torque 1600 N·m $= P \times d$.

$P_1 = 1600/0.25 = 6400$ N, $P_2 = 1600/0.2 = 8000$ N.

Moments about R_1:

$6400 \times 0.39 - R_{2H} \times 0.78 + 8000 \times 0.995 = 0.$ $R_{2H} = 13\,405.13$ N

Resolving horizontal loads:

$R_{1H} + R_{2H} = P_1 + P_2$

$R_{1H} = 6400 + 8000 - 13\,405.13 = 994.87$ N

Horizontal shear force diagram:

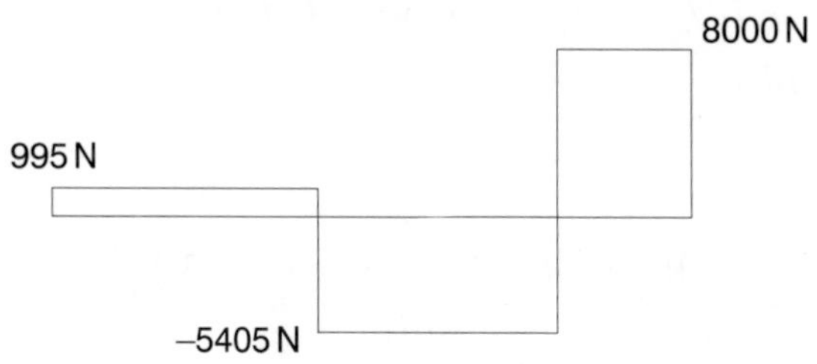

Horizontal bending moment diagram:
At Pulley 1, $995 \times 0.39 = 388\,\mathrm{N \cdot m}$; at Pulley 2, $8000 \times 0.215 = 1720\,\mathrm{N \cdot m}$.

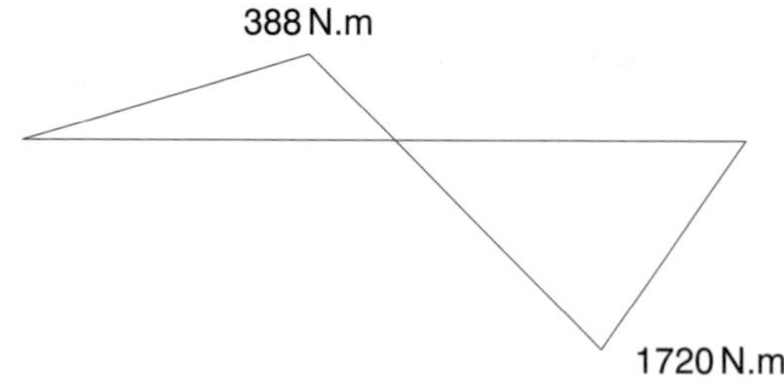

Vertical loads and bending moments:

$R_{1V} + R_{2V} = 882.9 + 1226.3$

Moments about R_1: $882.9 \times 0.39 - R_{2V} \times 0.78 + 1226.3 \times 0.995 = 0$. $R_{2V} = 2005.7\,\mathrm{N}$, $R_{1V} = 103.45\,\mathrm{N}$.

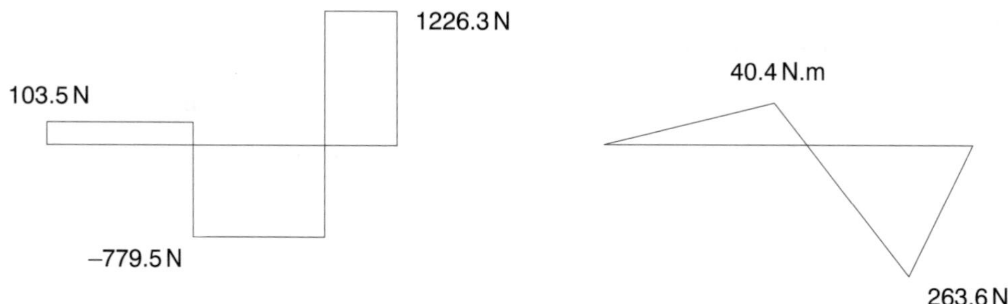

Vertical moment at pulley 1:

$103.45 \times 0.39 = 40.4\,\mathrm{N \cdot m}$

Vertical moment at pulley 2:

$1226.3 \times 0.215 = 263.6\,\mathrm{N \cdot m}$

Resultant maximum bending moment:

$M_{\max} = (263.6^2 + 1720^2)^{0.5} = 1740\,\mathrm{N \cdot m}$

Need factors for ASME equation:

$k_a = 4.51\sigma_{uts}^{-0.265} = 0.843$.
$k_b = 1.85d^{-0.19}$. Assume $d = 70\,\mathrm{mm}$, so $k_b = 0.83$.
$k_c = 0.9$.
$k_d = 1$.
$k_e = 1$.

$k_f = 0.5$. (The bending moment at the profiled keyway region is far greater. Also in the absence of information about the hardness assume worst case value.)
$k_g = 1$.

$$\sigma'_e = 0.504\sigma_{uts} = 0.504 \times 560 = 282\,\text{MPa}$$

$$\sigma_e = 0.843 \times 0.83 \times 0.9 \times 1 \times 1 \times 0.5 \times 1 \times 282 = 88.87\,\text{MPa}$$

Substituting values into the ASME equation for transmission shafts:

$$d = \left[\frac{32 \times 2}{\pi}\sqrt{\left(\frac{1740}{88.87 \times 10^6}\right)^2 + \frac{3}{4}\left(\frac{1600}{440 \times 10^6}\right)^2}\right]^{0.333} = 0.0739\,\text{m}$$

$d = 74\,\text{mm}$

As a standard diameter, the shaft could be specified as 75 mm.
Note: if d assumed for k_b is significantly different from the result, repeat calculation until reasonable convergence.

Chapter 4: Gears

Question 1
$d = mN$. $N = d/m$. $N_B = 67.5/1.5 = 45$, $N_C = 27/1.5 = 18$, $N_F = 56/2 = 28$.

$$n_H = \frac{N_G}{N_H} n_G$$

$$n_G = n_F$$

$$n_F = \frac{N_E}{N_F} n_E$$

$$n_E = n_D$$

$$n_D = \frac{N_C}{N_D} n_C$$

$$n_C = n_B$$

$$n_B = \frac{N_A}{N_B} n_A$$

$$n_H = \frac{N_G}{N_H} \times \frac{N_E}{N_F} \times \frac{N_C}{N_D} \times \frac{N_A}{N_B} n_A = \frac{18}{30} \times \frac{18}{28} \times \frac{18}{38} \times \frac{20}{45} \times 1490 = 121\,\text{rpm}$$

Gear	A	B	C	D	E	F	G	H
Direction	CW	ACW	ACW	CW	CW	ACW	ACW	CW

CW = clockwise; ACW = anticlockwise.

Question 2
Calculating the pinion pitch diameter:

$$d_P = mN_P = 2.5 \times 18 = 45\,\text{mm}$$

Calculating the pitch line velocity:

$$V = \frac{d_P}{2} \times 10^{-3} \times n\frac{2\pi}{60} = \frac{0.045}{2} \times 1100 \times 0.1047 = 2.592\,\text{m/s}$$

Calculating the velocity factor:

$$K_v = \frac{6.1}{6.1 + V} = \frac{6.1}{6.1 + 2.592} = 0.7018$$

Calculating the transmitted load:

$$W_t = \frac{60 \times 10^3 H}{\pi d n} = \frac{60 \times 10^3 \times 4}{\pi 45 \times 1100} + 1.543\,\text{kN}$$

From Table 4.6 for $N_P = 18$, the Lewis form factor $Y = 0.29327$.

The Lewis equation for bending stress gives

$$\sigma = \frac{W_t}{K_v F m Y} = \frac{1.543 \times 10^3}{0.7018 \times 0.03 \times 0.0025 \times 0.29327} = 9.998 \times 10^7\,\text{N/m}^2 \approx 100\,\text{MPa}$$

Question 3

Calculating the pinion pitch diameter:

$$d_P = m N_P = 2 \times 18 = 36\,\text{mm}$$

Calculating the pitch line velocity:

$$V = \frac{d_P}{2} \times 10^{-3} \times n \frac{2\pi}{60} = \frac{0.036}{2} \times 1200 \times 0.1047 = 2.262\,\text{m/s}$$

Calculating the velocity factor:

$$K_v = \frac{6.1}{6.1 + V} = \frac{6.1}{6.1 + 2.262} = 0.7295$$

Calculating the transmitted load:

$$W_t = \frac{60 \times 10^3 H}{\pi d n} = \frac{60 \times 10^3 \times 1.75}{\pi 36 \times 1200} = 0.7737\,\text{kN}$$

From Table 4.6 for $N_p = 18$, the Lewis form factor $Y = 0.29327$.

The Lewis equation for bending stress is

$$\sigma = \frac{W_t}{K_v F m Y}$$

If $\sigma \leq 75\,\text{MPa}$:

$$F = \frac{W_t}{K_v m Y \sigma} = \frac{0.7737 \times 10^3}{0.7295 \times 0.002 \times 0.29327 \times 75 \times 10^6} = 0.0241\,\text{m}$$

So $F = 25\,\text{mm}$.

Question 4

One answer could be for $m = 2.5$, $F = 26\,\text{mm}$. If these values are used: $d_P = 47.5\,\text{mm}$, $W_t = 1.005\,\text{kN}$, $V = 2.487\,\text{m/s}$, $K_v = 0.7104$, $Y = 0.30078$, the Lewis bending stress $\sigma = 72.38 \times 10^6\,\text{N/m}^2$.

Question 5

From Table 4.6 for $N_P = 21$, the Lewis form factor $Y = 0.31406$.

Calculating the pinion pitch diameter:

$$d_P = m N_P = 6 \times 21 = 126\,\text{mm}$$

Calculating the pitch line velocity:

$$V = \frac{d_P}{2} \times 10^{-3} \times n \frac{2\pi}{60} = \frac{0.126}{2} \times 850 \times 0.1047 = 5.608\,\text{m/s}$$

Calculating the velocity factor:

$$K_v = \frac{6.1}{6.1 + V} = \frac{6.1}{6.1 + 5.608} = 0.5210$$

Calculating the transmitted load from the Lewis equation for bending stress:

$$W_t = \sigma K_v F m Y = 117 \times 10^6 \times 0.5210 \times 0.05 \times 0.006 \times 0.31406 = 5.743\,\text{kN}$$

$$H = \frac{\pi d n W_t}{60 \times 10^3} = \frac{\pi 126 \times 850 \times 5.743}{60 \times 10^3} \approx 32.2\,\text{kW}$$

Questions 6, 7 and 8
No unique solution.

Question 9
For a cast iron pinion and a cast iron gear, $C_P = 163\sqrt{\text{MPa}}$.

$\phi = 20°, \quad m = 4, \quad N_P = 20$

$d_P = 4 \times 20 = 80\,\text{mm}, \quad N_G = 32$

$H = 10.5\,\text{kW}, \quad n = 950\,\text{rpm}, \quad F = 50\,\text{mm}$

$d_G = 4 \times 32 = 128\,\text{mm}$

$$\sigma_c = -C_P\left[\frac{W_t}{C_v F \cos\phi}\left(\frac{1}{r_1} + \frac{1}{r_2}\right)\right]^{0.5}$$

$$r_1 = \frac{d_P}{2}\sin\phi = \frac{0.08}{2}\sin 20 = 0.01368$$

$$r_2 = \frac{d_G}{2}\sin\phi = \frac{0.128}{2}\sin 20 = 0.02189$$

$$W_t = \frac{60 \times 10^3 H}{\pi d n} = \frac{60 \times 10^3 \times 10.5}{\pi 80 \times 950} = 2.639\,\text{kN}$$

$$V = \frac{d_P}{2} \times 10^{-3} n \frac{2\pi}{60} = 3.979\,\text{m/s}$$

$$C_v = \frac{6.1}{6.1 + V} = 0.6052$$

$$\sigma_c = -163\left[\frac{2639}{0.6052 \times 0.05 \times \cos 20}\left(\frac{1}{0.01368} + \frac{1}{0.02189}\right)\right]^{0.5}$$

$$= -163[92\,808 \times 118.8]^{0.5}$$

$$= -163 \times 3320\,\text{kPa} = 541\,200\,\text{kPa} = 541\,\text{MPa}$$

Question 10
$d_P = 30\,\text{mm}, \quad W_t = 0.9549\,\text{kN}, \quad V = 0.1571\,\text{m/s}, \quad C_P = 174\sqrt{\text{MPa}}, \quad C_m = 1.6, \quad Q_v = 6, \quad C_v = 0.9287,$
$I = 0.1033052, \sigma_c = 896.4\,\text{MPa}$.

Question 11
$N_P = 18, \phi = 20°, n = 1800\,\text{rpm}, H = 3\,\text{kW}, N_G = 52, m = 2.5, F = 30\,\text{mm}, Q_v = 6$.
Exceptional rigidity $\rightarrow C_m = 1.3$.
Steel pinion, cast iron gear $\rightarrow C_P = 174\sqrt{\text{MPa}}$.

$d_P = 2.5 \times 18 = 45\,\text{mm}$

$V = 4.241\,\mathrm{m/s}$

$$W_t = \frac{60 \times 10^3 \times 3}{\pi 45 \times 1800} = 0.7074\,\mathrm{kN}$$

$$B = \frac{(12 - Q_v)^{2/3}}{4} = 0.8255$$

$A = 59.77$, $C_v = 0.7206$

$$I = \frac{\cos\phi \sin\phi}{2} \frac{m_G}{m_G + 1}$$

$$m_G = \frac{N_G}{N_P} = \frac{52}{18} = 2.889$$

$I = 0.1194$

AGMA contact stress:

$$\sigma_c = C_P\left(\frac{W_t C_a}{C_v}\frac{C_s}{Fd}\frac{C_m C_f}{I}\right)^{0.5} = 174\left(\frac{707.4 \times 1}{0.7206} \times \frac{1}{0.03 \times 0.045} \times \frac{1.3 \times 1}{0.1194}\right)^{0.5} = 489\,600\,\mathrm{kPa} = 489.6\,\mathrm{MPa}$$

$$\sigma_{c,all} = \frac{S_c C_L C_H}{C_T C_R}$$

From Table 4.15, $C_R = 0.85$.
$C_L = 2.466\,\mathrm{N}^{-0.056} = 1.1376$.
AGMA number 30 cast iron: so from Table 4.17, BHN = 175.

$$C_H = 1 + A(m_G - 1)$$

$$A = 8.98 \times 10^{-3}\left(\frac{H_{BP}}{H_{BG}}\right) - 8.29 \times 10^{-3} = 8.98 \times 10^{-3}\left(\frac{240}{175}\right) - 8.29 \times 10^{-3} = 4.025 \times 10^{-3}$$

$C_H = 1.008$

For the gear (weaker material), from Table 4.17, S_c for no. 30 cast iron = 450 MPa.

$$\sigma_{c,all} = \frac{450 \times 1.138 \times 1.008}{1.0 \times 0.85} = 606.8\,\mathrm{MPa}$$

$$n_c = \frac{\sigma_{c,all}}{\sigma_c} = 1.239$$

Question 12

$N_P = 19$, $m = 3$, $d_P = 3 \times 19 = 57\,\mathrm{mm}$.
$N_G = 4.42 \times 19 = 84$ (whole number only).
For $N_P = 19$, $N_G = 84$, $J \approx 0.358$.
$K_a = 1$, $K_s = 1$, $K_m = 1.3$.
$Q_v = 6$, $\rightarrow B = 0.8255$, $A = 59.77$.

$$V = \frac{d}{2} \times 10^{-3} n \times \frac{2\pi}{60} = 2.238\,\mathrm{m/s}$$

$$K_v = \left(\frac{A}{A + \sqrt{200V}}\right)^B = 0.7787$$

$W_t = 1.675\,\mathrm{kN}$.

$$\sigma = \frac{W_t K_a}{K_v}\frac{1}{Fm}\frac{K_s K_m}{\mathrm{J}} = \frac{1675 \times 1}{0.7787} \times \frac{1}{0.04 \times 0.003} \times \frac{1 \times 1.3}{0.358} = 65.1\,\mathrm{MPa}$$

Question 13

Power $= 30\,\text{kW} = 30\,000\,\text{W}$, $n_{i/p} = 2900\,\text{rpm}$, 655M13 case hardened steel.
$\phi = 20°$, $Q_v = 8$, $N_G = 38$, $N_P = 20$. $N_G/N_P = 1.9$, $d_G = mN_G = 95\,\text{mm}$.
$m = 2.5$, $F = 30\,\text{mm}$. Accurate mounting.
$L > 1 \times 10^8$ cycles, $R = 0.999$.

$$V = \frac{d}{2} n \frac{2\pi}{60} = \frac{95 \times 10^{-3}}{2} \times 2900 \times \frac{2\pi}{60} = 14.43\,\text{m/s}$$

$$K_v = \left(\frac{A}{A + \sqrt{200V}}\right)^B$$

$$B = \frac{(12 - 8)^{2/3}}{4} = 0.6300$$

$$A = 50 + 56(1 - B) = 70.72$$

$$K_v = 0.7005$$

Accurate mounting and $F < 50\,\text{mm}$, so from Table 4.13, $K_m = 1.3$.
From Table 4.14:

$$J = 0.39170 + \frac{20 - 17}{25 - 17}(0.40466 - 0.39170) = 0.3965$$

W_t = transmitted load = power/pitch line velocity $= H/V = 30\,000/14.43 = 2079.7\,\text{N}$

K_a = application factor = 1 (assumption not excessive)

K_s = size factor = 1 (assumption small gears)

$$\sigma = \frac{2079.7 \times 1}{0.70052} \times \frac{1}{0.03 \times 0.0025} \times \frac{1 \times 1.3}{0.3965} = 1.29 \times 10^8\,\text{Pa} = 129.8\,\text{MPa}$$

$$\sigma_{all} = \frac{S_t K_L}{K_T K_R}$$

Figure 4.22: $K_L = 0.96$ (10^8 mid value).
Table 4.15: $K_R = 1.25$.
K_T = temperature factor = 1 (default value: no excessive temperatures envisaged).
$S_t = 280\,\text{MPa}$ (use low value).

$$\sigma_{all} = \frac{280 \times 0.96}{1 \times 1.25} = 215\,\text{MPa}$$

$$n_s = \frac{\sigma_{all}}{\sigma} = \frac{215}{129.8} = 1.658 \text{ for the gear}$$

$J_{pinion} = 0.353$, $\sigma = 145.8\,\text{MPa}$ for the pinion. $n_s = 215/145.8 = 1.47$ for the pinion.
With $n_s = 1.658$ for the gear the proposed design appears reasonable. If n_s is nearer unity the design could be suspect.

Question 14

$N_P = 19$, $\phi = 20°$, $H = 14.5\,\text{kW}$, $N_G = 77$, $n_p = 375\,\text{rpm}$, $F = 75\,\text{mm}$, $m = 6$, $Q_v = 6$, $K_m = 1.6$, $K_a = K_s = 1$, life 10^8 cycles, 99% reliability.

$$K_v = \left(\frac{A}{A + \sqrt{200V}}\right)^B$$

$$B = \frac{(12 - 6)^{2/3}}{4} = 0.8255$$

$$A = 50 + 56(1 - B) = 59.77$$

$d_P = mN_P = 6 \times 19 = 114\,\text{mm},\ V = 2.238\,\text{ms}^{-1},\ K_v = 0.7787$
$N_P = 19,\ N_G = 77$
From Table 4.14 $J = 0.35\,665$.

$$W_t = \frac{60 \times 10^3 H}{\pi dn} = 6.478\,\text{kN}$$

$$\sigma = \frac{6478 \times 1}{0.7787} \times \frac{1}{0.075 \times 0.006} \times \frac{1 \times 1.6}{0.35\,665} = 82.93\,\text{MPa}$$

$$\sigma_{all} = \frac{S_t K_L}{K_T K_R}$$

For a 300BHN steel pinion from Table 4.16, $S_t = 250\,\text{MPa}$.
From Fig. 4.22 for 10^8 cycles, $K_L = 0.93$.
From Table 4.15, $K_R = 1$. Assume $K_T = 1$ (no excessive temperatures).

$$\sigma_{all} = \frac{S_t K_L}{K_T K_R} = \frac{250 \times 0.93}{1 \times 1} = 232.5\,\text{MPa}$$

$n_s = \sigma_{all}/\sigma = 232.5/82.93 = 2.8$
For $N_G = 77$, $N_P = 19$, Table 4.14 gives $J = 0.43\,082$, i.e. $\sigma = 68.66\,\text{MPa}$.
For an AGMA class 30 cast iron gear from Table 4.16, $S_t = 69\,\text{MPa}$.

$$\sigma_{all} = \frac{S_t K_L}{K_T K_R} = \frac{69 \times 0.93}{1 \times 1} = 64.17\,\text{MPa}$$

$$n_s = \frac{64.17}{68.66} = 0.93.$$

This is less than one so there is no margin of safety. The gearset is likely to fail.

Chapter 5: Seals

Question 1
(a) Labyrinth seal.
(b) Mechanical face seal.
(c) Gasket.

Question 2
$d = 10.5\,\text{mm},\ D = 12.8\,\text{mm},\ B = 2.3\,\text{mm},\ R = 0.5\,\text{mm}.$

Question 3
$r_i = 75 + 4.5 = 79.5\,\text{mm}.\ r_o = 75 + 4.5 + 0.5 = 80\,\text{mm}.$
$A = \pi(r_o^2 - r_i^2) = \pi(0.08^2 - 0.0795^2) = 2.505 \times 10^{-4}\,\text{m}^2.$
$c = 0.5\,\text{mm},\ t = 0.35\,\text{mm},\ p = 7\,\text{mm}.$

$c/t = 0.5/0.35 = 1.429,\ \text{so}\ \alpha = 0.71$

$n = 8.\ c/p = 0.5/7 = 0.07143$

$\gamma = 1 + 10.2(0.5/7) = 1.729$

$$\varphi = \sqrt{\frac{1 - (1.01/2)}{8 + \ln(2/1.01)}} = 0.2388$$

$$\rho_o = \frac{p_o}{RT_o} = \frac{2 \times 10^5}{287 \times 323} = 2.157\,\text{kg/m}^3$$

$$\dot{m} = 2.505 \times 10^{-4} \times 0.71 \times 1.729 \times 0.2388\sqrt{2.157 \times 2 \times 10^5} = 0.04823\,\text{kg/s} \approx 0.048\,\text{kg/s}$$

Question 4

No unique solution. Try six fins, $h = 3.2\,\text{mm}$, $p = 4.5\,\text{mm}$, $t = 0.3\,\text{mm}$.

$$\varphi = \sqrt{\frac{1 - (1.01/1.2)}{6 + \ln(1.2/1.01)}} = 0.1602$$

Try $c = 0.4\,\text{mm}$.

$\gamma = 1.784$.

$$\rho_o = \frac{p_o}{RT_o} = \frac{1.2 \times 10^5}{287 \times 318} = 1.315\,\text{kg/m}^3$$

$\alpha = 0.71$

$$A = \pi\{[(30 + 3.2 + 0.4) \times 10^{-3}]^2 - 0.0332^2\} = 8.394 \times 10^{-5}\,\text{m}^2$$

$\dot{m} = 0.0068\,\text{kg/s}$

Question 5

$r_o = 75 + 3.2 + 0.4 = 78.6\,\text{mm}$, $r_i = 78.2\,\text{mm}$, $A = 1.97 \times 10^{-4}\,\text{m}^2$.
$n = 6$ fins, $c = 0.4\,\text{mm}$, $h = 3.2\,\text{mm}$, $p = 4.5\,\text{mm}$, $t = 0.3\,\text{mm}$.
$c/t = 0.4/0.3 = 1.333$, $\alpha = 0.71$.
$\gamma = 1 + 8.82(0.4/4.5) = 1.784$.

$$\varphi = \sqrt{\frac{1 - (2.2/2.5)}{6 + \ln(2.5/2.2)}} = 0.1399$$

$$\rho_o = \frac{p_o}{RT_o} = \frac{2.5 \times 10^5}{287 \times 443} = 1.966\,\text{kg/m}^3$$

$\dot{m} = 0.0244\,\text{kg/s}$

Question 6

$\phi = 0.05\,\text{m}$, $c = 0.0003\,\text{m}$, $L = 0.03\,\text{m}$, $\mu = 0.02\,\text{Pa} \cdot \text{s}$, $p_o = 6 \times 10^5\,\text{Pa}$, $p_a = 5.5 \times 10^5\,\text{Pa}$.

$$Q = \frac{\pi \times 0.05 \times 0.0003^3(6 \times 10^5 - 5.5 \times 10^5)}{12 \times 0.03 \times 0.02} = 2.945 \times 10^{-5}\,\text{m}^3/\text{s}$$

Question 7

$\phi = 0.1\,\text{m}$, $c = 0.0001\,\text{m}$, $L = 0.05\,\text{m}$.

$$Q = \frac{\pi \times 0.1 \times 0.0001^3[(1.4 \times 10^5)^2 - (1.01 \times 10^5)^2]}{24 \times 1.85 \times 10^{-5} \times 0.05 \times 1.01 \times 10^5} = 1.317 \times 10^{-3}\,\text{m}^3/\text{s}$$

Question 8

$a = 40\,\text{mm}$, $b = 20\,\text{mm}$, $c = 0.2\,\text{mm}$.

$$Q = \frac{\pi \times 0.0002^3(6 \times 10^5 - 5.2 \times 10^5)}{6 \times 0.022 \times \ln(0.04/0.02)} = 2.2 \times 10^{-5}\,\text{m}^3/\text{s}$$

Chapter 6: Belt and chain drives

Question 1

45 kW, $n_1 = 2000$ rpm, $n_2 = 890$ rpm. $C \approx 0.8\,\text{m}$.
The speed ratio is $2000/890 = 2.247$.
From Table 6.2 the service factor is 1.1.
Design power $= 45 \times 1.1 = 49.5\,\text{kW}$.
From Fig. 6.5 an SPB belt is suitable.
From Table 6.3, the minimum pulley diameter is approximately 160 mm.

From Table 6.4, selecting $D_1 = 280$ mm, $D_2 = 630$ mm, for a speed ratio of 2.25, $C = 662$ mm (SPB2800) and the combined arc and belt length correction factor is 0.9.

From Table 6.5 the rated power per belt for $n_1 = 2000$ rpm and $D_1 = 280$ mm is 25.72 kW and from Table 6.6 the additional power per belt is 1.62 kW. The corrected power per belt is

$(25.72 + 1.62) \times 0.9 = 24.6$ kW

$$\text{Number of belts} = \frac{49.5}{24.6} = 2.01$$

Rounding up, use three belts.

Question 2

18.5 kW, $n_1 = 1455$ rpm, $n_2 = 400$ rpm. $C \approx 1.4$ m.
The speed ratio is

$1455/400 = 3.6375$

From Table 6.2 the service factor is 1.1.

Design power $= 18.5 \times 1.1 = 20.35$ kW

From Fig. 6.5 an SPB belt is suitable.
From Table 6.3, the minimum pulley diameter is approximately 112 mm.
From Table 6.4, selecting the closest ratio as 3.57, $D_1 = 224$ mm, $D_2 = 800$ mm, $C = 1.416$ m (SPB4500). The combined arc and belt length correction factor is 1.0.

From Table 6.5 the rated power per belt is approximately 15.47 kW. From Table 6.6 the additional power per belt is 1.21 kW. The corrected power per belt is

$(15.47 + 1.21) \times 1.0 = 16.68$ kW

$$\text{Number of belts required} = \frac{20.35}{16.68} = 1.22$$

Rounding up, use two SPB4500 belts with pulley diameters of 224 and 800 mm and centre distance 1.416 m.

Question 3

11 kW, $n_1 = 720$ rpm, $n_2 = 140$ rpm. $C < 1$ m, $D_1 < 90$ cm, $D_2 < 90$ cm.
The speed ratio is

$720/140 = 5.14$

From Table 6.2 the service factor is 1.4.

Design power $= 11 \times 1.4 = 15.4$ kW

From Fig. 6.5, the speed and power requirements fall within the range of capability for SPB belts.
From Table 6.3, the minimum pulley diameter is approximately 132 mm.
From Table 6.4, select a speed ratio of 5, with $D_1 = 160$ mm and $D_2 = 800$ mm (<900 mm), and $C = 968$ mm (SPB3550). The combined arc and belt length correction factor is 0.9.

From Table 6.5 the rated power per belt is approximately 5.11 kW. From Table 6.6 the additional power per belt is 0.62 kW. The corrected power per belt is

$(5.11 + 0.62) \times 0.9 = 5.157$ kW

$$\text{Number of belts required} = \frac{15.4}{5.157} = 2.986$$

Rounding up, use three SPB3550 belts with pulley diameters of 160 and 800 mm and centre distance 968 mm.

Question 4

22 kW, $D_1 = 250$ mm, $D_2 = 355$ mm, $n_1 = 1450$ rpm. $\mu = 0.7$, $\rho = 1100$ kg/m^3, $\sigma_{max} = 7$ MPa, $C = 1.8$ m, $t = 3.5$ mm.

$$\theta_d = \pi - 2\sin^{-1}\frac{355 - 250}{2 \times 1800} = 3.083$$

$$F_1 = \sigma_{max}A = \sigma_{max}tw = 7 \times 10^6 \times 3.5 \times 10^{-3}w = 24\,500w$$

$$F_c = \rho AV^2 = \rho tw \times \left(1450 \times \frac{2\pi}{60} \times \frac{0.25}{2}\right)^2 = w \times 1100 \times 3.5 \times 10^{-3} \times \left(1450 \times \frac{2\pi}{60} \times \frac{0.25}{2}\right)^2 = 1387w$$

$$\frac{F_1 - 1387w}{F_2 - 1387w} = e^{2.1581} = 8.655$$

$$F_1 - 1387w = 8.655(F_2 - 1387w)$$

$$F_1 - F_2 = \frac{22\,000}{18.98} = 1159$$

$$F_2 = F_1 - 1159 = 24\,500w - 1159$$

$$\frac{24\,500w - 1387w}{8.655} = 24\,500w - 1159 - 1387w$$

$w = 0.0567$ m. $w = 56.7$ mm. Use $w = 60$ mm.

Question 5

5 kW, $d_1 = 100$ mm, $d_2 = 250$ mm, $n_1 = 1500$ rpm. $\mu = 0.75$, $\rho = 1100\,\text{kg/m}^3$, $\sigma_{max} = 9$ MPa, $C = 0.6$ m, $t = 3.5$ mm.

$$\theta_d = \pi - 2\sin^{-1}\frac{250 - 100}{2 \times 600} = 2.891$$

$$F_1 = \sigma_{max}A = \sigma_{max}tw = 9 \times 10^6 \times 3.5 \times 10^{-3}w = 31\,500w$$

$$F_c = \rho AV^2 = \rho tw \times (7.854)^2 = w \times 1100 \times 3.5 \times 10^{-3} \times (7.854)^2 = 237.5w$$

$$\frac{F_1 - 237.5w}{F_2 - 237.5w} = e^{\mu\theta} = e^{0.75 \times 2.891} = 8.743$$

$$(F_1 - F_2)V = 5000$$

$$F_1 - F_2 = 636.6$$

$$F_2 = F_1 - 636.6 = 31\,500w - 636.6$$

$$F_1 - 237.5w = 8.743F_2 - 2076w$$

$$31\,500w - 237.5w = 8.743(31\,500w - 636.6) - 2076w$$

$w = 0.02299$ m. $w \approx 23$ mm. Use a 30 mm wide belt.

Question 6

37 kW, $d_1 = 250$ mm, $d_2 = 700$ mm, $n_1 = 1470$ rpm. $\mu = 0.8$, $\rho = 1100\,\text{kg/m}^3$, $\sigma_{max} = 6$ MPa, $C = 1.2$ m, $t = 2.9$ mm.

$$F_1 = \sigma_{max}A = \sigma_{max}tw = 17\,400w$$

$$F_c = \rho AV^2, \quad V = 19.24\,\text{m/s}$$

$$F_c = 1181w$$

$$\frac{F_1 - F_c}{F_2 - F_c} = e^{\mu\theta}$$

$$\theta = 2.764\,\text{rad}$$

$$\frac{F_1 - 1181w}{F_2 - 1181w} = 9.129$$

$(F_1 - F_2)V = 37 \times 10^3$

$F_1 - F_2 = 1923$

$F_2 = 17\,400w - 1923$

$$\frac{17\,400w - 1181w}{17\,400w - 1923 - 1181w} = 9.129$$

$w = 0.1332\,\text{m}$, $w = 133.2\,\text{mm}$. Use $w = 140\,\text{mm}$ or more.

Question 7

$$\theta_d = \pi - 2\sin^{-1}\frac{0.25 - 0.2}{2 \times 1.4} = 3.106$$

$F_1 = \sigma_{max}A = \sigma_{max}tw = 6.6 \times 10^6 \times 0.15 \times 0.0042 = 4158\,\text{N}$

$$V = 3000 \times \frac{2\pi}{60} \times \frac{0.2}{2} = 31.42\,\text{m/s}$$

$F_c = \rho A V^2 = 1100 \times 0.15 \times 0.0042 \times (31.42)^2 = 684.0\,\text{N}$

$$\frac{F_1 - F_c}{F_2 - F_c} = \frac{4158 - 684}{F_2 - 684} = e^{\mu\theta} = e^{0.75 \times 3.106} = 10.27$$

$F_2 = 1022\,\text{N}$

Power capacity is

$(F_1 - F_2)V = (4158 - 1022) \times 31.42 = 98.5\,\text{kW}$

No, the drive is not suitable.

Question 8

75 kW, $d_1 = 200\,\text{mm}$, $d_2 = 300\,\text{mm}$, $n_1 = 2946\,\text{rpm}$. $\mu = 0.8$, $\rho = 1100\,\text{kg/m}^3$, $\sigma_{max} = 6\,\text{MPa}$, $C = 1.4\,\text{m}$, $t = 4.3\,\text{mm}$.

$F_1 = \sigma_{max}A = \sigma_{max}tw = 6 \times 10^6 \times 4.3 \times 10^{-3} \times w = 25\,800w$

$F_c = \rho A V^2$, $\quad V = \omega r = 2946(2\pi/60) \times 0.1 = 30.85\,\text{m/s}$

$F_c = 1100 \times 4.3 \times 10^{-3} \times w \times 30.85^2 = 4502w.$

$$\theta_d = \pi - 2\sin^{-1}\frac{D - d}{2C} = \pi - 2\sin^{-1}\left(\frac{300 - 200}{2 \times 1400}\right) = 3.070$$

$$\frac{25\,800w - 4502w}{F_2 - 4502w} = e^{0.8 \times 3.07}$$

Also $(F_1 - F_2)V = 75 \times 10^3$.

$F_1 - F_2 = 2431$

$F_2 = 25\,800w - 2431$

$$\frac{25\,800w - 4502w}{25\,800w - 2431 - 4502w} = e^{0.8 \times 3.07} = e^{2.456} = 11.66$$

$21\,298w = 11.66(21\,298w - 2431)$

$w = 0.1248\,\text{m}$ or $w = 124.8\,\text{mm}$. Use $w = 130\,\text{mm}$ or more.

Question 9

The desired reduction ratio is

$$\frac{725}{400} = 1.813$$

The nearest ratio available (*see* Table 6.9) using standard sized sprockets is 1.80 using $N_1 = 21$, $N_2 = 38$.

The application factor from Table 6.10 is 1.0.

The tooth factor is

$$f_2 = \frac{19}{N_1} = \frac{19}{21} = 0.9048$$

The selection power is

$$18.5 \times 1.0 \times 0.9048 = 16.74\,\text{kW}$$

Using the BS/ISO selection chart (Fig. 6.15) the following drives would be suitable:

25.4 mm pitch in oil bath (simple)
19.05 mm pitch in oil bath (duplex)
15.875 mm pitch in oil bath (triplex).

Selecting the simple chain, $p = 25.4$ mm.

The chain length is given by

$$L = \frac{21+38}{2} + \frac{2 \times 470}{25.4} + \left(\frac{38-21}{2\pi}\right)^2 \frac{25.4}{470} = 66.9 \text{ pitches}$$

Rounding up to the nearest even integer gives $L = 68$ pitches.

The exact centre distance is given by

$$C = \frac{25.4}{8}\left[2 \times 68 - 38 - 21 + \sqrt{(2 \times 68 - 38 - 21)^2 - \frac{\pi}{3.88}(38-21)^2}\right] = 484.1\,\text{mm}$$

Question 10

There is no unique solution to this question. What follows is one possible solution. The speed ratio 710/75 = 9.47 is too high for a single reduction and is outside the range listed in Table 6.4. A two-stage reduction might be feasible. Examination of Table 6.9 for a combination of reductions shows that choosing ratios of 3.8 and 2.48 might be suitable. As $3.8 \times 2.48 = 9.424$ is close to the target reduction, this combination is likely to be acceptable and would give an output speed of 75.3 rpm.

Use the larger reduction on the higher speed drive: so for a ratio of 3.8 from Table 6.9, $N_1 = 25$, $N_2 = 95$.

The application factor from Table 6.10 assuming moderate shocks is 1.4.

The tooth factor is

$$f_2 = \frac{19}{N_1} = \frac{19}{25} = 0.76$$

The selection power is

$$2.2 \times 1.4 \times 0.76 = 2.34\,\text{kW}$$

Using the BS/ISO selection chart (Fig. 6.15) the following drives would be suitable:

12.7 mm pitch in oil bath (simple)
9.525 mm drip feed (duplex).

Selecting the simple chain, $p = 12.7$ mm.

Take the distance between the sprocket centres as the minimum recommended, i.e. $30 \times p = 30 \times 12.7 = 381$ mm, to ensure the design fits in the space available. This constraint can be relaxed at a later stage if appropriate.

The chain length is given by

$$L = \frac{25+95}{2} + \frac{2 \times 381}{12.7} + \left(\frac{95-25}{2\pi}\right)^2 \frac{12.7}{381} = 124.1 \text{ pitches}$$

Rounding up to the nearest even integer gives $L = 126$ pitches.

The exact centre distance is given by

$$C = \frac{12.7}{8}\left[2 \times 126 - 95 - 25 + \sqrt{(132)^2 - \frac{\pi}{3.88}(70)^2}\right] = 393.7\,\text{mm}$$

For the second chain with a ratio of 2.48, $N_1 = 23$, $N_2 = 57$.

The application factor from Table 6.10 is 1.4.

The tooth factor is

$$f_2 = \frac{19}{N_1} = \frac{19}{23} = 0.8261$$

The selection power $= 2.2 \times 1.4 \times 0.8261 = 2.544\,\text{kW}$.

Using the BS/ISO selection chart (Fig. 6.15) the following drives would be suitable:

12.7 mm pitch in oil bath (simple)
9.525 mm drip feed (duplex).

Selecting the simple chain, $p = 12.7\,\text{mm}$.

Again the distance between the sprocket centres can be taken as the minimum recommended, i.e. $30 \times p = 30 \times 12.7 = 381\,\text{mm}$.

The chain length is given by

$$L = \frac{23 + 57}{2} + \frac{2 \times 381}{12.7} + \left(\frac{57 - 23}{2\pi}\right)^2 \frac{12.7}{381} = 101.0 \text{ pitches}$$

Rounding up to the nearest even integer gives $L = 102$ pitches.

The exact centre distance is given by

$$C = \frac{12.7}{8}\left[2 \times 102 - 57 - 23 + \sqrt{(124)^2 - \frac{\pi}{3.88}(34)^2}\right] = 387.6\,\text{mm}$$

The combination of the two centre distances fits in the space available.

Question 11

The desired reduction ratio is

$$\frac{728}{400} = 1.82$$

From Table 6.9 the nearest ratio using standard sized sprockets is 1.8. $N_1 = 21$, $N_2 = 38$. From Table 6.10, $f_1 = 1.0$.

The tooth factor is

$$f_2 = \frac{19}{N_1} = \frac{19}{21} = 0.905$$

The selection power $= 30 \times 1.0 \times 0.905 = 27.15\,\text{kW}$.

Using the BS/ISO selection chart (Fig. 6.15) a 25.4 mm pitch simple BS chain drive is suitable with oil bath lubrication:

$$L = \frac{21 + 38}{2} + \frac{2 \times 1000}{25.4} + \left(\frac{38 - 21}{2\pi}\right)^2 \frac{25.4}{1000} = 108.4 \text{ pitches}$$

Rounding up to the nearest even integer gives $L = 110$ pitches.

The exact centre distance is

$$C = \frac{25.4}{8}\left[2 \times 110 - 38 - 21 + \sqrt{(2 \times 110 - 38 - 21)^2 - \frac{\pi}{3.88}(38 - 21)^2}\right] = 1020\,\text{mm}$$

Question 12
3 kW, $n_2 = 400$ rpm, $n_1 = 2820$ rpm, $C = 300$ mm.
The desired reduction ratio is

$$\frac{2820}{470} = 6$$

The nearest ratio available (*see* Table 6.9) using standard sized sprockets is 6, using $N_1 = 19$, $N_2 = 114$.
The application factor from Table 6.10 is 1.0.
The tooth factor is

$$f_2 = \frac{19}{N_1} = \frac{19}{19} = 1$$

The selection power is

$$3 \times 1.0 \times 1 = 3\,\text{kW}$$

Using the BS/ISO selection chart (Fig. 6.15) the following drives would be suitable:

9.525 mm pitch in oil bath (simple)
8 mm pitch in oil bath (duplex).

Selecting the simple chain, $p = 9.525$ mm.
The chain length is given by

$$L = \frac{19 + 114}{2} + \frac{2 \times 300}{9.525} + \left(\frac{114 - 19}{2\pi}\right)^2 \frac{9.525}{300} = 136.7 \text{ pitches}$$

Rounding up to the nearest even integer gives $L = 138$ pitches.
The exact centre distance is given by

$$C = \frac{9.525}{8}\left[2 \times 138 - 114 - 19 + \sqrt{(2 \times 138 - 114 - 19)^2 - \frac{\pi}{3.88}(114 - 19)^2}\right] = 306.7\,\text{mm}$$

Chapter 7: Clutches and brakes

Question 1
$T = I\alpha$
$\alpha = \Delta n/t$
$I = 0.25\,\text{kg}\cdot\text{m}^2$

(a) $T = 0.25\left(\dfrac{500(2\pi/60)}{2.5}\right) = 5.236\,\text{N}\cdot\text{m}$

(b) $T = 0.25\left(\dfrac{1000(2\pi/60)}{2}\right) = 13.10\,\text{N}\cdot\text{m}$

Question 2

$$E = \frac{1}{2}m(V_i^2 - V_f^2) = \frac{1}{2} \times 100 \times 10^3\left[\left(\frac{250 \times 10^3}{3600}\right)^2 - 0\right] = 241.1\,\text{MJ}$$

Assuming that the aircraft brakes uniformly, then the power that must be dissipated is

$$\frac{241.1 \times 10^6}{40} = 6.028\,\text{MW}$$

This is a significant quantity of power, equivalent to six thousand 1 kW electric heaters, and evenly divided between multiple brakes gives an indication why aborted takeoffs can result in burnt out brakes.

Note that the effects of any thrust reversal and aerodynamic drag have been ignored. Both of these effects would reduce the heat dissipated within the brakes.

Question 3

$F = 2\pi p_{max} r_i(r_o - r_i)$

(a) $T = \frac{\mu F}{2}(r_o + r_i) = \frac{0.3 \times 4000}{2}(0.3 + 0.2) = 300\,\text{N} \cdot \text{m}$

(b) $T = \frac{2}{3}\mu F \frac{r_o^3 - r_i^3}{r_o^2 - r_i^2} = \frac{2}{3} \times 0.3 \times 4000 \left(\frac{0.3^3 - 0.2^3}{0.3^2 - 0.2^2}\right) = 304\,\text{N} \cdot \text{m}$

Question 7

The torque capacity per pad is $320/2 = 160\,\text{N} \cdot \text{m}$.

The effective radius is given by

$$r_e = \frac{0.1 + 0.14}{2} = 0.12\,\text{m}$$

The actuation force required is

$$F = \frac{T}{\mu r_e} = \frac{160}{0.35 \times 0.12} = 3810\,\text{N}$$

The maximum pressure is

$$p_{max} = \frac{F}{\theta r_i(r_o - r_i)} = \frac{3810}{40(2\pi/360) \times 0.1(0.14 - 0.1)} = 1.364 \times 10^6\,\text{N/m}^2$$

The average pressure is

$$p_{av} = \frac{2r_i/r_o}{1 + (r_i/r_o)} p_{max} = \frac{2 \times 0.1/0.14}{1 + (0.1/0.14)} 1.36 \times 10^6 = 1.137 \times 10^6\,\text{N/m}^2$$

The area of one of the hydraulic cylinders is

$\pi 0.0127^2 = 5.067 \times 10^{-4}\,\text{m}^2$

The hydraulic pressure required is given by

$$p_{hydraulic} = \frac{F}{A_{cylinder}} = \frac{3810}{5.067 \times 10^{-4}} = 7.519 \times 10^6\,\text{N/m}^2$$

i.e. $p_{hydraulic} \approx 75\,\text{bar}$.

Question 8

Taking moments about the pivot:

$$\sum M_{pivot} = 600 \times 0.4 - F_n \times 0.2 + \mu F_n 0.05 = 0$$

$$F_n = \frac{0.4 \times 600}{0.2 - 0.35 \times 0.05} = 1315\,\text{N}$$

$$T = \mu F_n r = 0.35 \times 1315 \times 0.15 = 69.04\,\text{N} \cdot \text{m}$$

Question 9

For the left hand shoe:

$$\sum M_{pivot} = F_a \times 0.3 - F_n \times 0.2 + \mu F_n 0.03 = 0$$

$$F_n = \frac{1000 \times 0.3}{0.2 - 0.3 \times 0.03} = 1571\,\text{N}$$

For the right hand shoe:

$$\sum M_{\text{pivot}} = -F_a \times 0.3 + F_n \times 0.2 + \mu F_n \times 0.03 = 0$$

$$F_n = \frac{1000 \times 0.3}{0.2 + 0.3 \times 0.03} = 1435\,\text{N}$$

$$T = \mu(F_{n\,\text{total}}) \times r = 0.3 \times (1571 + 1435) \times 0.15 = 135.4\,\text{N}\cdot\text{m}$$

Question 10

For the top shoe:

$$F \times 0.56 - F_n 0.3 + \mu F_n 0.05 = 0$$

$$F_n = \frac{2400 \times 0.56}{0.3 - 0.35 \times 0.05} = 4758\,\text{N}$$

For the bottom shoe:

$$-F \times 0.56 + F_n 0.3 + \mu F_n 0.05 = 0$$

$$F_n = \frac{2400 \times 0.56}{0.3 + 0.35 \times 0.05} = 4233\,\text{N}$$

$$\text{Torque} = \mu(F_{n\,\text{total}}) \times 0.125 = 0.35(4758 + 4233) \times 0.125 = 393.4\,\text{N}\cdot\text{m}$$

$$\text{Heat generation} = \omega \times T = 100 \times \frac{2\pi}{60} \times 393.4 = 4119\,\text{W}$$

Question 11

Distance from pivot to drum centre is

$$b = \sqrt{0.2^2 + 0.15^2} = 0.25\,\text{m}$$

$$\theta_{\text{pivot}} = \tan^{-1}\left(\frac{0.15}{0.2}\right) = 36.87^\circ$$

$$\theta_1 = 3.13^\circ, \quad \theta_2 = 103.1^\circ$$

$$(\sin\theta)_{\max} = \sin 90 = 1$$

$$\mu = 0.3, \quad r = 0.15\,\text{m}, \quad w = 0.04\,\text{m}$$

From equation 7.32 the moment of the normal force with respect to the shoe pivot is

$$M_n = \frac{0.15 \times 0.04 \times 0.25 p_{\max}}{\sin 90}\left[\frac{1}{2}\left((103.1 - 3.13) \times \frac{2\pi}{360}\right) - \frac{1}{4}(\sin 206.2 - \sin 6.26)\right]$$

$$= 1.5 \times 10^{-3} p_{\max}(0.8727 + 0.1376) = 1.515 \times 10^{-3} p_{\max}$$

From equation 7.33 the moment of the frictional forces with respect to the shoe pivot is

$$M_f = \frac{0.3 \times 0.04 \times 0.15 p_{\max}}{\sin 90}\left[0.15(\cos 3.13 - \cos 103.1) + \frac{0.25}{4}(\cos 206.2 - \cos 6.26)\right]$$

$$= 1.8 \times 10^{-3} p_{\max}(0.1838 - 0.1182) = 1.181 \times 10^{-4} p_{\max}$$

From equation 7.31 the actuation force is

$$F_a = \frac{M_n - M_f}{a} = \frac{1.515 \times 10^{-3} p_{\max} - 1.181 \times 10^{-4} p_{\max}}{0.5} = 500$$

$$p_{\max} = 179\,000\,\text{N/m}^2$$

$$T = \mu w r^2 \frac{p_{max}}{(\sin\theta)_{max}}(\cos\theta_1 - \cos\theta_2) = 0.3 \times 0.04 \times 0.15^2 \frac{0.179 \times 10^6}{1}(\cos 3.13 - \cos 103.1) = 59.2\,\mathrm{N \cdot m}$$

Question 12

First it is necessary to calculate values for θ_1 and θ_2 as these are not indicated directly on the diagram:

$$\theta_1 = 25^\circ - \tan^{-1}\left(\frac{90}{300}\right) = 8.301^\circ$$

$$\theta_2 = 25^\circ + 130^\circ - \tan^{-1}\left(\frac{90}{300}\right) = 138.3^\circ$$

The maximum value of $\sin\theta$ would be $\sin 90 = 1$.

The distance between the pivot and the drum centre is

$$b = \sqrt{0.09^2 + 0.3^2} = 0.3132\,\mathrm{m}$$

The normal moment is given by

$$M_n = \frac{wrbp_{max}}{(\sin\theta)_{max}}\left[\frac{1}{2}(\theta_2 - \theta_1) - \frac{1}{4}(\sin 2\theta_2 - \sin 2\theta_1)\right]$$

$$= \frac{0.05 \times 0.25 \times 0.3132 \times 1 \times 10^6}{\sin 90}\left[\frac{1}{2}\left((138.3 - 8.301) \times \frac{2\pi}{360}\right) - \frac{1}{4}(\sin 276.6 - \sin 16.6)\right]$$

$$= 5693\,\mathrm{N \cdot m}$$

$$M_f = \frac{\mu w r p_{max}}{(\sin\theta)_{max}}\left[r(\cos\theta_1 - \cos\theta_2) + \frac{b}{4}(\cos 2\theta_2 - \cos 2\theta_1)\right]$$

$$= \frac{0.32 \times 0.05 \times 0.25 \times 1 \times 10^6}{\sin 90}\left[0.25(\cos 8.301 - \cos 138.3) + \frac{0.3132}{4}(\cos 276.6 - \cos 16.6)\right]$$

$$= 1472\,\mathrm{N \cdot m}$$

The orthogonal distance between the actuation force and the pivot is $a = 0.7\,\mathrm{m}$.

The actuation load on the left hand shoe is given by

$$F_{a\,\text{left shoe}} = \frac{M_n - M_f}{a} = \frac{5693 - 1472}{0.7} = 6031\,\mathrm{N}$$

The torque contribution from the left hand shoe is given by

$$T_{\text{left shoe}} = \mu w r^2 \frac{p_{max}}{(\sin\theta)_{max}}(\cos\theta_1 - \cos\theta_2)$$

$$= 0.32 \times 0.05 \times 0.25^2 \times 1 \times 10^6(\cos 8.301 - \cos 138.3)$$

$$= 1736\,\mathrm{N \cdot m}$$

The actuation force on the right hand shoe can be determined by considering each member of the lever mechanism as a free body:

$$F - A_V + B_V = 0$$

$$A_H = B_H$$

$$B_H = C_H \quad A_H = C_H$$

$$0.35F = 0.1B_H, \quad F = 0.1B_H/0.35. \quad B_H = 6031, \quad F = 6031 \times 0.1/0.35 = 1723\,\mathrm{N}$$

So the limiting lever force is $F = 1723\,\mathrm{N}$.

$C_V = 0, \quad B_V = 0$

The actuating force for the right hand lever is the resultant of F and B_H. The resultant angle is given by $\tan^{-1}(0.1/0.35) = 15.95°$.

$$F_{\text{a right shoe}} = \frac{1723}{\cos 15.95} = 6272\,\mathrm{N}$$

The perpendicular distance between the actuation force vector and the pivot is given by

$a = 0.6 \times \cos 15.95 = 0.5769\,\mathrm{m}$

The normal and frictional moments for the right hand shoe can be determined using equations 7.39 and 7.40:

$$M'_n = \frac{M_n p'_{max}}{p_{max}} = \frac{5693 p'_{max}}{1 \times 10^6}$$

$$M'_f = \frac{M_f p'_{max}}{p_{max}} = \frac{1472 p'_{max}}{1 \times 10^6}$$

For the right hand shoe the maximum pressure can be determined from

$$F_{\text{a right shoe}} = \frac{M'_n + M'_f}{a} = 6272 = \frac{5693 p'_{max} - 1472 p'_{max}}{1 \times 10^6 \times 0.5769}$$

$p'_{max} = 0.8571 \times 10^6\,\mathrm{N/m^2}$

The torque contribution from the right hand shoe is

$$T_{\text{right shoe}} = \mu w r^2 \frac{p'_{max}}{(\sin\theta)_{max}}(\cos\theta_1 - \cos\theta_2)$$

$$= 0.32 \times 0.05 \times 0.25^2 \times 0.8571 \times 10^6(\cos 8.301 - \cos 138.3)$$

$$= 1488\,\mathrm{N \cdot m}$$

The total torque is given by

$$T_{total} = T_{\text{left shoe}} + T_{\text{right shoe}} = 1736 + 1488 = 3224\,\mathrm{N \cdot m}$$

Question 13

$b = 0.12\,\mathrm{mm}$.

$\theta_1 = 0°, \theta_2 = 120°$.

As $\theta_2 > 90°$, the maximum value of $\sin\theta$ is $\sin 90 = 1 = (\sin\theta)_{max}$.

For this brake with the direction of rotation as shown, the right hand shoe is self-energising.

For the right hand shoe:

$$M_n = \frac{0.04 \times 0.14 \times 0.12 \times p_{max}}{1}\left[\frac{1}{2}\left((120 - 0) \times \frac{2\pi}{360}\right) - \frac{1}{4}(\sin 240 - \sin 0)\right] 8.492 \times 10^{-4} p_{max}$$

$$M_f = \frac{0.25 \times 0.04 \times 0.14 \times p_{max}}{1}\left[0.14(\cos 0 - \cos 120) + \frac{0.12}{4}(\cos 240 - \cos 0)\right] 2.31 \times 10^{-4} p_{max}$$

$a = 120 \cos 30 + 120 = 0.2239$

$$F_a = \frac{M_n - M_f}{a} = \frac{8.492 \times 10^{-4} p_{max} - 2.31 \times 10^{-4} p_{max}}{0.2239} = 2000\,\mathrm{N}$$

$p_{max} = 0.7244 \times 10^6\,\mathrm{N/m^2}$

The torque applied by the right hand shoe is given by

$$T_{\text{right shoe}} = \frac{\mu w r^2 p_{max}}{(\sin\theta)_{max}}(\cos\theta_1 - \cos\theta_2)$$

$$= \frac{0.25 \times 0.04 \times 0.14^2 \times 0.7244 \times 10^6}{1}(\cos 0 - \cos 120) = 213.0\,\text{N}\cdot\text{m}$$

The left hand shoe is self-de-energising, so

$$F_a = \frac{M_n + M_f}{a}$$

$$2000 = \frac{8.492 \times 10^{-4} p'_{max} + 2.31 \times 10^{-4} p'_{max}}{0.2239}$$

$$p'_{max} = 0.4146 \times 10^6\,\text{N/m}^2$$

The torque applied by the left hand shoe is given by

$$T_{\text{left shoe}} = \frac{\mu w r^2 p'_{max}}{(\sin\theta)_{max}}(\cos\theta_1 - \cos\theta_2)$$

$$= \frac{0.25 \times 0.04 \times 0.14^2 \times 0.4146 \times 10^6}{1}(\cos 0 - \cos 120) = 121.9\,\text{N}\cdot\text{m}$$

The total torque applied by both shoes is

$$T_{\text{total}} = T_{\text{right shoe}} + T_{\text{left shoe}} = 213 + 121.9 = 334.9\,\text{N}\cdot\text{m}$$

Question 14

$b = 0.08$ m.

$\theta_1 = 0°$, $\theta_2 = 132°$.

As $\theta_2 > 90°$, the maximum value of $\sin\theta$ is $\sin 90 = 1 = (\sin\theta)_{max}$.

For this brake with the direction of rotation as shown the right hand shoe is self-energising.

For the right hand shoe:

$$M_n = \frac{0.03 \times 0.1 \times 0.08 \times 1.2 \times 10^6}{1}\left[\frac{1}{2}\left((132 - 0) \times \frac{2\pi}{360}\right) - \frac{1}{4}(\sin 264 - \sin 0)\right]$$

$$= 403.3\,\text{N}\cdot\text{m}$$

$$M_f = \frac{0.4 \times 0.03 \times 0.1 \times 1.2 \times 10^6}{1}\left[0.1(\cos 0 - \cos 132) + \frac{0.08}{4}(\cos 264 - \cos 0)\right]$$

$$= 208.5\,\text{N}\cdot\text{m}$$

$$a = 0.08\cos 24 + 0.078 = 0.1511\,\text{m}$$

$$F_a = \frac{M_n - M_f}{a} = \frac{403.3 - 208.5}{0.1511} = 1289\,\text{N}$$

The actuating force is 1289 N.

The torque applied by the right hand shoe is given by

$$T_{\text{right shoe}} = \frac{\mu w r^2 p_{max}}{(\sin\theta)_{max}}(\cos\theta_1 - \cos\theta_2) = \frac{0.4 \times 0.03 \times 0.1^2 \times 1.2 \times 10^6}{1}(\cos 0 - \cos 132) = 240.4\,\text{N}\cdot\text{m}$$

The torque applied by the left hand shoe cannot be determined until the maximum operating pressure p'_{max} for the left hand shoe has been calculated.

As the left hand shoe is self-de-energising, the normal and frictional moments can be determined using equations 7.39 and 7.40:

$$M'_n = \frac{M_n p'_{max}}{p_{max}} = \frac{403.3p'_{max}}{1.2 \times 10^6}$$

$$M'_f = \frac{M_f p'_{max}}{p_{max}} = \frac{208.5p'_{max}}{1.2 \times 10^6}$$

The left hand shoe is self-de-energising, so

$$F_a = \frac{M_n + M_f}{a}$$

$F_a = 1289\,\text{N}$ as calculated earlier.

$$1289 = \frac{403.3p'_{max} + 208.5p'_{max}}{1.2 \times 10^6 \times 0.1511}$$

$$p'_{max} = 0.382 \times 10^6\,\text{N/m}^2$$

The torque applied by the left hand shoe is given by

$$T_{\text{left shoe}} = \frac{\mu w r^2 p'_{max}}{(\sin\theta)_{max}}(\cos\theta_1 - \cos\theta_2)$$

$$= \frac{0.4 \times 0.03 \times 0.1^2 \times 0.382 \times 10^6}{1}(\cos 0 - \cos 132) = 76.5\,\text{N}\cdot\text{m}$$

The total torque applied by both shoes is

$$T_{\text{total}} = T_{\text{right shoe}} + T_{\text{left shoe}} = 240.4 + 76.5 = 316.9\,\text{N}\cdot\text{m}$$

Question 15

$F_1 = p_{max} r w.$

$F_1 = 0.6 \times 10^6 \times 0.18 \times 0.1 = 10\,800\,\text{N}.$

$$F_2 = \frac{F_1}{e^{\mu\theta}} = \frac{10\,800}{e^{0.3\times 270\pi/180}} = 2627\,\text{N}$$

$$T = (F_1 - F_2)r = (10\,800 - 2627)0.18 = 1471\,\text{N}\cdot\text{m}$$

Chapter 8: Engineering tolerancing

Question 1

Medium drive fit, H7/s6.

Hole $62.5^{+0.030}_{+0}$ mm, bushing $62.5^{+0.072}_{+0.053}$ mm.

Limits of interference 0.072 mm to 0.023 mm.

Maximum pressure will occur when $\delta = 0.072$ mm.

$$p = \frac{0.072 \times 10^{-3}}{2 \times 31.25 \times 10^{-3} \times \left[\frac{1}{200 \times 10^9}\left(\frac{43.75^2 + 31.25^2}{43.75^2 - 31.25^2} + 0.27\right) + \frac{1}{113 \times 10^9}\left(\frac{31.25^2 + 25^2}{31.25^2 - 25^2} - 0.27\right)\right]}$$

$$= \frac{0.072 \times 10^{-3}}{0.0625(1.677 \times 10^{-11} + 3.793 \times 10^{-11})} = 21.06\,\text{MPa}$$

$$\sigma_o = 21.06 \times 10^6(3.083) = 64.94\,\text{MPa}$$

$$\sigma_i = 21.06 \times 10^6 \times 4.556 = 95.94\,\text{MPa}$$

$\delta_o = 2.206 \times 10^{-5}\,\text{m}$

$\delta_i = 4.992 \times 10^{-5}\,\text{m}$

Question 2

$T = 2fp\pi b^2 L$. Power $= T\omega$.

$$T = \frac{6000}{30 \times 2\pi/60} = 1910\,\text{N}\cdot\text{m}$$

$$p = \frac{T}{2f\pi b^2 L} = \frac{1910}{2 \times 0.12\pi(0.02)^2 \times 0.08} = 79.16 \times 10^6\,\text{N/m}^2$$

The interference required to generate the pressure is given by

$$\delta = 2bp\left[\frac{1}{E_o}\left(\frac{c^2+b^2}{c^2-b^2}+\mu_o\right)+\frac{1}{E_i}\left(\frac{b^2+a^2}{b^2-a^2}-\mu_i\right)\right]$$

$$= 2 \times 0.02 \times 79.16 \times 10^6\left[\frac{1}{100 \times 10^9}\left(\frac{0.05^2+0.02^2}{0.05^2-0.02^2}+0.3\right)+\frac{1}{207 \times 10^9}(1-0.3)\right]$$

$$= 6.393 \times 10^{-5}\,\text{m}$$

So the required interference fit is approximately 0.064 mm or 64 μm.

Question 4

$\sigma_1 = 0.05/3$, $\sigma_2 = 0.05/3$, $\sigma_3 = 0.03/3$.

$\sigma_z^2 = \sigma_1^2 + \sigma_2^2 + \sigma_3^2$.

$\sigma_z = 0.0256$.

$\Delta z = 0.1536$.

$y = x_1 + x_2 + x_3$. $\partial y/\partial x = 1$.

$\Delta y = \Delta x_1 + \Delta x_2 + \Delta x_3 = 0.1 + 0.1 + 0.06 = 0.26$.

$\sigma_z = 0.07/3 = 0.02333$.

$\sigma_z^2 = \sigma_1^2 + \sigma_2^2 + (0.03/3)^2 = 2\sigma_1^2 + 0.01^2$.

$\sigma_1 = 0.0149$.

$3\sigma_1 = 0.0447$.

Dimensions: 40 ± 0.045, 20 ± 0.045, 35 ± 0.03 mm.

Question 5

$$\Delta z^2 = \left|\frac{\partial z}{\partial x_1}\right|^2 \Delta x_1^2 + \left|\frac{\partial z}{\partial x_2}\right|^2 \Delta x_2^2 + \left|\frac{\partial z}{\partial x_3}\right|^2 \Delta x_3^2$$

$\Delta x_1 = 6\sigma_{x_1}$, $\Delta z = 6\sigma_z$.

$$\frac{\partial z}{\partial x_1} = \frac{1}{4x_3},\quad \frac{\partial z}{\partial x_2} = \frac{1}{4x_3},\quad \frac{\partial z}{\partial x_3} = -\frac{x_1+x_2}{4x_3^2}$$

$$\Delta z^2 = \left|\frac{1}{4 \times 2.05}\right|^2 0.12^2 + \left|\frac{1}{4 \times 2.05}\right|^2 0.18^2 + \left|-\frac{2.49+3.05}{4 \times 2.05^2}\right|^2 0.24^2$$

$$= 2.142 \times 10^{-4} + 4.819 \times 10^{-4} + 6.256 \times 10^{-3} = 6.952 \times 10^{-3}$$

$\Delta z = 0.0834$

$z = 0.6756 \pm 0.0417$

$\sigma_z \approx 0.0139$

Question 6

$$Q = \frac{\pi\phi\,\Delta r^3\,\Delta P}{12\mu L}$$

$\phi = 0.15\,\text{m}$, $\Delta r = 0.001\,\text{m}$, $\Delta P = 0.5 \times 10^5\,\text{Pa}$, $\mu = 0.02$, $L = 0.01$.
$\Delta\phi = 10 \times 10^{-5} = 1 \times 10^{-4}$, $\Delta\Delta r = 0.12 \times 10^{-3}$, $\Delta\Delta P = 0.05 \times 10^5$, $\Delta\mu = 2 \times 10^{-4}$, $\Delta L = 0.2 \times 10^{-3}$.
(Twice tolerance limits: $\Delta z = 6\sigma_z$.)

$$\Delta Q_{\text{sure-fit}} \approx \left|\frac{\partial Q}{\partial\phi}\right|\Delta\phi + \left|\frac{\partial Q}{\partial\Delta r}\right|\Delta\Delta r + \left|\frac{\partial Q}{\partial\Delta P}\right|\Delta\Delta P + \left|\frac{\partial Q}{\partial\mu}\right|\Delta\mu + \left|\frac{\partial Q}{\partial L}\right|\Delta L$$

$$\frac{\partial Q}{\partial\phi} = \frac{\pi\Delta r^3\Delta P}{12\mu L} = \frac{\pi 0.001^3 \times 0.5 \times 10^5}{12 \times 0.02 \times 0.01} = 0.06545$$

$$\frac{\partial Q}{\partial\Delta r} = \frac{3\pi\phi\Delta r^2\Delta P}{12\mu L} = \frac{3\pi 0.15 \times 0.001^2 \times 0.5 \times 10^5}{12 \times 0.02 \times 0.01} = 29.45$$

$$\frac{\partial Q}{\partial\Delta P} = \frac{\pi\phi\Delta r^3}{12\mu L} = \frac{\pi 0.15 \times 0.001^3}{12 \times 0.02 \times 0.01} = 1.963 \times 10^{-7}$$

$$\frac{\partial Q}{\partial\mu} = -\frac{\pi\phi\Delta r^3\Delta P}{12\mu^2 L} = \frac{\pi 0.15 \times 0.001^3 \times 0.5 \times 10^5}{12 \times 0.02^2 \times 0.01} = -0.4909$$

$$\frac{\partial Q}{\partial L} = -\frac{\pi\phi\Delta r^3\Delta P}{12\mu L^2} = \frac{\pi 0.15 \times 0.001^3 \times 0.5 \times 10^5}{12 \times 0.02(0.01)^2} = 0.9817$$

$$\Delta Q_{\text{sure-fit}} = 6.545 \times 10^{-6} + 3.534 \times 10^{-3} + 9.815 \times 10^{-4} + 9.818 \times 10^{-5} + 1.963 \times 10^{-4} = 4.817 \times 10^{-3}$$

$\bar{Q} = 9.817 \times 10^{-3}\,\text{m}^3/\text{s} \quad (\pm 2.409 \times 10^{-3})$

$$\Delta Q_{\text{basic normal}}\sqrt{\left|\frac{\partial Q}{\partial\phi}\right|^2\Delta\phi^2 + \left|\frac{\partial Q}{\partial\Delta r}\right|^2\Delta\Delta r^2 + \left|\frac{\partial Q}{\partial\Delta p}\right|^2 + \Delta\Delta p^2 + \left|\frac{\partial Q}{\partial\mu}\right|^2\Delta\mu^2 + \left|\frac{\partial Q}{\partial L}\right|^2\Delta L^2}$$

$$= [(0.06545 \times 1 \times 10^{-4})^2 + (29.45 \times 0.12 \times 10^{-3})^2 + (1.963 \times 10^{-7} \times 0.05 \times 10^5)^2$$
$$+ (0.06545 \times 1 \times 10^{-4})^2 + (0.9817 \times 0.2 \times 10^{-3})^2]^{0.5}$$
$$= 3.674 \times 10^{-3}$$

$\pm 3\sigma$, so $\sigma_Q = 3.674 \times 10^{-3}/6 = 6.124 \times 10^{-4}$.
$\bar{Q} = 9.817 \times 10^{-3} \pm 1.837 \times 10^{-3}\,\text{m}^3/\text{s}$.

Question 8

$$k = \frac{4E}{(1-\mu^2)} \times \frac{t^3}{K_1 D_e^2}$$

$K_1 = \text{constant}$

Assuming quantities are subject to variability, are uncorrelated and random:

$$\Delta k_{\text{sure-fit}} = \left|\frac{\partial k}{\partial E}\right|\Delta E + \left|\frac{\partial k}{\partial \mu}\right|\Delta\mu + \left|\frac{\partial k}{\partial t}\right|\Delta t + \left|\frac{\partial k}{\partial D_e}\right|\Delta D_e$$

$$\Delta k^2_{\text{basic normal}} = \left|\frac{\partial k}{\partial E}\right|^2\Delta E^2 + \left|\frac{\partial k}{\partial \mu}\right|^2\Delta\mu^2 + \left|\frac{\partial k}{\partial t}\right|^2\Delta t^2 + \left|\frac{\partial k}{\partial D_e}\right|^2\Delta D_e^2$$

$$\frac{\partial k}{\partial E} = \frac{4t^3}{(1-\mu^2)K_1D_e^2} = \frac{4 \times (2.22 \times 10^{-3})^3}{(1-0.3^2) \times 0.69 \times 0.04^2} = 4.356 \times 10^{-5}$$

$$\frac{\partial k}{\partial \mu} = -\frac{2\mu}{(1-\mu^2)^2} \times \frac{4Et^3}{K_1D_e^2} = -\frac{2 \times 0.3}{(1-0.09)^2} \times \frac{4 \times 207 \times 10^9 \times (2.22 \times 10^{-3})^3}{0.69 \times 0.04^2} = -5.946 \times 10^6$$

Quotient rule: $\left(\frac{u}{v}\right)' = \frac{uv' - vu'}{v^2}$

$$\frac{\partial k}{\partial t} = \frac{12Et^2}{(1-\mu^2)K_1D_e^2} = \frac{12 \times 207 \times 10^9 \times (2.22 \times 10^{-3})^2}{(1-0.09) \times 0.69 \times 0.04^2} = 1.219 \times 10^{10}$$

$$\frac{\partial k}{\partial D_e} = -\frac{8Et^3}{(1-\mu^2)K_1D_e^3} = -\frac{8 \times 207 \times 10^9 \times (0.00222)^3}{(1-0.09) \times 0.69 \times (0.04)^3} = -4.509 \times 10^8$$

Assume natural tolerance limits $\pm 3\sigma$.

$\Delta E = 2 \times 2 \times 10^9 = 4 \times 10^9$

$\Delta\mu = 2 \times 0.003 = 0.006$

$\Delta t = 2 \times 0.03 \times 10^{-3} = 6 \times 10^{-5}$

$\Delta D_e = 2 \times 0.08 \times 10^{-3} = 1.6 \times 10^{-4}$

$$\begin{aligned}\Delta k_{\text{sure-fit}} &= (4.356 \times 10^{-5} \times 4 \times 10^9) + (5.946 \times 10^6 \times 0.006) + (1.219 \times 10^{10} \times 6 \times 10^{-5}) \\ &\quad + (4.509 \times 10^8 \times 1.6 \times 10^{-4}) \\ &= 174\,240 + 35\,676 + 731\,400 + 72\,144 = 1\,013\,460\end{aligned}$$

$$\bar{k} = \frac{4 \times 207 \times 10^9}{1-0.3^2} \times \frac{(2.22 \times 10^{-3})^3}{0.69 \times 0.04^2} = 9\,017\,347.3\ \text{N/m}$$

$k_{\text{sure-fit}} = 9\,017\,000 \pm 506\,700\ \text{N/m}$

$\Delta k^2_{\text{basic normal}} = 174\,240^2 + 35\,676^2 + 731\,400^2 + 72\,144^2 = 5.718 \times 10^{11}$

$\Delta k_{\text{basic normal}} = 756\,200$

$k_{\text{basic normal}} = 9\,017\,000 \pm 378\,100\ \text{N/m}$

To reduce the variability tighten tolerances on t, as this has the highest sensitivity coefficient.

Index

Page numbers in **bold** refer to figures; *italic* page numbers refer to tables.